SARAH STEIGERWALD

W9-AUA-648

U. Tietze · Ch. Schenk

Advanced Electronic Circuits

With the Assistance of
E. Schmid

With 570 Figures

Springer-Verlag Berlin Heidelberg New York 1978

Dr.-Ing. Ulrich Tietze · Dr.-Ing. Christoph Schenk
Universität Erlangen-Nürnberg

Dipl.-Ing. Eberhard Schmid
Ph.D. (University of Southampton, UK)
Siemens AG, Erlangen

The book is based on
Tietze, U.; Schenk, Ch.: Halbleiter-Schaltungstechnik.
4. Aufl. Berlin, Heidelberg, New York: Springer 1978

ISBN 3-540-08750-8 Springer-Verlag Berlin Heidelberg New York
ISBN 0-387-08750-8 Springer-Verlag New York Heidelberg Berlin

Library of Congress Cataloging in Publication Data. Tietze, Ulrich, 1946–. Advanced electronic circuits. Based on the 4th ed. (1978) of the authors' Halbleiter-Schaltungstechnik. Bibliography: p.   Includes index. 1. Integrated circuits. 2. Electronic circuits. I. Schenk, Christoph, 1945– joint author. II. Schmid, Eberhard, 1946– joint author. III. Title TK7874.T53 621.3815'3 78-13342.

Typesetting, printing and binding: Universitätsdruckerei H. Stürtz AG, Würzburg

2362/3020-543210

Preface

In the earlier stages of integrated circuit design, analog circuits consisted simply of type 741 operational amplifiers, and digital circuits of 7400-type gates. Today's designers must choose from a much larger and rapidly increasing variety of special integrated circuits marketed by a dynamic and creative industry. Only by a proper selection from this wide range can an economical and competitive solution be found to a given problem. For each individual case the designer must decide which parts of a circuit are best implemented by analog circuitry, which by conventional digital circuitry and which sections could be microprocessor controlled.

In order to facilitate this decision for the designer who is not familiar with all these subjects, we have arranged the book so as to group the different circuits according to their field of application. Each chapter is thus written to stand on its own, with a minimum of cross-references.

To enable the reader to proceed quickly from an idea to a working circuit, we discuss, for a large variety of problems, typical solutions, the applicability of which has been proved by thorough experimental investigation. Our thanks are here due to Prof. Dr. D. Seitzer for the provision of excellent laboratory facilities.

The subject is extensive and the material presented has had to be limited. For this reason, we have omitted elementary circuit design, so that the book addresses the advanced student who has some background in electronics, and the practising engineer and scientist.

The book is based on the fourth edition (1978) of the text book "Tietze/Schenk, Halbleiter-Schaltungstechnik". Chapter 17 has been added to help the reader settle questions arising from definitions and nomenclature.

The English manuscript was prepared by Eberhard Schmid in close collaboration with the authors. We should like to express our gratitude to Patricia Schmid for her careful revision and to Springer-Verlag for their continued close co-operation.

Erlangen, September 1978   U. Tietze   Ch. Schenk

Contents

1 Linear and non-linear operational circuitry

Digital computers allow mathematical operations to be performed to a very high standard of accuracy. The quantities involved are often continuous signals, for instance in the form of voltages, which in turn may be the analogs for some other measured quantities. In such cases the digital computer requires two additional units, an analog-to-digital converter (ADC) and a digital-to-analog converter (DAC). Such expenditure is only justified if the demand for accuracy is too high to be met by analog computer circuitry involving operational amplifiers. An upper limit for the accuracy of this kind of circuit is in the order of 0.1%.

In the following sections the most important families of operational circuits are classified and described. They are circuits for the four fundamental arithmetic operations, for differential and integral operations with respect to time, and for the synthesis of transcendental or any other chosen functions. In order to illustrate as clearly as possible the operating principles of these circuits, we initially assume ideal characteristics of the operational amplifiers involved. When using real operational amplifiers, restrictions and additional conditions must be observed in the choice of the circuit parameters and, since they are treated thoroughly in the literature, they are not repeated here. We want to discuss in more detail only those effects that play a special role in the performance of the particular circuit.

1.1 Summing amplifier

For the addition of several voltages, an operational amplifier can be used when it is connected as an inverting amplifier. As Fig. 1.1

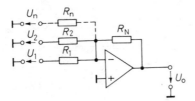

Fig. 1.1 Inverting summing amplifier.

Output voltage: $\quad -U_o = \dfrac{R_N}{R_1} U_1 + \dfrac{R_N}{R_2} U_2 + \cdots + \dfrac{R_N}{R_n} U_n$

indicates, the input voltages are connected via series resistors to the N-input of the operational amplifier. Since this node represents virtual ground, Kirchhoff's current law (KCL) immediately renders the relation for the output voltage:

$$\frac{U_1}{R_1}+\frac{U_2}{R_2}+\cdots+\frac{U_n}{R_n}+\frac{U_o}{R_N}=0.$$

The inverting summing amplifier can also be used as an amplifier with a wide-range zero adjustment if a direct voltage is added to the signal voltage in the manner described.

1.2   Subtracting circuits

1.2.1   Reduction to an addition

A subtraction can be reduced to the problem of an addition by inverting the signal to be subtracted. This requires the circuit shown in Fig. 1.2. The operational amplifier OA 1 inverts the input voltage U_2; the output voltage is then

$$U_o=A_P\,U_2-A_N\,U_1. \tag{1.1}$$

An evaluation of the actual difference based on the equation

$$U_o=A_D(U_2-U_1)$$

can be carried out if both gains, A_P and A_N, are made equal to the desired differential gain A_D. The deviation of this voltage from the result of an ideal subtraction, is determined by the common-mode rejection ratio (CMRR) which is defined as $G=A_D/A_{CM}$. The deviation can be calculated by allowing

$$U_2=U_{CM}+\tfrac{1}{2}U_D$$

and

$$U_1=U_{CM}-\tfrac{1}{2}U_D. \tag{1.2}$$

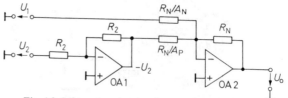

Fig. 1.2 Subtracting circuit using summing amplifier.

Output voltage: $U_o=A_D(U_2-U_1)$
Condition for coefficients: $A_N=A_P=A_D$

Equation (1.1) will then give

$$U_o = \underbrace{(A_P - A_N)}_{A_{CM}} U_{CM} + \underbrace{\tfrac{1}{2}(A_P + A_N)}_{A_D} U_D, \tag{1.3}$$

where U_{CM} is the common-mode voltage and U_D is the differential signal voltage. From Eq. (1.3) the common-mode rejection ratio can be determined as

$$G = \frac{A_D}{A_{CM}} = \frac{1}{2} \cdot \frac{A_P + A_N}{A_P - A_N}. \tag{1.4}$$

We can now write:

$$A_N = A - \tfrac{1}{2}\Delta A,$$

$$A_P = A + \tfrac{1}{2}\Delta A,$$

where A is the arithmetic mean of A_N and A_P. Introducing these expressions into Eq. (1.4) yields the result

$$G = \frac{A}{\Delta A}. \tag{1.5}$$

The common mode rejection ratio thus equals the reciprocal of the relative matching of the individual gains.

1.2.2 Subtraction using a single operational amplifier

For the calculation of the output voltage of the subtracting amplifier in Fig. 1.3, we may use the principle of superposition. We therefore write

$$U_o = k_1 U_1 + k_2 U_2.$$

For $U_2 = 0$, the circuit is an inverting amplifier where $U_o = -\alpha_N U_1$. It follows that $k_1 = -\alpha_N$. For $U_1 = 0$, the circuit represents a non-inverting amplifier having a voltage divider connected at its input. The potential

$$V_P = \frac{R_P}{R_P + R_P/\alpha_P}$$

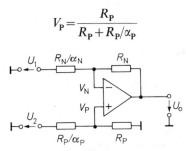

Fig. 1.3 Subtracting circuit using a single amplifier (subtracting amplifier).

Output voltage: $U_o = \alpha(U_2 - U_1)$
Condition for coefficients: $\alpha_N = \alpha_P = \alpha$

is thus amplified by the factor $(1+\alpha_N)$, this resulting in the output voltage

$$U_o = \frac{\alpha_P}{1+\alpha_P}(1+\alpha_N)\,U_2.$$

If both resistor ratios are the same, i.e. if $\alpha_P = \alpha_N = \alpha$, it follows that

$$U_o = \alpha\,U_2$$

and that $k_2 = \alpha$. We now apply superposition and obtain the output voltage for the general case as

$$U_o = \alpha(U_2 - U_1).$$

Should the ratio of the resistors connected to the P-terminal not be the same as that of the resistors connected to the N-terminal, the circuit does not evaluate the precise difference of the input voltages. In this case

$$U_o = \frac{1+\alpha_N}{1+\alpha_P}\,\alpha_P\,U_2 - \alpha_N\,U_1.$$

For the calculation of the common-mode rejection ratio we use the formulation of Eq. (1.2) again and obtain

$$G = \frac{A_D}{A_{CM}} = \frac{1}{2}\cdot\frac{(1+\alpha_N)\,\alpha_P + (1+\alpha_P)\,\alpha_N}{(1+\alpha_N)\,\alpha_P - (1+\alpha_P)\,\alpha_N}.$$

With $\alpha_N = \alpha - \frac{1}{2}\Delta\alpha$ and $\alpha_P = \alpha + \frac{1}{2}\Delta\alpha$, the expression may be rewritten and expanded into a series. Neglecting higher-order terms one obtains

$$G \approx (1+\alpha)\frac{\alpha}{\Delta\alpha}. \tag{1.6}$$

For constant α, the common-mode rejection ratio is inversely proportional to the tolerance of the resistor ratios. If the resistor ratios are identical, $G = \infty$, although this applies to ideal operational amplifiers only. In order to obtain a particularly high common-mode rejection ratio under real conditions, R_P may be varied slightly. In this way, $\Delta\alpha$ can be adjusted and the finite common-mode rejection ratio of the operational amplifier can be compensated for.

Equation (1.6) also shows that the common-mode rejection ratio for a given resistor matching tolerance $\Delta\alpha/\alpha$ is approximately proportional to the chosen differential gain $A_D = \alpha$. This is a great improvement over the previous circuit.

An example may best illustrate this: two voltages of about 10 V are to be subtracted one from the other. Their difference is 100 mV maximum. This value is to be amplified and to appear at the output of the subtraction amplifier as a voltage of 5 V, with an accuracy of 1 %. The differential gain must therefore be set as $A_D = 50$. The absolute error at the output must be smaller than $5\,V \cdot 1\% = 50\,mV$. If we assume the favourable case of the common-mode gain representing the only source of error, we then find it necessary to limit the common-mode gain to

$$A_{CM} \leqq \frac{50\,mV}{10\,V} = 5 \cdot 10^{-3},$$

i.e.

$$G \geqq \frac{50}{5 \cdot 10^{-3}} = 10^4 \triangleq 80\,dB.$$

For the subtracting amplifier in Fig. 1.3, this demand can be met by a relative resistor matching tolerance of $\Delta\alpha/\alpha = 0.5\%$, as follows from Eq. (1.6). For the subtraction circuit of Fig. 1.2 however, Eq. (1.5) yields a maximum tolerable mismatch of 0.01%.

Figure 1.4 shows an expansion of the subtracting amplifier for any number of additional summing and/or subtracting inputs. The determining factor for the proper functioning of the circuit is that the coefficient condition mentioned below Fig. 1.4 is satisfied. In the case of all coefficients α_i and α_i' being fixed but Eq. (1.7) not being satisfied, zero voltage must be added or subtracted with that coefficient required to satisfy the equation.

In order to deduce the relationships given below Fig. 1.4, we apply Kirchhoff's current law to the N-terminal:

$$\sum_{i=1}^{m} \frac{U_i - V_N}{\left(\dfrac{R_N}{\alpha_i}\right)} + \frac{U_o - V_N}{R_N} = 0.$$

Hence

$$\sum_{i=1}^{m} \alpha_i U_i - V_N \left[\sum_{i=1}^{m} \alpha_i + 1\right] + U_o = 0.$$

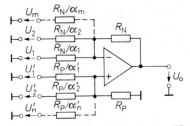

Fig. 1.4 Multiple-input subtracting amplifier.

Output voltage: $\qquad U_o = \sum_{i=1}^{n} \alpha_i' U_i' - \sum_{i=1}^{m} \alpha_i U_i$

Condition for coefficients: $\qquad \sum_{i=1}^{n} \alpha_i' = \sum_{i=1}^{m} \alpha_i$

Similarly, one obtains for the P-terminal:

$$\sum_{i=1}^{n} \alpha_i' U_i' - V_P \left[\sum_{i=1}^{n} \alpha_i' + 1 \right] = 0.$$

With $V_N = V_P$ and the additional condition

$$\sum_{i=1}^{m} \alpha_i = \sum_{i=1}^{n} \alpha_i', \tag{1.7}$$

the subtraction of the two equations results in

$$U_o = \sum_{i=1}^{n} \alpha_i' U_i' - \sum_{i=1}^{m} \alpha_i U_i.$$

For $n = m = 1$, the multiple-input subtracting amplifier becomes the basic circuit of Fig. 1.3.

The inputs of computing circuits represent loads to the signal voltage sources. The output resistances of the latter have to be sufficiently low to keep the computing errors small. Often the sources themselves are feedback amplifiers and they inherently have low-impedance outputs. When using other signal sources it may become necessary to employ impedance converters connected to the inputs. These converters often take the form of non-inverting amplifiers, and the resulting subtracting circuits are then called instrumentation amplifiers. They are commonly used in the field of measurement, and they are dealt with extensively in Chapter 15.

1.3　Bipolar-coefficient circuit

The circuit in Fig. 1.5 allows the multiplication of the input voltage by a constant factor, the value of which can be set between the limits $+n$ and $-n$ by the potentiometer R_2. If the slider of the potentiometer is positioned as far to the right as possible, then $q = 0$ and the circuit operates as an inverting amplifier with the gain $A = -n$. In this case, the resistor $R_1/(n-1)$ is ineffective since there is no voltage across it.

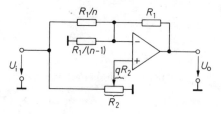

Fig. 1.5 Bipolar-coefficient circuit.

Output voltage:　$U_o = n(2q-1) U_i$

For $q=1$, the full input voltage U_i is at the P-terminal. The voltage across R_1/n is therefore zero, and the circuit operates as a non-inverting amplifier having the gain

$$A=1+\frac{R_1}{R_1/(n-1)}=+n.$$

For intermediate positions the gain is

$$A=n(2q-1).$$

It is thus linearly dependent on q and can be easily adjusted, for instance by means of a calibrated helical potentiometer. The factor n determines the range of the coefficient. The smallest value is $n=1$; the resistor $R_1/(n-1)$ may in this case be omitted.

1.4 Integrators

In analog computer application, the operational amplifier as an integrator is particularly important. Its output voltage can be expressed by the general form

$$U_o(t)=K\int_0^t U_i(\tilde{t})\,d\tilde{t}+U_o(t=0),$$

where $\tilde{t}$ is a dummy variable of integration.

1.4.1 Inverting integrator

The inverting integrator in Fig. 1.6 differs from the inverting amplifier in that the feedback resistor R_N is replaced by the capacitor C. The output voltage is then expressed by

$$U_o=\frac{Q}{C}=\frac{1}{C}\left[\int_0^t I_C(\tilde{t})\,d\tilde{t}+Q_0\right],$$

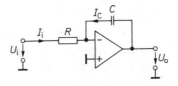

Fig. 1.6 Inverting integrator.

Output voltage: $U_o=-\dfrac{1}{RC}\displaystyle\int_0^t U_i(\tilde{t})\,d\tilde{t}+U_{oo}$

where Q_0 is the charge on the capacitor at the beginning of the integration $(t=0)$. As $I_C = -U_i/R$, it follows that

$$U_o = -\frac{1}{RC} \int_0^t U_i(\tilde{t}) \, d\tilde{t} + U_{oo}.$$

The constant U_{oo} represents the initial condition: $U_{oo} = U_o(t=0) = Q_0/C$. It has to be set to a defined value by the additional measures described in the next section.

Let us now look at two special cases. If the input voltage U_i is constant in time, one obtains the output voltage

$$U_o = -\frac{U_i}{RC} t + U_{oo},$$

which increases linearly with time. The circuit is thus very well suited to the generation of triangular and sawtooth voltages.

If U_i is a cosinusoidal alternating voltage, $u_i = \hat{U}_i \cos \omega t$, the output voltage becomes

$$U_o(t) = -\frac{1}{RC} \int_0^t \hat{U}_i \cos \omega \tilde{t} \, d\tilde{t} + U_{oo} = -\frac{\hat{U}_i}{\omega RC} \sin \omega t + U_{oo}.$$

The amplitude of the alternating output voltage is therefore inversely proportional to the angular frequency. When the amplitude-frequency response is plotted in log-log coordinates, the result is a straight line having the slope $-6\,dB/octave$. This characteristic is a simple criterion for determining whether a circuit behaves as an integrator.

The behaviour in the frequency domain can also be determined directly with the help of complex calculus:

$$\underline{A} = \frac{U_o}{U_i} = -\frac{Z_C}{R} = -\frac{1}{j\omega RC}. \tag{1.8}$$

Hence, it follows that for the ratio of the amplitudes

$$\frac{\hat{U}_o}{\hat{U}_i} = |\underline{A}| = \frac{1}{\omega RC},$$

as shown before.

As regards the frequency compensation, it must be noted that the feedback network causes a phase shift, contrary to all circuits previously dealt with. This means that the feedback factor becomes complex:

$$\underline{k} = \frac{U_N}{U_o}\bigg|_{U_i=0} = \frac{j\omega RC}{1+j\omega RC}. \tag{1.9}$$

For high frequency, $\underline{k}$ approaches $\underline{k}=1$ and the phase shift becomes zero. Therefore, in this range of frequency the same conditions obtain as for a unity-gain inverting amplifier. The frequency compensation necessary for the latter case must therefore also be used for the integrator circuit. Internally compensated amplifiers are normally designed for this application, and are therefore suitable for integration.

The frequency range usable for integration can be seen in Fig. 1.7 for a typical example. As integration time constant, $\tau = RC = 100\,\mu s$, is chosen. It is apparent that by doing so, a maximal loop gain of $|g| = |\underline{k}\underline{A}_D| \approx 600$ is attained, this corresponding to an output accuracy of about $1/|g| \approx 0.2\%$. In contrast to that of the inverting amplifier, the output accuracy falls not only at high, but also at low frequencies.

For the real operational amplifier, the input bias current I_B and the offset voltage U_O may be very troublesome, as their effects add up with time. If the input voltage U_i is reduced to nil, the capacitor carries the error current

$$\frac{U_O}{R} + I_B.$$

This results in a change in output voltage

$$\frac{dU_o}{dt} = \frac{1}{C}\left(\frac{U_O}{R} + I_B\right). \tag{1.10}$$

An error current of $I_B = 1\,\mu A$ causes the output voltage to rise at a rate of $1\,V/s$ if $C = 1\,\mu F$. Equation (1.10) indicates that for a given time constant, the contribution of the input bias current is smaller, the larger

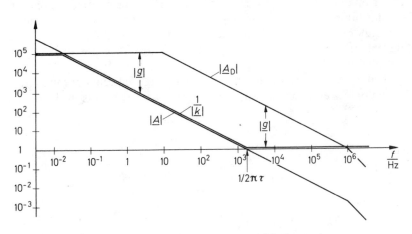

Fig. 1.7 Amplitude-frequency response of the loop gain $\underline{g}$

the value of C chosen. The contribution of the offset voltage remains constant. Because there is a limit to the size of C, one should at least make certain that the influence of I_B does not prevail over that of U_O. This is the case when

$$I_B < \frac{U_O}{R} = \frac{U_O C}{\tau}.$$

If, for instance, a time constant of $\tau = 1\,\mathrm{s}$ has to be achieved with a capacitance of $C = 1\,\mu\mathrm{F}$, an operational amplifier having an offset voltage of $U_O = 1\,\mathrm{mV}$ should possess an input bias current smaller than

$$I_B = \frac{1\,\mu\mathrm{F} \cdot 1\,\mathrm{mV}}{1\,\mathrm{s}} = 1\,\mathrm{nA}.$$

Operational amplifiers with bipolar input transistors rarely have such low input bias currents. The only solution to this problem is to compensate for this current, as is shown in Fig. 1.8. Resistor R_1 is chosen to have the same order of magnitude as R. The voltage across it is thus $I_B R$. Because $V_N \approx V_P$, for $U_i = 0$, a current of the magnitude

$$I = \frac{V_N}{R} = \frac{I_B R}{R} = I_B$$

flows through resistor R. The error current through capacitor C is therefore zero. Small offset currents and offset voltages can also be compensated for, and this is achieved by slightly varying resistor R_1. Unfortunately, the offset current drift, which may be relatively large for bipolar-input operational amplifiers, cannot be compensated for, and it is therefore better to use amplifiers with FET input. Their input bias current may be so small that only the offset voltage has to be compensated for.

A further source of error may be leakage currents through the capacitor. As electrolytic capacitors have leakage currents in the order

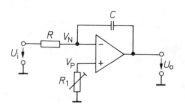

Fig. 1.8 Integrator having input bias current compensation

of μA, their use as integration capacitors is out of the question. One is therefore dependent on Polystyrene or Teflon dielectrics which make capacitors of over $10\,\mu F$ very unwieldy.

1.4.2 Initial condition

An integrator is only fit for use if its output voltage $U_o(t=0)$ can be set independently of the input voltage. With the circuit in Fig. 1.9 it is possible to stop integration and set the initial condition.

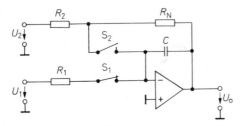

Fig. 1.9 Integrator having three modes of operation: integrate, hold, set initial condition.

Initial condition: $\quad U_o(t=0) = -\dfrac{R_N}{R_2} U_2$

If the switch S_1 is closed and S_2 open, the circuit behaves as that in Fig. 1.6; the voltage U_1 is integrated. On opening switch S_1, the charging current becomes zero in the case of an ideal integrator, and the output voltage remains at the value it had at the time of switching. This may be of use if one wants to interrupt computation, e.g. in order to read the output voltage at leisure. To set the initial condition, S_1 is left open and S_2 is closed. The integrator becomes an inverting amplifier with an output voltage of

$$U_o = -\frac{R_N}{R_2} U_2.$$

The output assumes this voltage only after a certain delay determined by the time constant $R_N C$.

Figure 1.10 shows one possibility of replacing the switches by electronic components. The two FETs T_1 and T_2 substitute the switches S_1 and S_2 of Fig. 1.9. They are conducting (ON) if the corresponding mode control signal voltage is greater than zero. For sufficiently negative control voltages they are in the OFF-mode. The precise operation of the FET switches and of the diodes D_1 to D_6 is described in more detail in Chapter 7.

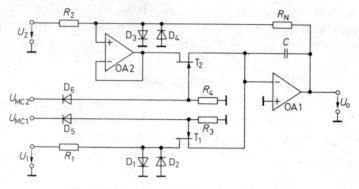

Fig. 1.10 Integrator with electronic mode control.

Initial condition: $U_o(t=0) = -\dfrac{R_N}{R_2} U_2$

The voltage follower OA 2 reduces the delay time constant for setting the initial condition, from the value $R_N C$ to the small value of $R_{DS\ ON} \cdot C$.

1.4.3   Summing integrator

Just as the inverting amplifier can be extended to become a summing amplifier, so an integrator can be developed into a summing integrator. The relationship given for the output voltage can be directly derived by the application of KCL to the summing point.

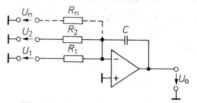

Fig. 1.11 Summing integrator.

Output voltage: $U_o = -\dfrac{1}{RC} \displaystyle\int_0^t \left(\dfrac{U_1}{R_1} + \dfrac{U_2}{R_2} + \cdots + \dfrac{U_n}{R_n} \right) d\tilde{t} + U_{o0}$

1.4.4   Non-inverting integrator

The integrators previously described produce an output voltage opposite in polarity to the input voltage. For an integration without polarity reversal, an inverting amplifier can be added to the integrator. Another solution is shown in Fig. 1.12. In principle, the circuit consists of a lowpass filter as integrating element. A NIC, having the internal

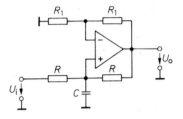

Fig. 1.12 Non-inverting integrator.

Output voltage: $U_o = \dfrac{2}{RC} \displaystyle\int_0^t U_i(\tilde{t})\, d\tilde{t} + U_{oo}$

resistance $-R$, is connected in parallel to the filter and simultaneously acts as an impedance converter (see Chapter 2). For the calculation of the output voltage, we apply KCL to the P-input and obtain

$$\frac{U_o - V_P}{R} + \frac{U_i - V_P}{R} - C\frac{dV_P}{dt} = 0.$$

Hence, with $V_P = V_N = \tfrac{1}{2}U_o$, we arrive at the result

$$U_o = \frac{2}{RC}\int_0^t U_i(\tilde{t})\, d\tilde{t}.$$

It must be noted that the input voltage source must have a very low impedance otherwise the stability condition for the NIC is not fulfilled. The operational amplifier evaluates differences between large quantities, and therefore this integrator does not have the same precision as the basic circuit in Fig. 1.6.

1.5 Differentiators

1.5.1 Basic circuit

If resistor and capacitor of the integrator in Fig. 1.6 are interchanged, one obtains the differentiator in Fig. 1.13. The application of KCL to the summing point yields the relationship

$$C\frac{dU_i}{dt} + \frac{U_o}{R} = 0, \qquad U_o = -RC\frac{dU_i}{dt}. \tag{1.11}$$

Thus, for sinusoidal alternating voltages, $u_i = \hat{U}_i \sin\omega t$, we obtain the output voltage

$$u_o = -\omega RC\,\hat{U}_i \cos\omega t.$$

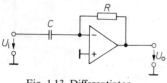

Fig. 1.13 Differentiator.

Output voltage: $U_o = - RC \dfrac{dU_i}{dt}$

For the ratio of amplitudes, it follows that

$$\frac{\hat{U}_o}{\hat{U}_i} = |\underline{A}| = \omega RC. \tag{1.12}$$

When the frequency response of the gain is plotted in log-log co-ordinates (Bode plot), the result is a straight line with the slope $+6\,\text{dB/octave}$. In general, a circuit is said to behave as a differentiator in a particular frequency range if, in that range, the amplitude-frequency response rises at a rate of $6\,\text{dB/octave}$.

The behaviour in the frequency domain can also be determined directly with the help of complex calculus:

$$\underline{A} = \frac{U_o}{\underline{U}_i} = -\frac{R}{\underline{Z}_C} = -j\omega RC. \tag{1.13}$$

Hence,

$$|\underline{A}| = \omega RC,$$

in accordance with Eq. (1.12).

1.5.2 Practical construction

The practical construction of the differentiator circuit in Fig. 1.13 presents certain problems since the circuit is prone to oscillations. These are caused by the feedback network which, at higher frequencies, gives rise to a phase lag of 90°, as the feedback factor is

$$\underline{k} = \frac{1}{1 + j\omega RC}. \tag{1.14}$$

The lag adds to the phase shift of the operational amplifier which in the most favourable case is already $-90°$. The remaining phase margin is zero, the circuit being therefore unstable. The instability can be overcome if one reduces the phase shift of the feedback network at high frequencies by connecting a resistor R_1 in series with the differentiating capacitor, as in Fig. 1.14. This measure need not necessarily reduce the usable frequency range, since the reduction in loop gain limits the satisfactory function of the differentiator at higher frequencies.

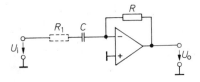

Fig. 1.14 Practical design of a differentiator.

$$Output\ voltage:\quad U_o = -RC\frac{dU_i}{dt}\quad for\ f \ll \frac{1}{2\pi R_1 C}$$

For the cutoff frequency f_1 of the RC element $R_1 C$, it is appropriate to choose the value for which the loop gain becomes unity. To find this value, one considers a fully compensated amplifier, the amplitude-frequency response of which is shown in the example of Fig. 1.15 as a broken line. The phase margin at the frequency f_1 is then approximately 45°. Since, in the vicinity of the frequency f_1 the amplifier has a feedback factor of less than unity, one can obtain an increase in the phase margin by a reduction of the frequency compensation, and hence approach a transient behaviour of near critical damping.

To optimize the compensation capacitor C_k, one applies a triangular voltage to the input of the differentiator and reduces C_k so that the rectangular output voltage is optimally damped.

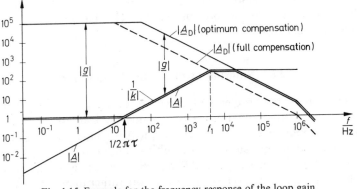

Fig. 1.15 Example for the frequency response of the loop gain.
$f_1 = \sqrt{f_T/2\pi\tau}$ where $\tau = RC$ and f_T is the unity-gain bandwidth

1.5.3 Differentiator with high input impedance

The input impedance of the differentiator described shows capacitive behaviour and this can lead to difficulties in some cases; for example, an operational amplifier circuit used as an input voltage

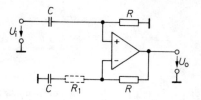

Fig. 1.16 Differentiator with high input impedance.

Output voltage: $U_o = RC \dfrac{dU_i}{dt}$

Input impedance: $|Z_i| \geqq R$

source can easily become unstable. The differentiator in Fig. 1.16 is better in this respect. Its input impedance does not fall below the value of R, even at high frequency.

The operation of the circuit is best illustrated as follows. Alternating voltages of low frequency are differentiated by the RC network at the input. In this frequency range, the operational amplifier corresponds to a non-inverting amplifier, having the gain $\underline{A} = 1$.

Alternating voltages of high frequency pass the input RC element unchanged and are differentiated by the feedback amplifier. If both time constants are equal, the effects of differentiation at low frequencies and at high frequencies overlap and make a smooth change-over.

As regards the stabilization against likely oscillations, the same principles apply as for the previous circuit. The damping resistor R_1 is drawn in Fig. 1.16 with broken lines.

1.6   Solution of differential equations

There are many problems which can be described most easily in the form of differential equations. One obtains the solution by using the analog computing circuits described above to model the differential equation and by measuring the resulting output voltage. In order to avoid stability problems, the differential equation is transformed in such a way to require exclusively integrators rather than differentiators.

We shall illustrate this method with the example of a linear second-order differential equation

$$y'' + k_1 y' + k_0 y = f(x). \qquad (1.15)$$

In the first step, the independent variable x is replaced by the time variable t:

$$x = \frac{t}{\tau}.$$

Since

$$y' = \frac{dy}{dt} \cdot \frac{dt}{dx} = \tau \dot{y} \quad \text{and} \quad y'' = \tau^2 \ddot{y},$$

the differential equation (1.15) becomes

$$\tau^2 \ddot{y} + k_1 \tau \dot{y} + k_0 y = f(t/\tau). \tag{1.16}$$

In the second step, the equation is solved for the undifferentiated quantities:

$$k_0 y - f(t/\tau) = -\tau^2 \ddot{y} - k_1 \tau \dot{y}.$$

Thirdly, the equation is multiplied throughout by the factor $(-1/\tau)$ and integrated

$$-\frac{1}{\tau} \int [k_0 y - f(t/\tau)]\, dt = \tau \dot{y} + k_1 y. \tag{1.17}$$

In this way, an expression is formed on the left side of Eq. (1.17), which can be computed by a simple summing integrator. Its output voltage is called a state variable, z_n, where n is the order of the differential equation; here $n = 2$. Therefore

$$z_2 = -\frac{1}{\tau} \int [k_0 y - f(t/\tau)]\, dt. \tag{1.18}$$

In this equation, the output variable y is initially taken as known.

By inserting Eq. (1.18) in Eq. (1.17), we arrive at

$$z_2 = \tau \dot{y} + k_1 y. \tag{1.19}$$

This differential equation is now treated in the same way as Eq. (1.16) and we obtain therefore

$$z_2 - k_1 y = \tau \dot{y},$$
$$-\frac{1}{\tau} \int [z_2 - k_1 y]\, dt = -y. \tag{1.20}$$

The left-hand side represents the state variable z_1:

$$z_1 = -\frac{1}{\tau} \int [z_2 - k_1 y]\, dt. \tag{1.21}$$

This expression is formed by a second summing integrator. Substitution in Eq. (1.20) gives the equation for the output

$$y = -z_1. \tag{1.22}$$

Since there are no longer any derivatives, the procedure is ended. The last equation, (1.22), provides the missing relation for the output variable y which had initially been taken as known.

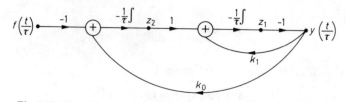

Fig. 1.17 Signal flow graph for the solution of the differential equation

$$y'' + k_1 y' + k_0 y = f(x)$$

The operations necessary for the solution of the differential equation (Eqs. (1.18), (1.21) and (1.22)) can be represented clearly with the aid of a signal flow chart as in Fig. 1.17. The appropriate analog computing circuit is shown in Fig. 1.18. In order to save an additional inverting amplifier for the computation of the expression $-k_1 y$ in Eq. (1.21), the fact is made use of that, from Eq. (1.22), $z_1 = -y$.

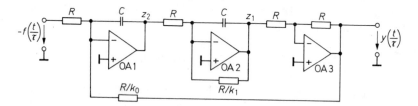

Fig. 1.18 Practical analog computing circuit

1.7   Function networks

The problem often arises that two voltages U_1 and $U_2 = f(U_1)$ have to be assigned one to the other, where f is a given function, so that for example

$$U_2 = U_A \log \frac{U_1}{U_B},$$

or

$$U_2 = U_A \sin \frac{U_1}{U_B}.$$

The correlation can also be given in the form of a diagram or table.

There are three possibilities of realizing such relationships. One can either make use of a physical effect that complies with the correlation required, or one can approximate the function by a series of straight lines or a power series. Below, we give some examples for these methods.

1.7.1   Logarithm

A logarithmic amplifier must give an output voltage that is proportional to the logarithm of the input voltage. It is therefore best to make use of the diode characteristic

$$I_A = I_S(e^{\frac{U_F}{mU_T}} - 1). \tag{1.23}$$

In this equation, I_S is the saturation leakage current, U_T is the thermal voltage, where $U_T = kT/e_0$, and m is a correction factor $(1 \leq m \leq 2)$. For the forward biased diode, when $I_A \gg I_S$, Eq. (1.23) can be approximated with good accuracy to

$$I_A = I_S e^{\frac{U_F}{mU_T}}. \tag{1.24}$$

Hence, the forward voltage

$$U_F = m U_T \ln \frac{I_A}{I_S}, \tag{1.25}$$

which is the required logarithmic function. The simplest way of using this relationship for computation of the logarithm is shown in Fig. 1.19, where a diode is incorporated in the feedback loop of an operational amplifier. This amplifier converts the input voltage U_i to a proportional current $I_A = U_i/R_1$. At the same time, the voltage $U_o = -U_F$ appears on its low-impedance output. Therefore

$$U_o = -m U_T \ln \frac{U_i}{I_S R_1} = -m U_T \ln 10 \lg \frac{U_i}{I_S R_1}, \tag{1.26}$$

or, at room temperature,

$$U_o = -(1 \dots 2) \cdot 60 \, \text{mV} \lg \frac{U_i}{I_S R_1}.$$

The usable range is limited by two effects. The diode possesses a parasitic series resistance, across which a considerable voltage occurs at high currents leading to errors in the computation of the logarithm. In addition, the correction factor m is current dependent. A satisfactory

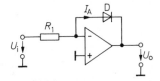

Fig. 1.19 Diode logarithmic amplifier.

Output voltage:

$$U_o = -m U_T \ln \frac{U_i}{I_S R_1} \quad \text{for } U_i > 0$$

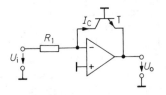

Fig. 1.20 Transistor logarithmic amplifier.

Output voltage:

$$U_o = -U_T \ln \frac{U_i}{I_{ES} R_1} \quad \text{for } U_i > 0$$

accuracy can therefore only be achieved over an input voltage range of 1 or 2 decades.

The unfavourable effect of the varying correction factor m can be eliminated by replacing the diode D by a transistor T, as shown in Fig. 1.20. For the collector current at $U_{CB}=0$, the following relationship holds:

$$I_C = \alpha I_E = \alpha I_{ES}(e^{\frac{U_{BE}}{mU_T}} - 1).$$

According to reference [1.1] the current dependence of α and m cancel so that

$$I_C = \gamma I_{ES}(e^{\frac{U_{BE}}{U_T}} - 1).$$

In this equation, the factor γ is only very slightly dependent on the current and has a value close to unity. Therefore, for $U_{BE}>0$,

$$I_C \approx I_{ES} e^{\frac{U_{BE}}{U_T}}, \tag{1.27}$$

hence

$$U_{BE} = U_T \ln \frac{I_C}{I_{ES}}. \tag{1.28}$$

It follows that the output voltage of the transistor logarithmic amplifier in Fig. 1.20 is

$$U_o = -U_T \ln \frac{U_i}{I_{ES} R_1}.$$

As the current-dependent correction factor m no longer influences the result, the transistor logarithmic amplifier can be used over a considerably larger range of input voltages than the diode log-amplifier. When suitable transistors are employed one has control over a variation of the collector current from the pA to the mA region, i.e. over nine decades. However, operational amplifiers with very low input currents are needed to exploit this range to the full.

Since the transistor T increases the loop gain by its own voltage gain, the circuit is prone to oscillations. The voltage gain of the transistor stage can be reduced quite simply by connecting a resistor R_E between emitter and amplifier output, as in Fig. 1.21. The resistance R_E must not, of course, give rise to voltage saturation of the operational amplifier at the largest possible output current. The capacitor C can further improve stability of the circuit by derivative action in the feedback. It must be noted, however, that the upper cutoff frequency decreases proportionally to the current because of the non-linear transistor characteristic.

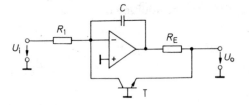

Fig. 1.21 Additional frequency compensation of the logarithmic amplifier

One disadvantage of the logarithmic amplifier described is its strong temperature dependence. The reason for this is that U_T and I_{ES} change drastically with temperature. For a temperature rise from 20 °C to 50 °C, U_T increases by 10 % while the reverse current is multiplied tenfold. The influence of the reverse current can be eliminated by computing the difference between two logarithms. We use this principle in Fig. 1.22, where the differential amplifier stage T_1, T_2 serves to find the logarithm. In order to examine the operation of the circuit, we determine the current sharing in the differential amplifier stage. From Kirchhoff's voltage law (KVL) it follows that

$$U_1 + U_{BE\,2} - U_{BE\,1} = 0.$$

The transfer characteristics of the transistors may be written as

$$I_{C1} = I_{ES}\, e^{\frac{U_{BE1}}{U_T}},$$

$$I_{C2} = I_{ES}\, e^{\frac{U_{BE2}}{U_T}};$$

and therefore

$$\frac{I_{C1}}{I_{C2}} = e^{\frac{U_1}{U_T}}. \tag{1.29}$$

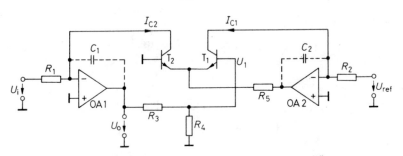

Fig. 1.22 Temperature-compensated logarithmic amplifier.

Output voltage: $\quad U_o = -U_T \cdot \dfrac{R_3 + R_4}{R_4}\, \ln \dfrac{R_2\, U_i}{R_1\, U_{ref}}\quad$ for $U_i, U_{ref} > 0$

From Fig. 1.22, we infer the additional equations

$$I_{C2} = \frac{U_i}{R_1},$$

$$I_{C1} = \frac{U_{ref}}{R_2},$$

$$U_1 = \frac{R_4}{R_3 + R_4} U_o,$$

if R_4 is not chosen to be too large. By substitution, we get the output voltage

$$U_o = -U_T \frac{R_3 + R_4}{R_4} \ln \frac{R_2 U_i}{R_1 U_{ref}}.$$

The value of R_5 does not appear in the result. Its resistance is chosen so that the voltage across it is smaller than the maximum possible output voltage swing of the amplifier OA 2.

As regards the frequency compensation of both amplifiers, the same argument holds as for the previous circuit. C_1 and C_2 are the additional compensation capacitors. The temperature influence of U_T can be offset by letting resistor R_4 have a positive temperature coefficient of 0.3 %/K.

1.7.2   Exponential function

Figure 1.23 shows an exponential function network the construction of which is analogous to that of the logarithmic amplifier in Fig. 1.20. When applying a negative voltage to the input, the current flowing through the transistor is given by Eq. (1.27):

$$I_C = I_{ES} e^{\frac{U_{BE}}{U_T}} = I_{ES} e^{-\frac{U_i}{U_T}},$$

and the output voltage is therefore

$$U_o = I_C R_1 = I_{ES} R_1 e^{-\frac{U_i}{U_T}}.$$

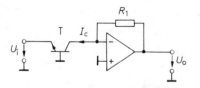

Fig. 1.23 Simple exponential function network.

$$U_o = I_{ES} R_1 e^{-\frac{U_i}{U_T}} \quad \text{for} \quad U_i < 0$$

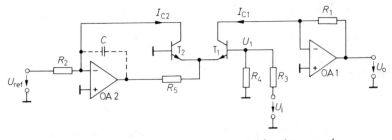

Fig. 1.24 Temperature compensated exponential function network.

$$U_o = \frac{U_{ref}R_1}{R_2} e^{\frac{R_4}{R_3+R_4} \cdot \frac{U_i}{U_T}} \quad \text{for } U_{ref} > 0$$

As with the logarithmic amplifier in Fig. 1.22, the temperature stability can be improved by using a differential amplifier. The appropriate circuit is represented in Fig. 1.24. Again, from Eq. (1.29)

$$\frac{I_{C1}}{I_{C2}} = e^{\frac{U_1}{U_T}}.$$

From Fig. 1.24 we deduce the following equations

$$I_{C1} = \frac{U_o}{R_1},$$

$$I_{C2} = \frac{U_{ref}}{R_2},$$

$$U_1 = \frac{R_4}{R_3+R_4} U_i.$$

By substitution we obtain the output voltage

$$U_o = \frac{U_{ref}R_1}{R_2} e^{\frac{R_4}{R_3+R_4} \cdot \frac{U_i}{U_T}}.$$

It can be seen that I_{ES} no longer appears in the result if the transistors are well matched. The resistor R_5 limits the current through the transistors T_1 and T_2, and its resistance does not affect the result as long as the operational amplifier OA 2 is not saturated.

The exponential function networks described enable the computation of expressions of the form

$$y = e^{ax}.$$

Since

$$b^{ax} = (e^{\ln b})^{ax} = e^{ax \ln b},$$

exponential functions to any base b can be computed according to

$$y = b^{ax}$$

by amplifying the input signal x by the factor $\ln b$, and by applying the result to the input of an exponential function network.

1.7.3 Computation of power functions using logarithms

The computation of power expressions

$$y = x^a$$

can be carried out for $x > 0$ by means of logarithmic amplifiers and exponential function networks because

$$x^a = (e^{\ln x})^a = e^{a \ln x}.$$

The basic arrangement for such a circuit is shown in Fig. 1.25. The equations mentioned apply to the logarithmic-function network of Fig. 1.22 and the exponential-function network of Fig. 1.24, where $R_3 = 0$, $R_4 = \infty$ and $R_1 = R_2$. We therefore obtain the output voltage

$$U_o = U_{ref} \, e^{\frac{a U_T \ln \frac{U_i}{U_{ref}}}{U_T}}$$

$$U_o = U_{ref} \left(\frac{U_o}{U_{ref}} \right)^a.$$

The output voltage shows an excellent temperature stability as the thermal voltage U_T-cancels out.

Involution (raising to the power) by means of logarithms is in principle defined for positive input voltages only. However, from a mathematical point of view, bipolar input signals are also permitted for whole-number exponents a. This case can be realized by using the multipliers described in Section 1.8.

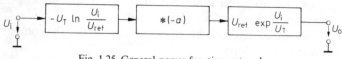

Fig. 1.25 General power-function network.

$$U_o = U_{ref} \left(\frac{U_i}{U_{ref}} \right)^a$$

1.7.4 Sine and cosine functions

The output of a sine-function network should approximate the expression

$$U_o = \hat{U}_o \sin\left(\frac{\pi}{2} \cdot \frac{U_i}{\hat{U}_i}\right) \tag{1.30}$$

within the range $-\hat{U}_i \leq U_i \leq +\hat{U}_i$. For small input voltages,

$$U_o = \hat{U}_o \cdot \frac{\pi}{2} \cdot \frac{U_i}{\hat{U}_i}.$$

It is suitable to choose a value for $\hat{U}_o$ so that near the origin, $U_o = U_i$. This is the case for

$$\hat{U}_o = \frac{2}{\pi} \cdot \hat{U}_i. \tag{1.31}$$

For small input voltages, the sine-function network must possess unity gain, whereas at higher voltages the gain must fall. Figure 1.26 represents a circuit fulfilling these conditions, based on the principle of *piecewise approximation*.

For small input voltages, all diodes are reverse biased, and $U_o = U_i$, as required. When U_o rises above U_1, diode D_1 becomes forward biased. U_o then increases more slowly than U_i because of the voltage divider formed by R_v and R_4. When U_o becomes larger than U_2, the output of the network

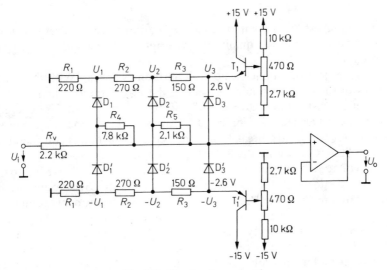

Fig. 1.26 Sine function network.

$$U_o \approx \frac{2}{\pi} \cdot \hat{U}_i \sin\left(\frac{\pi}{2} \frac{U_i}{\hat{U}_i}\right) \quad \text{for } \hat{U}_i = 5.0 \text{ V}$$

is additionally loaded with R_5, so that the rise in voltage is slowed down even more. Diode D_3 finally produces the horizontal tangent at the crest of the sine curve. The diodes D_1' to D_3' have the corresponding effects at negative input voltages, i.e. for the negative section of the sine curve. Considering that diodes do not become conducting suddenly, but have an exponential characteristic, one can obtain low distortion factors with only a small number of diodes.

In order to determine the parameters of the network, one begins by choosing the breakpoints of the approximation curve. It can be shown that the first n odd harmonics disappear if $2n$ breakpoints are assigned to the following values of the input voltage:

$$U_{ik} = \pm \frac{2k}{2n+1} \hat{U}_i, \quad \text{for } 0 < k \leq n, \qquad (1.32)$$

k being a whole number. According to Eqs. (1.30) and (1.31), the corresponding output voltages are

$$U_{ok} = \pm \frac{2}{\pi} \hat{U}_i \sin \frac{\pi k}{2n+1}, \quad \text{for } 0 < k \leq n. \qquad (1.33)$$

Therefore, the slope of the line segment above the k-th breakpoint is given as

$$m_k = \frac{U_{o(k+1)} - U_{ok}}{U_{i(k+1)} - U_{ik}} = \frac{2n+1}{\pi} \left[\sin \frac{\pi(k+1)}{2n+1} - \sin \frac{\pi k}{2n+1} \right]. \qquad (1.34)$$

For the highest breakpoint, when $k=n$, the slope becomes zero, as was stipulated earlier in the qualitative description. The slope m_0 must be chosen as unity.

For reasons of symmetry, no even harmonics appear. With the r.m.s. values of the odd harmonics present in the waveform, one arrives at a theoretical distortion factor of 1.8%, if $2n=6$ breakpoints are chosen, this being reduced to 0.8% for $2n=12$. However, as real diode characteristics do not have sharp breakpoints, the actual distortion is considerably lower. This is shown by the following example.

A voltage of triangular waveshape with a peak value of $\hat{U}_i = 5\,\text{V}$ is to be converted into a sinusoidal voltage. According to Eq. (1.31), the amplitude of the latter must be 3.18 V, so that the slope of the line segment around the origin is unity. For the approximation, we want to use $2n=6$ breakpoints. Following Eq. (1.33), they must appear at the output voltages $\pm 1.4\,\text{V}$, $\pm 2.5\,\text{V}$ and $\pm 3.1\,\text{V}$. For real diodes we assume that a sizeable current flows only for forward voltages of more than 0.5 V. The diode bias voltages must then be reduced by this amount. We thus obtain the voltages $U_1 = 0.9\,\text{V}$, $U_2 = 2.0\,\text{V}$ and $U_3 = 2.6\,\text{V}$ which define the values for the voltage divider chain R_1, R_2, R_3 shown in Fig. 1.26. The emitter-fol-

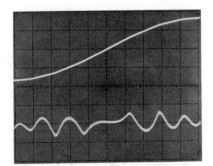

Fig. 1.27 Output voltage and error voltage (amplified 50 times) as a function of the input
voltage. Vertical: 2 V/Div; Horizontal: 1 V/Div

lower amplifiers T_1 and T_1', serve as low-impedance sources for U_3 and
$-U_3$, and simultaneously as temperature compensation for the forward
voltages of the diodes.

From Eq. (1.34), we obtain for the slopes of the three segments:
$m_1 = 0.78$, $m_2 = 0.43$ and $m_3 = 0$. We choose $R_v = 2.2 \, k\Omega$. From

$$m_1 = \frac{R_4}{R_v + R_4},$$

thereby neglecting the internal resistance of the divider chain, we obtain
$R_4 = 7.8 \, k\Omega$. The slope of the second segment is

$$m_2 = \frac{(R_5 \| R_4)}{R_v + (R_5 \| R_4)},$$

thus $R_5 = 2.1 \, k\Omega$.

For the fine adjustment of the network, the use of a notch filter for
the fundamental (see Chapter 3.9) and the display of the remaining error
voltage on the screen of an oscilloscope is recommended. The optimum is
reached when the peaks of the deviation curve have the same height, as can
be seen in the oscillogram in Fig. 1.27. The distortion factor measured for
this case was 0.42 % and therefore clearly below the theoretical value for
ideal diodes.

Power series expansion

Another method for the approximation of a sine function is the use of a
power series since

$$\sin x = x - \frac{x^3}{3!} + \frac{x^5}{5!} - + \cdots.$$

To keep the number of components low, the series is broken off after the second term and this results in an error. When limiting the range of values of the argument to $-\dfrac{\pi}{2} \leq x \leq \dfrac{\pi}{2}$, one can minimize the error by slightly changing the coefficients [1.3]. If one chooses

$$\sin x \approx y = 0.9825 x - 0.1402 x^3, \tag{1.35}$$

the error becomes zero for $x = 0$, ± 0.96 and $\pm \pi/2$. Between these values, the absolute error is less than 0.57% of the amplitude. The distortion factor is 0.6%. It can be reduced to 0.25% by a slight variation of the coefficients, and is therefore somewhat smaller than for the piecewise approximation method using 2×3 breakpoints. The lack of breakpoints is particularly advantageous when the signal is to be differentiated.

For a practical circuit, we define

$$x = \frac{\pi}{2} \cdot \frac{U_i}{\hat{U}_i}$$

and

$$y = \frac{U_o}{\hat{U}_o}.$$

Furthermore, we choose $\hat{U}_i = \hat{U}_o$ and thus obtain from Eq. (1.35)

$$U_o = 1.543 \, U_i - 0.543 \, \frac{U_i^3}{\hat{U}_i^2} \approx \hat{U}_i \sin\left(\frac{\pi}{2} \frac{U_i}{\hat{U}_i}\right).$$

The block diagram for this operation is represented in Fig. 1.28 where the input voltage amplitude $\hat{U}_i$ is equal to the computing unit E for the multipliers. We shall discuss the analog multipliers used, in the next section.

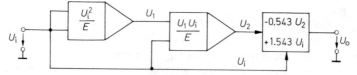

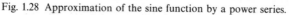

Fig. 1.28 Approximation of the sine function by a power series.

$$U_o \approx \hat{U}_i \sin\left(\frac{\pi}{2} \cdot \frac{U_i}{\hat{U}_i}\right) \quad \text{for } \hat{U}_i = E$$

Differential amplifier stage

Another way of approximating a sine wave is based on the fact that the function $\tanh x$ has a similar shape for small x. This function can be easily formed with the help of a differential amplifier stage, as in Fig. 1.29.

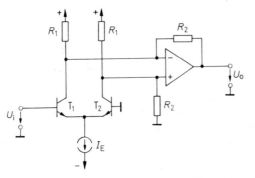

Fig. 1.29 Approximation of the sine function, using a differential amplifier stage.

$$U_0 \approx I_E R_2 \cdot \sin\left(\frac{\pi}{2}\frac{U_i}{\hat{U}_i}\right) \quad \text{for} \ \hat{U}_i = 2.3\,U_T$$

It is shown in Section 1.7.1, that for a differential amplifier, using Eq. (1.29)

$$\frac{I_{C1}}{I_{C2}} = e^{\frac{U_i}{U_T}}$$

and

$$I_{C1} + I_{C2} \approx I_E.$$

Therefore

$$I_{C1} - I_{C2} = \frac{e^{\frac{U_i}{U_T}} - 1}{e^{\frac{U_i}{U_T}} + 1} I_E = I_E \tanh\frac{U_i}{2U_T}. \tag{1.36}$$

The operational amplifier computes the difference between the two collector currents such that

$$U_0 = R_2 (I_{C1} - I_{C2}).$$

It follows that

$$U_0 = I_E R_2 \tanh\frac{U_i}{2U_T}. \tag{1.37}$$

This function can be interpreted as approximating the sine function

$$U_0 = \hat{U}_0 \sin\left(\frac{\pi}{2}\cdot\frac{U_i}{\hat{U}_i}\right) \quad \text{for} \ -\frac{\pi}{2} \leq x \leq \frac{\pi}{2}.$$

The quality of the sine approximation is dependent on the peak value $\hat{U}_i$ chosen. For $\hat{U}_i = 2.8\,U_T \approx 72\,\text{mV}$, the error is smallest and $\hat{U}_0$ is then $\hat{U}_0 = 0.86\,I_E R_2$. The value of the absolute error remains below 3% of the amplitude, and is greatest at the peak values. If two diodes are used to clamp the peaks of the approximating function, the distortion factor is reduced from 1.3% to approx. 0.4%.

Cosine function

The cosine function can be formed for values $0 \leqq x \leqq \pi$ by means of the sine function networks previously described. The input voltage U_i, which should lie between zero and $U_{i\,max}$, is converted to an auxiliary voltage

$$U_1 = U_{i\,max} - 2U_i. \tag{1.38}$$

As can be seen in Fig. 1.30, this equation is already a linear approximation of the cosine function. For the necessary rounding-off of the curve near

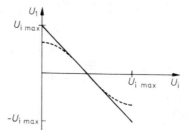

Fig. 1.30 Shape of the auxiliary voltage for the generation of the cosine function (broken line)

the maximum and minimum, one applies U_1 to the input of a sine function network. As is obvious from Fig. 1.31, the simple addition of a summing amplifier is all that is needed for a conversion of a sine function network to a cosine function network.

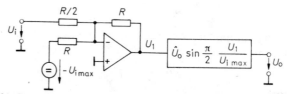

Fig. 1.31 Generation of a cosine function by means of a sine function network.

$$U_o = \hat{U}_o \cos\left(\pi \frac{U_i}{U_{i\,max}}\right) \quad \text{for } 0 \leqq U_i \leqq U_{i\,max}$$

Simultaneous generation of the sine and cosine function
for arguments $-\pi \leqq x \leqq \pi$

With the networks described so far, sine and cosine functions can be generated over a half-period. In cases where the range of argument has to be a full period or more, one initially generates triangular functions as a linear approximation, and uses the described circuits to round off the peaks. The shape of the required triangular voltages is represented in Fig. 1.32.

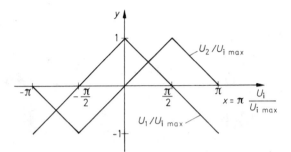

Fig. 1.32 Shape of the auxiliary voltages for the generation of the sine and cosine function for
$$-\pi \leqq x \leqq \pi$$

The voltage U_1 approximates the cosine function. For $U_i > 0$, it is identical to the voltage U_1 in Fig. 1.30. For $U_i < 0$, it is symmetrical about the y-axis. We can therefore use Eq. (1.38), and write, replacing U_i by $|U_i|$,

$$U_1 = U_{i\,max} - 2|U_i|. \tag{1.39}$$

The relationships for the sine function are rather more complicated, since we must differentiate between three cases:

$$U_2 = \begin{cases} -2(U_i + U_{i\,max}) & \text{for} \quad -U_{i\,max} \leqq U_i \leqq -\tfrac{1}{2}U_{i\,max}, & \text{(1.40a)} \\ 2\,U_i & \text{for} \quad -\tfrac{1}{2}U_{i\,max} \leqq U_i \leqq \tfrac{1}{2}U_{i\,max}, & \text{(1.40b)} \\ -2(U_i - U_{i\,max}) & \text{for} \quad \tfrac{1}{2}U_{i\,max} \leqq U_i \leqq U_{i\,max}. & \text{(1.40c)} \end{cases}$$

Such functions are best put into practice by using the general precision function network which is described below.

1.7.5 Variable function network

In Fig. 1.26 a diode network is shown for the piecewise linear approximation of functions. The calculation of the circuit parameters is possible to only a low accuracy because the forward voltage of the diodes and the loading of the voltage divider chains must be considered. Furthermore, the sign of the slope of each linear segment is determined by the structure of the network. Therefore, such a circuit can be optimized for one particular function only, and its parameters cannot be changed easily.

Figure 1.33, on the other hand, represents a circuit that allows the precise setting of the breakpoint and slope of each individual segment with a separate potentiometer. The part of the circuit formed by the operational amplifiers OA 1 and OA 2 permits a segment for positive input voltages to be formed, while the operational amplifiers OA 5 and OA 6 are effective for negative input voltages. The amplifier OA 4 determines the slope

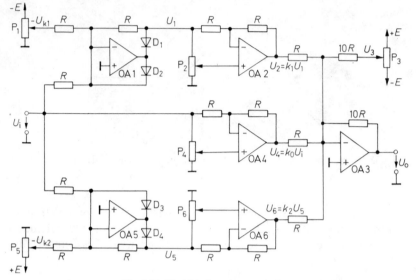

Fig. 1.33 Variable function network

about the origin. The circuit can be extended for any number of segments by adding further sections identical to those mentioned.

The amplifiers OA 2, OA 4 and OA 6 are connected as bipolar-coefficient circuits, as in Fig. 1.5 for $n=1$. Their gain can be adjusted to values between $-1 \leq k \leq +1$ by the corresponding potentiometers; and their output voltages are added by the summing amplifier OA 3. A constant voltage can be added by means of potentiometer P_3.

Near zero input voltage, only amplifier OA 4 contributes to the output voltage:

$$U_4 = k_0 U_i.$$

Both voltages U_1 and U_5 are nil in this case because the diodes D_1 and D_4 are reverse biased, and the amplifiers OA 1 and OA 5 have zero potential at the N-input since the loop is closed by the forward-biased diodes D_2 und D_3.

When the input voltage becomes greater than U_{k1}, diode D_1 is forward biased, and we obtain

$$U_1 = -(U_i - U_{k1}) \quad \text{for} \quad U_i \geq U_{k1} \geq 0.$$

The amplifier OA 1 therefore operates as a half-wave rectifier, with a positive bias voltage U_{k1}. Correspondingly, the operational amplifier OA 5 performs for negative input voltages according to

$$U_5 = -(U_i - U_{k2}) \quad \text{for} \quad U_i \leq U_{k2} \leq 0.$$

Hence, we obtain the general relationship for the slope of the output voltage U_o as

$$m = \frac{\Delta U_o}{\Delta U_i} = 10 \cdot \begin{cases} -k_0 + k_1 + \cdots + k_m & \text{for} \quad U_i > U_{km} > 0 \\ -k_0 + k_1 & \text{for} \quad U_i > U_{k1} > 0 \\ -k_0 & \text{for} \quad U_{k2} < U_i < U_{k1}. \\ -k_0 + k_2 & \text{for} \quad U_i < U_{k2} < 0 \\ -k_0 + k_2 + \cdots + k_n & \text{for} \quad U_i < U_{kn} < 0 \end{cases} \quad (1.41)$$

As an example, we shall demonstrate the implementation of the voltage waveshape $U_2/U_{i\,max}$ in Fig. 1.32. A positive breakpoint at $U_{k1} = \frac{1}{2} U_{i\,max}$ and a negative breakpoint at $U_{k2} = -\frac{1}{2} U_{i\,max}$ are needed. According to Eq. (1.40b) the slope of the segment through the origin must have the value $m = +2$, therefore $k_0 = -0.2$. Above the positive breakpoint, the slope must be -2. For this region, we take from Eq. (1.41)

$$m = 10(-k_0 + k_1)$$

and therefore obtain $k_1 = -0.4$, and correspondingly $k_2 = -0.4$. The shape of the output voltage functions of the operational amplifiers OA 2, OA 4 and OA 6 as resulting from this process, are shown in Fig. 1.34.

Even if no calibrated potentiometers are available, the network output can be given the desired shape in a simple way, using the following procedure. Initially, all breakpoint voltages and slopes are set to their maximum value and the input voltage is made zero. This ensures that $|U_i| < |U_{ki}|$. Only the zeroing potentiometer P_3 affects the output; it is used to adjust the output voltage $U_o(U_i = 0)$ to the desired value. In the next step, U_i is made equal to U_{k1} and P_4 is set so that $U_o(U_i = U_{k1})$ assumes the level required. The factor k_0 is now defined. Then, P_1 is adjusted to a point where the output voltage just begins to change; this happens when the

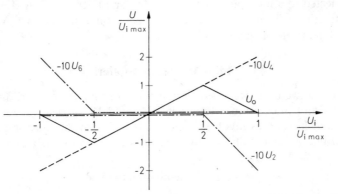

Fig. 1.34 Shape of the voltages in the circuit of Fig. 1.33 when generating the voltage U_2 in Fig. 1.32

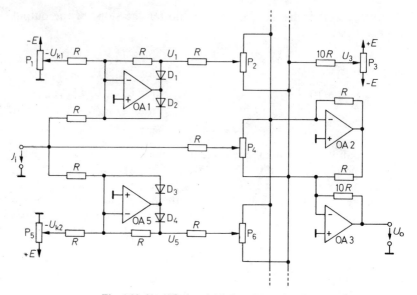

Fig. 1.35 Simplified variable function network

setting of P_1 corresponds to U_{k1}. Now U_i is set to the value of the next breakpoint up (or to the end of the range if there is no higher breakpoint), and P_2 is adjusted so that U_o attains the desired value. In this way, k_1 is defined. The remaining breakpoints and slopes are dealt with in the same manner.

In cases where no calibrated potentiometers are needed for the adjustment of the segment slopes, the circuit may be simplified. One can replace the bipolar coefficient circuits by simple potentiometers that are connected to a multiple-input subtracting amplifier, as in Fig. 1.35. The subtracting amplifier consists of the operational amplifiers OA 2 and OA 3, its principle being that of Fig. 1.2.

1.8   Analog multipliers

So far, we have described circuits for addition, subtraction, differentiation and integration. Multiplication could be carried out only if a constant factor was involved. Below, we deal with the most important principles for the multiplication and division of two variable voltages.

1.8.1   Time-division method

With the time-division method, a rectangular waveform of constant frequency f is generated, the amplitude of which is proportional to one

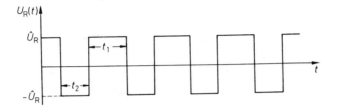

Fig. 1.36 Rectangular voltage of the time-division method

input voltage. The difference between the ON and OFF time, (t_1-t_2), is made proportional to the second input voltage. Figure 1.36 shows the shape of the rectangular voltage.

A lowpass filter is used to evaluate the arithmetic mean of this voltage. If

$$\hat{U}_R = U_1; \quad t_1 - t_2 = k\,U_2 \quad \text{and} \quad t_1 + t_2 = \frac{1}{f},$$

one obtains

$$\overline{U}_R = \frac{1}{t_1 + t_2} \int_0^{t_1+t_2} U_R(t)\,dt = f\left[\int_0^{t_1} U_1\,dt + \int_{t_1}^{t_1+t_2} (-U_1)\,dt\right]$$

$$\overline{U}_R = f\,U_1(t_1 - t_2) = f\,U_1\,k\,U_2 = K\,U_1\,U_2.$$

A problem inherent in this method is the determination of the parameters for the lowpass filter. On the one hand, it has to be designed in such a way that the ripple of the output caused by the rectangular input voltage, is low. On the other hand, the bandwidth of the resulting product should be as large as possible.

Figure 1.37 represents the block diagram of a time-division multiplier. The variable duty cycle is produced by a comparator which compares the input voltage U_2 with the output voltage of a triangular voltage generator. The intervals t_1 and t_2 are shown in Fig. 1.38. With the equation for the triangular voltage

$$U_D(t) = \frac{4\hat{U}_D}{T}t \quad \text{for } 0 \leqq t \leqq \frac{T}{4}$$

we obtain

$$t_2 = 2\left(\frac{T}{4} - \frac{U_2 T}{4\hat{U}_D}\right)$$

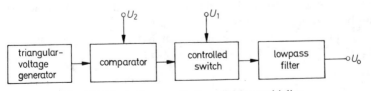

Fig. 1.37 Block diagram of a time-division multiplier

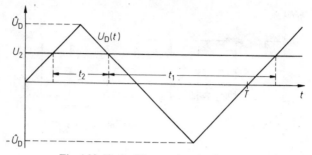

Fig. 1.38 Method for varying the duty cycle

and

$$t_1 - t_2 = \frac{T}{\hat{U}_D} U_2.$$

Hence,

$$\overline{U}_R = f\, U_1 \frac{T}{\hat{U}_D} U_2 = \frac{U_1 U_2}{\hat{U}_D}.$$

The construction of the individual modules is described in the following sections:

Triangular-voltage generator	Section 8.4,
Comparator	Section 7.5,
Controlled switch	Section 7.3.2
Lowpass filter	Chapter 3.

1.8.2   Multipliers with logarithmic function networks

Multiplication and division can be reduced to an addition and subtraction of logarithms:

$$\frac{x\,y}{z} = \exp[\ln x + \ln y - \ln z].$$

The function can be implemented by using three logarithmic amplifiers, one exponential function network and one summing amplifier. However, the problem can be solved in a more elegant way by using the logarithmic function network of Fig. 1.22 and the exponential function network of Fig. 1.24, and by considering the fact that the terminal for the reference voltage can be used as an additional signal input.

If for the logarithmic amplifier in Fig. 1.22, one chooses $R_1 = R_2$, $R_4 = \infty$, $R_3 = 0$, $U_i = U_z$ and $U_{ref} = U_y$, then

$$U_1 = U_{o\,\ln} = - U_T \ln \frac{U_z}{U_y}.$$

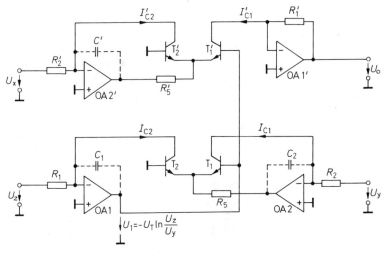

Fig. 1.39 Multiplication by means of logarithms.

$$U_o = \frac{U_x U_y}{U_z} \quad \text{for} \quad U_x, U_y, U_z > 0$$

We apply this voltage to the input of the exponential function network in Fig. 1.24 and in this circuit choose $U_{ref} = U_x$. With the same resistances as for the logarithmic amplifier, we obtain the output voltage

$$U_{o\,exp} = U_x\, e^{\frac{U_1}{U_T}} = \frac{U_x U_y}{U_z}.$$

No additional temperature compensation is necessary because the voltage U_T cancels out.

An inherent disadvantage of this method is that all input voltages must be positive and may not even fall to zero. Such multipliers are called one-quadrant multipliers.

The resulting circuit is shown complete in Fig. 1.39. The lower part represents the logarithmic amplifier and the upper section the exponential function network.

1.8.3   Transconductance multipliers

The transconductance of a transistor is defined as

$$g_m = \frac{dI_C}{dU_{BE}} = \frac{I_C}{U_T},$$

and is therefore proportional to the collector current. The variation of the collector current is then proportional to the product of the input voltage

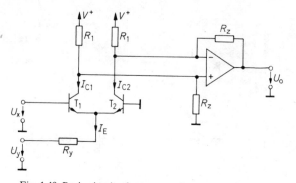

Fig. 1.40 Basic circuit of a transconductance multiplier.

$$U_o \approx \frac{R_z}{R_y} \frac{U_x U_y}{2U_T} \quad \text{for} \ U_y < 0$$

variation and the quiescent collector current. This property is made use of for multiplication, in the differential amplifier stage of Fig. 1.40.

The operational amplifier evaluates the difference between the collector currents:

$$U_o = R_z(I_{C2} - I_{C1}). \tag{1.42}$$

When applying a negative voltage U_y and setting U_x to zero, the currents through both transistors are equal, and the output voltage remains zero. If U_x is made positive, the collector current through T_1 rises and that of T_2 falls; the output voltage is negative. Correspondingly, U_o becomes positive when U_x is negative. The resulting difference in collector currents is greater, the larger the emitter current, i.e. the higher the value $|U_y|$. It can therefore be expected that U_o is at least approximately proportional to $U_x \cdot U_y$. For a more precise calculation we determine the current sharing within the differential amplifier stage. As is shown in Section 1.7.4, Eq. (1.36),

$$I_{C1} - I_{C2} = I_E \tanh \frac{U_x}{2U_T}. \tag{1.43}$$

A power series expansion up to the fourth order gives

$$I_{C1} - I_{C2} = I_E \left(\frac{U_x}{2U_T} - \frac{U_x^3}{24U_T^3} \right). \tag{1.44}$$

Hence,

$$I_{C1} - I_{C2} \approx I_E \cdot \frac{U_x}{2U_T} \quad \text{for} \ |U_x| \ll U_T. \tag{1.45}$$

When $|U_y| \gg U_{BE}$, then

$$I_E \approx -\frac{U_y}{R_y}.$$

Inserting this in Eq. (1.45) gives, in connection with Eq. (1.42), the result

$$U_o \approx \frac{R_z}{R_y} \cdot \frac{U_x U_y}{2 U_T}. \qquad (1.46)$$

As can be seen from Eq.' (1.44), the voltage U_x must be $|U_x| < 0.35 U_T \approx 9$ mV, if the error is not to exceed 1 %. Because of the small value of U_x, the transistors T_1 and T_2 must be closely matched to avoid drift of the offset voltage affecting the result.

For the correct operation of the circuit, it is necessary that U_y is always negative while the voltage U_x may have either sign. Such a multiplier is called a two-quadrant multiplier.

There are several properties of the transconductance multiplier in Fig. 1.40, which can be improved. In the deduction of the output equation (1.46), we had to use the approximation that $|U_y| \gg U_{BE} \approx 0.6$ V. This condition can be dropped if resistor R_y is replaced by a controlled current source for which I_E is proportional to U_y.

A further disadvantage of the circuit in Fig. 1.40 is that $|U_x|$ must be limited to small values in order to keep the error low. This can be avoided by not applying U_x directly, but rather its logarithm.

An expansion to a four-quadrant multiplier, i.e. a multiplier for input voltages of either polarity, is possible if a second differential amplifier stage is connected in parallel, the emitter current of which is controlled by U_y in opposition to that of the first transistor pair.

All these aspects are considered in the four-quadrant transconductance multiplier in Fig. 1.41. The differential amplifier stage T_1, T_2 is that of Fig. 1.40. It is supplemented symmetrically by the differential amplifier T_1', T_2'. The transistors T_5, T_6 form a differential amplifier with internal current feedback [1]. The collectors represent the outputs of two current sources which are controlled by U_y simultaneously but in opposition, as required:

$$I_5 = I_8 + \frac{U_y}{R_y}, \qquad I_6 = I_8 - \frac{U_y}{R_y}. \qquad (1.47)$$

For the difference of the collector currents in the two differential amplifier stages T_1, T_2 and T_1', T_2', we obtain, in analogy to the previous circuit,

$$I_1 - I_2 = I_5 \tanh \frac{U_1}{2 U_T} = \left(I_8 + \frac{U_y}{R_y} \right) \tanh \frac{U_1}{2 U_T}, \qquad (1.48)$$

$$I_1' - I_2' = I_6 \tanh \frac{U_1}{2 U_T} = \left(I_8 - \frac{U_y}{R_y} \right) \tanh \frac{U_1}{2 U_T}. \qquad (1.49)$$

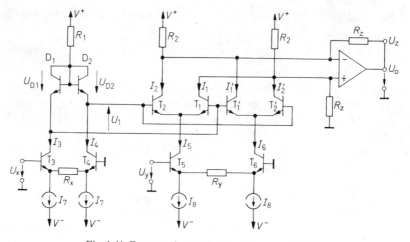

Fig. 1.41 Four-quadrant transconductance multiplier.

$$U_o = \frac{2R_z}{R_x R_y} \cdot \frac{U_x U_y}{I_7} \quad \text{for } I_7 > 0$$

As before, the operational amplifier evaluates the difference of the collector currents according to

$$U_o = R_z \Delta I = R_z (I_2 + I_1' - I_1 - I_2'). \tag{1.50}$$

By subtracting Eq. (1.48) from Eq. (1.49) and substituting, it follows that

$$U_o = -\frac{2R_z U_y}{R_y} \tanh \frac{U_1}{2U_T}, \tag{1.51}$$

where U_y may now have either polarity. When expanding this expression into a series, one can see that the same approximation of the multiplication is involved as for the previous circuit.

We now want to examine the relationship between U_1 and U_x. Two transistors are connected as diodes (transdiodes), D_1 and D_2, and they serve to form the logarithm of the input signals:

$$U_1 = U_{D2} - U_{D1} = U_T \ln \frac{I_4}{I_{ES}} - U_T \ln \frac{I_3}{I_{ES}}.$$

Hence,

$$U_1 = U_T \ln \frac{I_4}{I_3} = U_T \ln \frac{I_7 - \dfrac{U_x}{R_x}}{I_7 + \dfrac{U_x}{R_x}}. \tag{1.52}$$

Substitution in Eq. (1.51) gives the final result

$$U_o = \frac{2R_z}{R_x R_y} \cdot \frac{U_x U_y}{I_7} = \frac{U_x U_y}{E}, \tag{1.53}$$

where $E = R_x R_y I_7 / 2R_z$ is the computing unit. This is usually chosen to be 10 V. Good temperature compensation is attained as U_T cancels out. Equation (1.53) is obtained without recourse to power expansion, and therefore a considerably larger range of input voltages U_x is permissible. The limits of the input range are reached when one of the transistors in the controlled current source is turned off. Therefore,

$$|U_x| < R_x I_7 \quad \text{and} \quad |U_y| < R_y I_8.$$

If the currents I_7 are controlled by a further input voltage U_7, simultaneous division and multiplication is possible. In practice, however, it is difficult to keep both currents I_7 exactly the same over a wide input voltage range.

A more simple way of dividing is to open the connection between U_o and U_z and to link the voltages U_y and U_o instead. Because of the resulting feedback, the output voltage assumes a value so that $\Delta I = U_z / R_z$. Therefore, from Eqs. (1.50) and (1.53),

$$\Delta I = \frac{2U_x U_y}{R_x R_y I_7} = \frac{U_z}{R_z}.$$

Thus the new output voltage is

$$U_o = U_y = \frac{R_x R_y I_7}{2R_z} \cdot \frac{U_z}{U_x} = E \frac{U_z}{U_x}. \tag{1.54}$$

However, stability is only guaranteed when U_x is negative, otherwise the negative feedback becomes positive. The signal U_z, on the other hand, can have either polarity, and therefore the circuit is a two-quadrant divider. The limitation of the sign of the denominator is not a special feature of this arrangement, but is common to all divider circuits.

Transconductance multipliers operating on the principle of that shown in Fig. 1.41, are available as monolithic integrated circuits (e.g. Analog Devices AD 532, Intersil ICL 8013). A 3 dB bandwidth of approximately 1 MHz can be attained.

Transconductance divider with improved accuracy

We have mentioned two methods of division, one using the multiplication based on logarithms (Fig. 1.39) and the other using the transconductance multiplier described above. For a division, a basic problem arises in the region of zero input, as the output voltage is then chiefly determined by the input offset error. This error is particularly large for the transconductance multiplier, since in the input log-amplifier a positive constant (i.e. I_7 in Eq. (1.52)) is added to the input signal to avoid a change of polarity in the argument. The conditions are considerably more favourable if the circuit in Fig. 1.39 is used for the division although only one quadrant is available.

The advantages of the two methods, i.e. two-quadrant division and good accuracy near zero input, can be combined. This is attained not by adding a constant to the argument of the logarithm (to avoid the change in sign) but by adding to the numerator a quantity proportional to the denominator.

The divider output should conform to the expression

$$U_o = E\frac{U_x}{U_z}.$$

Assuming that $U_z > 0$ and $|U_x| < U_z$, two auxiliary voltages

$$U_1 = U_z - \tfrac{1}{2}U_x, \qquad U_2 = U_z + \tfrac{1}{2}U_x, \tag{1.55}$$

can be generated which are always positive. The logarithms of these two voltages are computed according to the block diagram of Fig. 1.42, each by means of the simple logarithmic amplifier in Fig. 1.20. With a differential amplifier stage, as in Fig. 1.40, the hyperbolic tangent of the difference of the output voltages U_3 and U_4 is calculated so that

$$U_o = R_z I_E \tanh\frac{U_T \ln(U_2/U_1)}{2U_T}. \tag{1.56}$$

Therefore, with Eq. (1.55),

$$U_o = \frac{R_z I_E}{2} \cdot \frac{U_x}{U_z}.$$

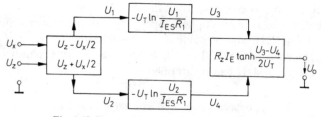

Fig. 1.42 Two-quadrant transconductance divider.

$$U_o = \frac{R_z I_E}{2} \cdot \frac{U_x}{U_z} \quad \text{for } U_z > 0 \text{ and } |U_x| \le U_z$$

With this method, an accuracy of 0.1 % of the computing unit E is accomplished over three decades of the output voltage (e.g. with model 436 from Analog Devices).

1.8.4 Multiplier using electrically isolated couplers

A voltage can be multiplied by a constant, using a simple voltage divider. Analog multiplication is possible if, by employing closed-loop control, one ensures that the constant is proportional to a second input voltage.

The principle of such a circuit is represented in Fig. 1.43. The arrangement contains two identical coefficient elements K_x and K_z, the output voltages of which are proportional to their input voltages. Their constant of proportionality k can be controlled by the voltage U_1. Due to the feedback via K_z, the output voltage U_1 of the operational amplifier assumes a level such that $k U_z = U_y$, this resulting in $k = U_y/U_z$. If the voltage U_x is applied to the second coefficient element K_x, its output voltage becomes

$$U_o = k U_x = \frac{U_x U_y}{U_z}.$$

Voltage U_z must be larger than zero so that the negative feedback does not become positive. The voltages U_x and U_y may have either polarity.

A FET can be employed as an electrically controlled resistor, and this is made use of in the circuit in Fig. 1.44. Amplifier OA1 operates as a controller for the adjustment of the coefficients. Its output voltage causes the resistance R_{DS} to vary so that

$$\frac{\alpha U_z}{R_{DS}} + \frac{U_y}{R_4} = 0.$$

Hence,

$$R_{DS} = -\alpha R_4 \frac{U_z}{U_y}.$$

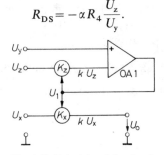

Fig. 1.43 Principle of the circuit.

$$U_o = \frac{U_x U_y}{U_z} \quad \text{for} \quad U_z > 0$$

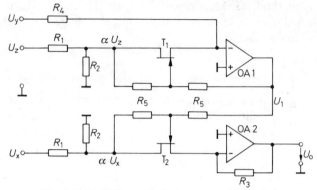

Fig. 1.44 Multiplier with FETs as controlled resistors.

$$U_o = \frac{R_3}{R_4} \frac{U_x U_y}{U_z} \quad \text{for } U_z > 0, \ U_y < 0$$

The output voltage of the operational amplifier OA 2 is

$$U_o = -\alpha \frac{R_3}{R_{DS}} U_x = \frac{R_3}{R_4} \cdot \frac{U_x U_y}{U_z}.$$

In order that FETs may be operated as resistors, the voltage across them must be kept below approx. 0.5 V. The voltage dividers R_1, R_2 have the necessary attenuation effect. The resistors R_5 make the voltage-current characteristic of the FET more linear, as is described in reference [1]. Their resistance must be kept high in comparison to the resistance of the input voltage divider.

Voltage U_z must be positive to ensure negative feedback for the closed-loop circuit. The FETs do not allow the implementation of bipolar coefficients. Therefore, U_y must always be negative so that the closed loop around OA 1 can balance, but voltage U_x may have either sign. In order to accomplish good accuracy, the FETs should be well matched over a wide range of resistances. The types VCR 10N ... VCR 20N from Siliconix are suited for this application.

1.8.5   Adjustment of multipliers

A multiplier should conform to the expression

$$U_o = \frac{U_x U_y}{E},$$

where E is the computing unit, e. g. $E = 10\,\text{V}$. In practice, there is a small offset voltage superposed on any terminal voltage. Therefore, in general

$$U_o + U_{oo} = \frac{1}{E}(U_x + U_{xo})(U_y + U_{yo}).$$

Thus

$$U_o = \frac{U_x U_y}{E} + \frac{U_y U_{xo} + U_x U_{yo} + U_{xo} U_{yo}}{E} - U_{oo}. \tag{1.57}$$

The product $U_x U_y$ must be nil whenever U_x or U_y is nil. This is only possible if the parameters U_{xo}, U_{yo} and U_{oo} become zero independently. Therefore, three trimmers are essential for the compensation of the offset voltages. A suitable trimming procedure is as follows. Firstly, U_x is made zero. Then, according to Eq. (1.57)

$$U_o = \frac{U_y U_{xo} + U_{xo} U_{yo}}{E} - U_{oo}.$$

When varying the voltage U_y, the output voltage also changes because of the term $U_y U_{xo}$. The zero trimmer for U_x is adjusted in such a way that, despite variation of U_y, a constant output voltage is obtained; U_{xo} is then zero.

In a second step, U_y is set to nil and U_x is varied. In the same way as above, the offset of U_y can now be compensated. Thirdly, U_x and U_y are made zero and the third trimmer is adjusted so that the output offset U_{oo} becomes zero.

A fourth trimmer potentiometer may often be necessary for adjusting the constant of proportionality, E, to the desired value.

1.8.6   Expansion of one- and two-quadrant multipliers to four-quadrant multipliers

There are cases where one- and two-quadrant multipliers have to be operated with input voltages of a polarity for which they are not designed. The most obvious remedy would then be to invert the polarity of the input and output of the multiplier whenever the prohibited polarity combination occurs. However, this method involves a large number of components and is also not particularly fast. It is more convenient to add constant voltages U_{xk} and U_{yk} to the input voltages U_x and U_y, so that the resulting input voltages remain within the permitted limits under all conditions. Then, for the output voltage,

$$U_o = \frac{(U_x + U_{xk})(U_y + U_{yk})}{E}.$$

Hence,

$$\frac{U_x U_y}{E} = U_o - \frac{U_{xk}}{E} U_y - \frac{U_{yk}}{E} U_x - \frac{U_{xk} U_{yk}}{E}.$$

It follows that a constant voltage, and also two voltages each proportional to an input voltage, must be subtracted from the output voltage of the

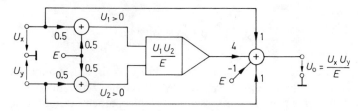

Fig. 1.45 Expansion of a one-quadrant multiplier to a four-quadrant multiplier

multiplier. The circuits necessary for these operations are described in the first sections of this chapter.

The block diagram of the resulting arrangement is represented in Fig. 1.45. The constant voltages and coefficients are chosen so that the range of control is fully exploited. If the input voltage U_x is within $-E \leq U_x \leq +E$, the range for the voltage $U_1 = 0.5 U_x + 0.5 E$ is $0 \leq U_1 \leq E$. Therefore, the output voltage

$$U_o = 4 \frac{\frac{1}{2}(U_x + E) \cdot \frac{1}{2}(U_y + E)}{E} - U_x - U_y - E = \frac{U_x U_y}{E}.$$

1.8.7 Multiplier as a divider or square-rooter

Figure 1.46 illustrates a method by which a multiplier without division input can be used as a divider. Because of negative feedback, the output voltage of the operational amplifier finds a level so that

$$\frac{U_o U_z}{E} = U_x.$$

Thus, the circuit computes the quotient $U_o = E U_x / U_z$, but only as long as $U_z > 0$. For negative denominators, the feedback is positive.

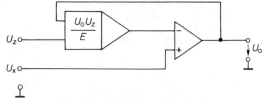

Fig. 1.46 Multiplier used as divider.

$$U_o = E \frac{U_x}{U_z} \quad \text{for } U_z > 0$$

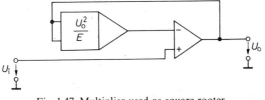

Fig. 1.47 Multiplier used as square-rooter.

$$U_o = \sqrt{E U_i} \quad \text{for} \quad U_i > 0$$

A multiplier can be employed as a square-rooter if operated as a squarer and inserted in the feedback loop of an operational amplifier, as shown in Fig. 1.47. The output voltage finds a level such that

$$\frac{U_o^2}{E} = U_i, \quad \text{hence} \quad U_o = \sqrt{E U_i}.$$

Correct operation is ensured only for positive input and output voltages. Difficulties may arise if the output becomes momentarily negative, e.g. at switching on. In such a case, the squarer causes a phase inversion in the feedback loop so that a positive feedback arises, and the output voltage becomes more negative until it reaches the negative level of output saturation. The circuit is then said to be in "Latch-up" and is inoperable. Therefore, additional measures in the circuit must ensure that the output voltage cannot become negative.

1.9 Transformation of coordinates

Cartesian as well as polar coordinates play an important role in many technical applications. We therefore discuss in this section some circuits which allow transformation from one coordinate system to the other.

1.9.1 Transformation from polar to Cartesian coordinates

In order to carry out the transformation equations

$$x = r \cos \varphi, \quad y = r \sin \varphi \tag{1.58}$$

by means of an analog computing circuit, the coordinates must be expressed as voltages. We let

$$\varphi = \pi \frac{U_\varphi}{E} \quad \text{for} \quad -E \leq U_\varphi \leq +E.$$

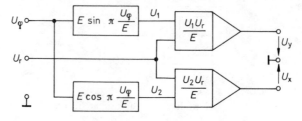

Fig. 1.48 Transformation from polar to Cartesian coordinates.

$$U_x = U_r \cos\left(\pi \frac{U_\varphi}{E}\right); \qquad U_y = U_r \sin\left(\pi \frac{U_\varphi}{E}\right)$$

The range of argument is therefore defined between $\pm\pi$. We define the remaining coordinates as

$$x = \frac{U_x}{E}; \qquad y = \frac{U_y}{E}; \qquad r = \frac{U_r}{E}.$$

Thus, Eq. (1.58) can be rewritten

$$U_x = U_r \cos\left(\pi \frac{U_\varphi}{E}\right), \qquad U_y = U_r \sin\left(\pi \frac{U_\varphi}{E}\right). \tag{1.59}$$

For the generation of the sine and cosine functions for the range of argument $\pm\pi$, we employ the network described in Section 1.7.4 and, in addition, two multipliers. The complete circuit is represented in the block diagram of Fig. 1.48.

1.9.2   Transformation from Cartesian to polar coordinates

The inversion of the transformation equation (1.58) yields

$$r = \sqrt{x^2 + y^2} \quad \text{or} \quad U_r = \sqrt{U_x^2 + U_y^2}, \tag{1.60}$$

$$\varphi = \tan^{-1} \frac{y}{x} \quad \text{or} \quad U_\varphi = \frac{E}{\pi} \tan^{-1} \frac{U_y}{U_x} \tag{1.61}$$

respectively. The magnitude, U_r, of the vector can be computed according to the block diagram in Fig. 1.49, using two squarers and one square-rooter. A more simple circuit which also has a larger range of input voltage, can be deduced by applying a few more mathematical operations. From Eq. (1.60)

$$U_r^2 - U_y^2 = U_x^2,$$
$$(U_r - U_y)(U_r + U_y) = U_x^2.$$

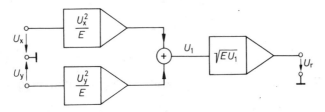

Fig. 1.49 Computation of the magnitude of a vector (vector voltmeter).

$$U_r = \sqrt{U_x^2 + U_y^2}$$

Hence,

$$U_r = \frac{U_x^2}{U_r + U_y} + U_y.$$

The implicit equation for U_r can be put into practice by means of a multiplier with division input, as in Fig. 1.50. The summing amplifier S_1 computes the expression

$$U_1 = U_r + U_y,$$

and therefore

$$U_2 = \frac{U_x^2}{U_r + U_y}.$$

In order to obtain U_r, this voltage U_2 is added to the input voltage U_y using the summing amplifier S_2.

According to Eq. (1.61), the computation of the angle is possible with the aid of a divider and a function network for the $\tan^{-1}$.

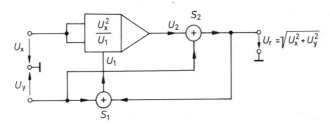

Fig. 1.50 Simplified circuit for the computation of the vector magnitude.

$$U_r = \sqrt{U_x^2 + U_y^2}$$

2 Controlled sources and impedance converters

In linear network synthesis, not only passive components are used, but also idealized active elements such as controlled current and voltage sources. In addition, idealized converter circuitry such as the negative impedance converter (NIC), the gyrator and the circulator is often employed. In the following sections we describe the most important ways of implementing these circuits.

2.1 Voltage-controlled voltage sources

A voltage-controlled voltage source is characterized by having an output voltage U_2 proportional to the input voltage U_1. It is therefore nothing more than a voltage amplifier. Ideally, the output voltage should be independent of the output current and the input current should be nil. Hence, the transfer characteristics are

$$I_1 = 0 \cdot U_1 + 0 \cdot I_2 = 0,$$
$$U_2 = A_v U_1 + 0 \cdot I_2 = A_v U_1.$$

In practice, the ideal source can only be approximated. Considering that the reaction of the output on the input is usually negligibly small, the equivalent circuit of a real source is that in Fig. 2.1, the transfer characteristics of which are

$$I_1 = \frac{1}{r_i} U_1 + 0 \cdot I_2,$$
$$U_2 = A_v U_1 - r_o I_2. \tag{2.1}$$

The internal voltage source shown must be assumed to be ideal. The input resistance is r_i, and the output resistance is r_o.

Voltage-controlled voltage sources of low output resistance and of defined, but adjustable gain are well known in the form of inverting

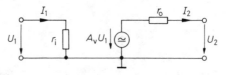

Fig. 2.1 Low-frequency equivalent circuit of a voltage-controlled voltage source

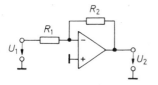

Fig. 2.2 Inverting amplifier as voltage-controlled voltage source.	Fig. 2.3 Non-inverting amplifier as voltage-controlled voltage source.

Ideal transfer characteristic:

$$U_2 = -\frac{R_2}{R_1} U_1$$

Input impedance: $\underline{Z}_i = R_1$

Ideal transfer characteristic:

$$U_2 = \left(1 + \frac{R_2}{R_1}\right) U_1$$

Input impedance: $\underline{Z}_i = r_{CM} \left\| \dfrac{1}{j\omega C_{CM}}\right.$

where r_{CM} is the common-mode input resistance of the operational amplifier

Output impedance: $\underline{Z}_0 = \dfrac{r_0}{\underline{g}}$

Output impedance: $\underline{Z}_0 = \dfrac{r_0}{\underline{g}}$

amplifiers and non-inverting amplifiers. They are shown for completeness in Figs. 2.2 and 2.3. It is easy to obtain output resistances of far less than $1\,\Omega$ and therefore to approach the ideal behaviour fairly closely. It should be noted, however, that the output impedance is somewhat inductive, i.e. it rises with increasing frequency.

The input resistance of a non-inverting amplifier is very high. At low frequencies, one easily attains values in the $G\Omega$-range and hence very nearly ideal conditions. The high (incremental) input resistance must not be reason for overlooking the additional errors that may arise due to the constant input bias current I_B, particularly when the output resistance of the signal source is high. In critical cases an amplifier with FET input should be employed.

For low-resistance signal sources, the inverting amplifier circuit in Fig. 2.2 can be used, as its low input resistance R_1 then causes no error. The advantage is that no inaccuracies due to common-mode signals can arise.

2.2 Current-controlled voltage sources

The equivalent circuit of a current-controlled voltage source, represented in Fig. 2.4, is identical to that of the voltage-controlled voltage source in Fig. 2.1. The only difference between the two sources is that the input current is now the controlling signal which should be influenced by the circuit as little as possible, a condition fulfilled in the

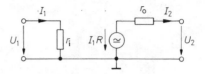

Fig. 2.4 Low-frequency equivalent circuit of a current-controlled voltage source

ideal case when $r_i=0$. The transfer characteristics, when neglecting the effect of the output on the input, are

$$U_1 = r_i I_1 + 0 \cdot I_2 \qquad U_1 = 0$$
$$U_2 = R I_1 - r_o I_2 \quad \Rightarrow \quad U_2 = R I_1 \qquad (2.2)$$
$$\text{(real)} \qquad\qquad \text{(ideal, } r_i = r_o = 0\text{).}$$

When implementing this circuit as in Fig. 2.5, we use the fact that the summing point of an inverting amplifier represents virtual ground. Because of this, the input resistance is low, as required. The output voltage becomes $U_2 = -RI_1$ if the input bias current of the amplifier can be neglected as against I_1. If very small currents I_1 are to be used as control signals, an amplifier with FET input should be employed. Errors may arise due to the offset voltage and these will increase if the output resistance R_g of the signal source is reduced, since the offset voltage is amplified by a factor of $(1 + R/R_g)$.

For the output impedance of the circuit, the same conditions hold as for the previous circuit where the loop gain g is dependent on the output resistance R_g of the signal source and is

$$g = k \underline{A}_D = \frac{R_g}{R + R_g} \underline{A}_D.$$

A current-controlled voltage source with a floating input is discussed in Chapter 15.2.1.

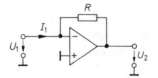

Fig. 2.5 Current-controlled voltage source.

Ideal transfer characteristic: $U_2 = -RI_1$

Input impedance: $\underline{Z}_i = \dfrac{R}{\underline{A}_D}$

Output impedance: $\underline{Z}_o = \dfrac{r_o}{g}$

2.3 Voltage-controlled current sources

The purpose of voltage-controlled current sources is to impress on a load a current I_2 which is independent of the output voltage U_2 and is determined only by the control voltage U_1. Therefore

$$I_1 = 0 \cdot U_1 + 0 \cdot U_2,$$
$$I_2 = g_f U_1 + 0 \cdot U_2.$$
(2.3)

In practice, these conditions can be fulfilled only approximately. Taking into account that the effect of the output on the input is normally very small indeed, the equivalent circuit of a real current source becomes that in Fig. 2.6. The transfer characteristics are then

$$I_1 = \frac{1}{r_i} U_1 + 0 \cdot U_2,$$
$$I_2 = g_f U_1 - \frac{1}{r_o} U_2.$$
(2.4)

For $r_i \to \infty$ and $r_o \to \infty$, one obtains the ideal current source. The parameter g_{fs} is known as forward transconductance or transfer conductance.

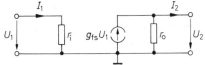

Fig. 2.6 Low-frequency equivalent circuit of a voltage-controlled current source

2.3.1 Current sources for floating loads

In inverting and non-inverting amplifiers, the current through the feedback resistor is $I_2 = U_1/R_1$ and is therefore independent of the voltage across the feedback resistor. Both circuits can thus be used as current sources if the load R_L is inserted in place of the feedback resistor, as shown in Figs. 2.7 and 2.8.

For the input impedance the same conditions obtain as for the corresponding voltage-controlled voltage sources in Figs. 2.2 and 2.3.

For a finite open-loop gain A_D of the operational amplifier, the output resistance assumes finite values only, as the potential difference $U_D = V_P - V_N$ does not remain exactly at zero. For the determination of the output resistance, we take the following relationships from Fig. 2.7

$$I_1 = I_2 = \frac{U_1 - V_N}{R_1}, \quad V_N = -\frac{V_o}{A_D}, \quad U_2 = V_N - V_o$$

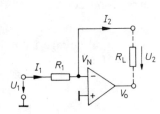

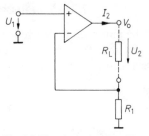

Fig. 2.7 Inverting amplifier
as voltage-controlled current source.

Fig. 2.8 Non-inverting amplifier
as voltage-controlled current source.

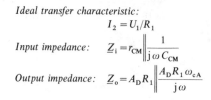

Ideal transfer characteristic:
$$I_2 = U_1/R_1$$

Ideal transfer characteristic:
$$I_2 = U_1/R_1$$

Input impedance: $\underline{Z}_i = R_1$

Input impedance: $\underline{Z}_i = r_{CM} \left\| \dfrac{1}{j\omega C_{CM}} \right.$

Output impedance: $\underline{Z}_o = A_D R_1 \left\| \dfrac{A_D R_1 \omega_{cA}}{j\omega} \right.$

Output impedance: $\underline{Z}_o = A_D R_1 \left\| \dfrac{A_D R_1 \omega_{cA}}{j\omega} \right.$

and obtain

$$I_2 = \frac{U_1}{R_1} - \frac{U_2}{R_1(1+A_D)} \approx \frac{U_1}{R_1} - \frac{U_2}{A_D R_1},$$

and therefore for the output resistance

$$r_o = -\frac{\partial U_2}{\partial I_2} = A_D R_1. \tag{2.5}$$

The output resistance is thus proportional to the differential gain of the operational amplifier.

Since the open-loop gain A_D of a frequency-compensated operational amplifier possesses a fairly low cutoff frequency (e.g. $f_{cA} \approx 10\,\text{Hz}$ for the 741 type), one must take into account that A_D is complex even at low frequencies. Equation (2.5) must then be rewritten and is in its complex form

$$\underline{Z}_o = \underline{A}_D R_1 = \frac{A_D}{1 + j\dfrac{\omega}{\omega_{cA}}} R_1. \tag{2.6}$$

This output impedance may be represented by a parallel connection of the resistor R_o and a capacitor C_o, as the following rearrangement of Eq. (2.6) shows:

$$\underline{Z}_o = \frac{1}{\dfrac{1}{A_D R_1} + \dfrac{j\omega}{A_D R_1 \omega_{cA}}} = R_o \left\| \dfrac{1}{j\omega C_o} \right., \tag{2.7}$$

where $R_o = A_D R_1$ and $C_o = \dfrac{1}{A_D R_1 \omega_{cA}}$. With an operational amplifier having $A_D = 10^5$ and $f_{cA} = 10\,\text{Hz}$, one obtains for $R_1 = 1\,\text{k}\Omega$

$$R_o = 100\,\text{M}\Omega \quad \text{and} \quad C_o = 159\,\text{pF}.$$

For a frequency of $10\,\text{kHz}$, the value of the output resistance $|\underline{Z}_o|$ is reduced to $100\,\text{k}\Omega$.

The same considerations apply for the output impedance of the circuit in Fig. 2.8.

As far as their electrical properties are concerned, the two current sources in Fig. 2.7 and Fig. 2.8 are well suited for many applications. However, they have a great technical disadvantage: The load R_L must be floating, i.e. it must not be connected to a fixed potential, otherwise the amplifier output or the N-input is short-circuited. The following circuits overcome this restriction.

2.3.2   Current sources for grounded loads

The principle of the current source in Fig. 2.9 is based on the fact that the output current is monitored by the voltage across R_1. The output voltage of the operational amplifier finds a value so that this voltage is equal to a given input voltage. In order to determine the output current, we apply KCL to the N-input, the P-input and to the output. Thus

$$\frac{V_o - V_N}{R_2} - \frac{V_N}{R_3} = 0,$$

$$\frac{U_1 - V_P}{R_2} + \frac{U_2 - V_P}{R_2} = 0,$$

$$\frac{V_o - U_2}{R_1} + \frac{V_P - U_2}{R_2} - I_2 = 0.$$

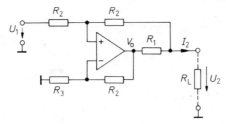

Fig. 2.9 Voltage-controlled current source for grounded loads.

Output current: $I_o = \dfrac{U_1}{R_1 \| R_2}$ for $R_3 = \dfrac{R_2^2}{R_1 + R_2}$

Since $V_N = V_P$, we obtain the output current as

$$I_2 = \left(\frac{1}{2R_2} + \frac{R_2+R_3}{2R_1R_3} \right) U_1 + \left(\frac{R_2+R_3}{2R_1R_3} - \frac{R_1+2R_2}{2R_1R_2} \right) U_2.$$

For one value of R_3, the output current becomes independent of the output voltage; this is the case when the second term is nil. The condition for R_3 is then

$$R_3 = \frac{R_2^2}{R_1+R_2}.$$

For this value the output current is

$$I_2 = \frac{U_1}{R_1 \| R_2}.$$

In practice, the resistance R_1 is made low so that the voltage across it remains in the order of a few volts. The resistance R_2 is chosen to be large in comparison with R_1, so that the operational amplifier and the voltage source U_1 are not unnecessarily loaded. On condition that $R_2 \gg R_1$, we approximate

$$R_3 \approx R_2 \quad \text{and} \quad I_2 \approx \frac{U_1}{R_1}.$$

The output resistance of the current source can, at low frequencies, be adjusted to infinity even for a real operation amplifier, by slightly varying R_3. However, there is a disadvantage in that the output resistance R_g of the controlling voltage source U_1 influences the adjustment. This is because it is connected in series with the resistor R_2. In addition, the current supplied by the control-voltage source is dependent on the load resistance. Hence, if R_g is load dependent, as for Zener diodes, a general adjustment is not possible.

　　The circuit in Fig. 2.10 is more favourable in this respect, as the resistor R_2 is connected to virtual ground. A further advantage is that there is no common-mode voltage.

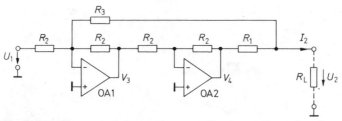

Fig. 2.10 Voltage-controlled current source without common-mode voltage.

Output current: $\quad I_2 = \dfrac{U_1}{R_1} \quad$ for $\quad R_3 = R_2 - R_1$

To determine the output current, we use KVL for the circuit and obtain

$$V_4 = -V_3 = U_1 + \frac{R_2}{R_3}U_2.$$

The application of KCL to the output gives

$$\frac{V_4 - U_2}{R_1} - \frac{U_2}{R_3} - I_2 = 0.$$

Eliminating V_4, we obtain

$$I_2 = \frac{U_1}{R_1} + \frac{R_2 - R_3 - R_1}{R_1 R_3}U_2.$$

The output current becomes independent of the output voltage when the condition

$$R_3 = R_2 - R_1$$

is fulfilled.

2.3.3 Precision current sources using transistors

Simple single-ended current sources employ a bipolar transistor or a field-effect transistor (FET) to feed loads having one terminal connected to a constant potential. Such circuits are described elsewhere [1]. Their disadvantage is that the output current is influenced by U_{BE} or U_{GS}, respectively, and therefore cannot be precisely defined. An operational amplifier can be used to eliminate this influence. Figure 2.11 shows such circuits for a bipolar transistor and for a field-

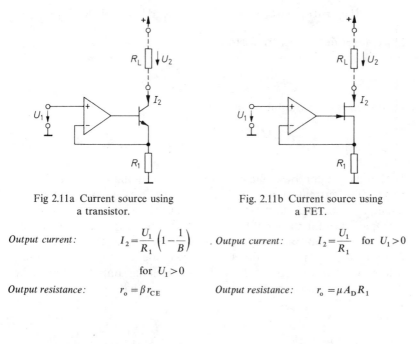

Fig 2.11a Current source using a transistor.

Fig. 2.11b Current source using a FET.

Output current: $\quad I_2 = \frac{U_1}{R_1}\left(1 - \frac{1}{B}\right)$ $\qquad$ *Output current:* $\quad I_2 = \frac{U_1}{R_1}$ for $U_1 > 0$

$\qquad\qquad\qquad\qquad$ for $U_1 > 0$

Output resistance: $\quad r_o = \beta r_{CE}$ $\qquad$ *Output resistance:* $\quad r_o = \mu A_D R_1$

effect transistor. The output voltage of the operational amplifier finds a value such that the voltage across the resistance R_1 equals U_1. (Obviously, this holds for positive voltages only, as otherwise the transistors are OFF). Since the current through R_1 is U_1/R_1, the load current is

for the bipolar transistor: $$I_2 = \frac{U_1}{R_1}\left(1 - \frac{1}{B}\right),$$

and for the FET: $$I_2 = \frac{U_1}{R_1}.$$

The difference between these currents is due to the fact that in the bipolar transistor, part of the emitter current flows through the base. As the current transfer ratio B is dependent on U_{CE}, the current I_B also changes with the output voltage U_2. It is shown elsewhere [1] that this effect limits the output resistance to the value $\beta \cdot r_{CE}$, even if the operational amplifier is assumed to be ideal.

The influence of the finite current transfer ratio can be reduced if the bipolar transistor is replaced by a Darlington circuit. It can be virtually eliminated by using a FET because the gate current is extremely small. The output resistance of the circuit in Fig. 2.11b is limited only by the finite gain of the operational amplifier. It can be determined by the following relationships obtained directly from the circuit for $U_1 = \text{const.}$

$$dU_{DS} \approx -dU_2,$$

$$dU_{GS} = dU_G - dU_S = -A_D R_1 dI_2 - R_1 dI_2 \approx -A_D R_1 dI_2.$$

Since, for a field-effect transistor,

$$dI_2 = g_{fs} dU_{GS} + \frac{1}{r_{DS}} dU_{DS},$$

we obtain for the output resistance

$$r_o = -\frac{dU_2}{dI_2} = r_{DS}(1 + A_D g_{fs} R_1) \approx \mu A_D R_1. \tag{2.8}$$

It is thus greater by a factor $\mu = g_{fs} \cdot r_{DS} \approx 150$ than that of the corresponding current source in Fig. 2.8 which uses an operational amplifier but no field-effect transistor. Using the same values as in the example given for the circuit in Fig. 2.8, we obtain the very high output resistance of approximately $15\,\text{G}\Omega$. Because of the frequency dependence of the open-loop gain $\underline{A}_D$, this value holds only for frequencies below the cutoff frequency f_{cA} of the operational amplifier. For higher frequencies, we have to take into consideration that the differential gain is complex and obtain, instead of Eq. (2.8), the output impedance

$$Z_0 = A_\mathrm{D} \mu R_1 = \frac{A_\mathrm{D}}{1 + \mathrm{j}\,\dfrac{\omega}{\omega_{\mathrm{c}\,\mathrm{A}}}} \mu R_1. \qquad (2.9)$$

A comparison with Eqs. (2.6) and (2.7) shows that this impedance is equivalent to a parallel connection of a resistor $R_0 = \mu A_\mathrm{D} R_1$ and a capacitor $C_0 = 1/\mu A_\mathrm{D} R_1 \omega_{\mathrm{c}\,\mathrm{A}}$. For the practical example mentioned, we obtain $C_0 = 1\,\mathrm{pF}$. The capacitance of the FET of the order of a few pF will appear in parallel.

When larger output currents are needed, a power FET may be used or the field-effect transistor may be replaced by a Darlington pair, consisting of a FET and a power transistor, as shown in Fig. 2.12. This measure does not change the properties of the circuit since, as before, no current flows into the gate.

The circuit in Fig. 2.11b can be modified by connecting the input voltage directly to R_1 and by grounding the P-input terminal. Figure 2.13 shows this version. U_1 must always be negative to ensure that the FET is not turned off. In contrast to the circuit in Fig. 2.11b, the control voltage source is loaded by the current $I_1 = -I_2$.

When a current source is needed, the output current of which flows in the opposite direction to that in the circuit of Fig. 2.11b, the n-channel FET is simply replaced by a p-channel FET, this being shown in the circuit of Fig. 2.14. If no p-channel FET is available, the

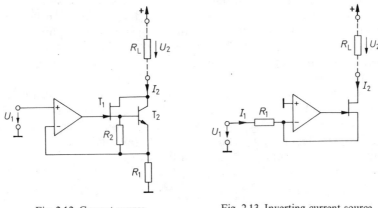

Fig. 2.12 Current source
for large output currents.

Fig. 2.13 Inverting current source
using a FET.

Output current:	$I_2 = \dfrac{U_1}{R_1}$ for $U_1 > 0$	*Output current:*	$I_2 = -\dfrac{U_1}{R_1}$ for $U_1 < 0$
Output resistance:	$r_0 = \mu_1 A_\mathrm{D} R_1$	*Output resistance:*	$r_0 = \mu A_\mathrm{D} R_1$

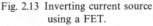

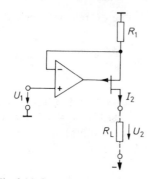

Fig. 2.14 Current source using a
p-channel FET.

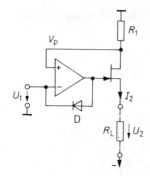

Fig. 2.15 Current source with a
quasi-p-channel FET.

Output current: $I_2 = -\dfrac{U_1}{R_1}$ for $U_1 < 0$ *Output current:* $I_2 = -\dfrac{U_1}{R_1}$ for $U_1 < 0$

Output resistance: $r_o = \mu A_D R_1$ *Output resistance:* $r_o = A_D R_1$

arrangement in Fig. 2.15 can also be used. Contrary to the previous circuits, the source terminal here serves as the output. However, this does not influence the output current, as it is controlled, as before, by the voltage across R_1. Negative feedback is established in the following manner. If the output current falls, V_p rises, resulting in an amplified increase of the gate potential, and U_{GS} is therefore reduced. This counteracts the reduction in current. When compared with those of the previous circuits, the output resistance is considerably smaller.

If, because of too large an input signal the gate junction becomes forward biased, the output voltage of the operational amplifier reacts directly on the P-input. Positive feedback then occurs, and the output reaches positive saturation. To avoid this latch-up, a diode D is inserted as in Fig. 2.15.

One disadvantage of all current sources previously described is that they can supply only unidirectional output currents. By combining the circuits of Figs. 2.11 and 2.14 one arrives at the current source in Fig. 2.16 which can supply currents of either polarity. For zero control voltage, $V_{P1} = \frac{3}{4}V^+$ and $V_{P2} = \frac{3}{4}V^-$. In this case,

$$I_2 = I_{D1} - I_{D2} = \frac{V^+}{4R_1} + \frac{V^-}{4R_1} = 0 \quad \text{for } V^+ = -V^-.$$

For positive input voltages U_1, current I_{D2} increases by $U_1/4R_1$, whereas I_{D1} decreases by the same amount. Therefore, one obtains a negative output current

$$I_2 = -\frac{U_1}{2R_1}.$$

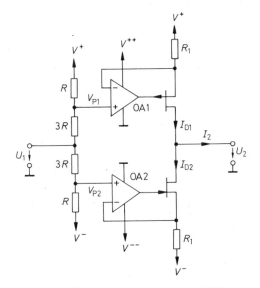

Fig. 2.16 Bipolar-current source using FETs.

$$\text{Output current:}\quad I_2 = -\frac{U_1}{2R_1}$$

For negative input voltages, I_{D2} decreases while I_{D1} becomes larger, this resulting in a positive output current. The limit for the control voltage is reached when one of the FETs is turned off. This is the case for $U_1 = \pm V^+$. In order to turn off the FETs, the absolute value of the gate potential must be higher than the supply voltage V^+. Therefore, the operational amplifier OA1 and OA2 need higher supply voltages, and these are labelled in Fig. 2.16 as V^{++} or V^{--}.

The circuit has a rather poor zero stability. This is because the output current is itself the difference between the two relatively large currents, I_{D1} and I_{D2}, which are influenced by changes in the supply voltages.

The circuit in Fig. 2.17 is considerably better in this respect. It differs from the former circuit in that it uses a different kind of control [2.1]. The two output stages are controlled by the currents I_3 and I_4, flowing in the supply terminals of the amplifier OA1. For the drain currents

$$I_{D1} = \frac{U_3}{R_1} = \frac{R_2}{R_1} I_3,$$

$$I_{D2} = \frac{U_4}{R_1} = \frac{R_2}{R_1} I_4.$$

(2.10)

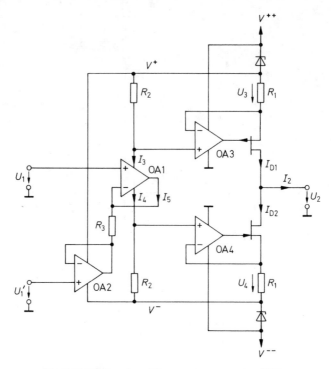

Fig. 2.17 Bipolar class-AB current source using FETs.

Output current: $I_2 = \dfrac{R_2}{R_1 R_3}(U_1 - U_1')$

Hence, the output current

$$I_2 = I_{D1} - I_{D2} = \frac{R_2}{R_1}(I_3 - I_4).$$ (2.11)

The two amplifiers OA1 and OA2 are connected as voltage followers. Therefore, the voltage across resistor R_3 is equivalent to the input voltage difference $U_1 - U_1'$. Thus, the output current of OA1 is

$$I_5 = \frac{U_1 - U_1'}{R_3}.$$ (2.12)

For further processing of this signal, use is made of the fact that KCL holds for any node in a circuit, and that an operational amplifier can be thought of as such a node if its supply currents are included in the analysis. As the input currents are negligible and as there is usually no ground connection, the following relationship holds with very good accuracy:

$$I_5 = I_3 - I_4.$$ (2.13)

Substitution in Eqs. (2.12) and (2.11) yields the output current

$$I_2 = \frac{R_2}{R_1 R_3}(U_1 - U_1').$$
(2.14)

If no differential input is required, amplifier OA 2 may be omitted and resistor R_3 connected to ground instead.

For zero input, $I_5 = 0$ and $I_3 = I_4 = I_Q$, where I_Q is the quiescent current flowing in the supply leads of the amplifier OA 1. It is small in comparison with the maximum possible output current, I_5, of the amplifier. For a positive differential input voltage, $I_3 \approx I_5 \gg I_4$. The output current I_2 is then supplied practically only from the upper output stage whereas the lower one is not in operation. The inverse is true for a negative input voltage difference. The circuit is therefore of the class-AB push-pull type. Since the quiescent current in the output stage is

$$I_{D1Q} = I_{D2Q} = \frac{R_2}{R_1} I_Q$$

and is small relative to the maximum output current, the output current at zero input signal is now determined by the difference of small quantities. This results in a very good zero-current stability. A further advantage is that of high efficiency, this being of special interest if the circuit is to be designed for high output currents.

The quiescent current can be made adjustable, if OA 1 is replaced by an amplifier (e.g. type μA 776), the quiescent current of which can be controlled by an external resistor. The quiescent current is chosen so that no crossover distortion occurs, even for higher frequencies.

It is possible to connect, as for OA 1, an additional output stage to the supply voltage terminals of amplifier OA 2. In this way, the currents I_2 and $-I_2$ are available simultaneously, and the circuit can be used as a floating current source, as will be shown in the next section.

2.3.4 Floating current sources

In the previous sections we mention two types of current sources. Neither load terminal in the circuits of Figs. 2.7 and 2.8 may be connected to a fixed potential. Such a load is called off-ground or floating and is illustrated in Fig. 2.18a. For this kind of operation, the load may in practice consist only of passive elements, as for active loads, there is normally a connection to ground via the supply.

Grounded loads can be supplied by a single-ended current source based on the principles in Fig. 2.18b. Its practical design is shown in Figs. 2.9 to 2.17.

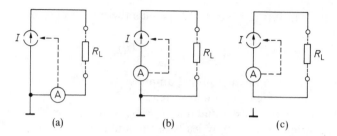

Fig. 2.18 (a) Current source for floating loads. (b) Current source for grounded loads, (single-ended current source). (c) Floating current source for any load

If one or the other load terminal is to be connected to any desired potential without the current being affected, a floating current source is needed. It can be constructed using two grounded current sources which supply equal but opposite currents, as shown in Fig. 2.19.

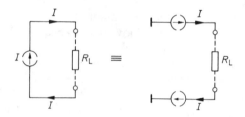

Fig. 2.19 Realization of a floating current source by two single-ended current sources

2.4   Current-controlled current sources

The equivalent circuit of the current-controlled current source is identical to that of the voltage-controlled current source in Fig. 2.6. The only difference is that the input current is now the controlling signal and should be influenced by the circuit as little as possible. This is so for the ideal condition where $r_i = 0$. When the effect of the output on the input is neglected, the transfer characteristics are

$$U_1 = r_i I_1 + 0 \cdot U_2 \qquad\qquad U_1 = 0$$
$$\qquad\qquad\qquad\qquad\qquad \Rightarrow$$
$$I_2 = A_1 I_1 - \frac{1}{r_o} \cdot U_2 \qquad\qquad I_2 = A_1 I_1 \qquad\qquad (2.15)$$

$$\text{(real)} \qquad\qquad\qquad\qquad \text{(ideal, } r_i = 0, \ r_o = \infty\text{).}$$

In Figs 2.7 and 2.13, we show two voltage-controlled current sources of finite input resistance. They can be operated as current-controlled

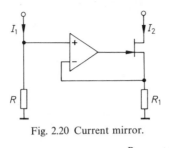

Fig. 2.20 Current mirror.

$$\textit{Output current:} \quad I_2 = \frac{R}{R_1} I_1$$

current sources having virtually ideal characteristics if the resistor R_1 is made zero; then $I_2 = I_1$.

Current-controlled current sources allowing reversal of the output currents are of particular interest. They are called current mirrors, and one example is shown in Fig. 2.20. It is based on the voltage-controlled current source in Fig. 2.11b, and the current-to-voltage conversion is effected by the additional resistor R. However, this results in non-ideal conditions for the input resistance.

The maximum freedom in specifying the circuit parameters is attained if a circuit from Section 2.2 is used for the current-to-voltage conversion and when one of the voltage-controlled current sources described is connected in series.

2.5 NIC (negative impedance converter)

There are cases when negative resistances or voltage sources having negative internal resistances are needed. By definition, the resistance $R = + U/I$, if the arrows for current and voltage have the same direction. When the voltage U on and the current I through a two-terminal network then have opposite signs, the quotient U/I becomes negative. Such a network is said to have a negative resistance. Negative resistances can, in principle, be realized only by the active circuits known as NIC's. There are two types: the UNIC which reverses the polarity of the voltage without affecting the direction of the current, and the INIC which reverses the current without changing the polarity of the voltage. The implementation of the INIC is particularly simple. Its ideal transfer characteristics are

$$U_1 = U_2 + 0 \cdot I_2, \qquad I_1 = 0 \cdot U_2 - I_2. \tag{2.16}$$

These equations can be implemented as in Fig. 2.21, by a voltage-controlled voltage source and a current-controlled current source.

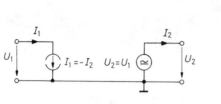

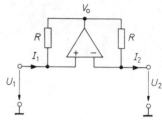

Fig. 2.21 Circuit of an INIC
using controlled sources

Fig. 2.22 INIC using a single
amplifier

However, both functions can be carried out by a single operational
amplifier, as represented in Fig. 2.22.

For the ideal operational amplifier, $V_P = V_N$, and therefore $U_1 = U_2$,
as required. The output potential of the amplifier has the value

$$V_o = U_2 + I_2 R.$$

Hence, the current at port 1 is, as required,

$$I_1 = \frac{V_o - U_2}{R} = -I_2.$$

For this deduction we have tacitly assumed stability of the circuit.
However, since it simultaneously employs positive and negative feedback, the validity of this assumption must be examined separately. To
do so, we determine what proportion of the output voltage affects the
P-input and N-input, respectively. Figure 2.23 shows the INIC in a
general application, where R_1 and R_2 are the internal resistances of the
circuits connected to it. The feedback of the voltage

$$V_P = V_o \frac{R_1}{R_1 + R} \quad \text{is positive,}$$

and that of

$$V_N = V_o \frac{R_2}{R_2 + R} \quad \text{is negative.}$$

The circuit is stable if the positive-feedback voltage V_P is smaller than
V_N, i.e. if

$$R_1 < R_2.$$

The circuit in Fig. 2.24 illustrates the use of the INIC for the simulation of negative resistances. When applying a positive voltage to port 1,

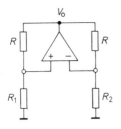

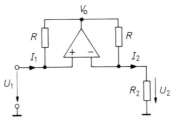

Fig. 2.23 INIC in a general
application

Fig. 2.24 Simulation of negative
resistances.

Negative resistance: $\dfrac{U_1}{I_1} = -R_2$

$U_2 = U_1$ becomes positive according to Eq. (2.16), and hence I_2 is also positive. From Eq. (2.16)

$$I_1 = -I_2 = -\frac{U_1}{R_2}.$$

A negative current thus flows into port 1 although a positive voltage is applied. Port 1 therefore behaves as if it were a negative resistance with the value

$$\frac{U_1}{I_1} = -R_2. \tag{2.17}$$

This arrangement is stable as long as the internal resistance R_1 of the circuit connected to port 1 is smaller than R_2. For this reason, the arrangement is stable even at short-circuit. It is also possible to have a negative resistance which is stable at open-circuit, by reversing the INIC, i.e. by connecting the resistance R_2 to port 1.

Since Eq. (2.16) also holds for alternating currents, one can replace the resistance R_2 by a complex impedance $\underline{Z}_2$ and in this way obtain any desired negative impedance.

The INIC can also be operated as a voltage source of negative output resistance. A voltage source, having the no-load voltage U_0 and the internal resistance r_0, yields at the load I the output voltage $U = U_0 - I r_0$. For normal voltage sources, r_0 is positive, this resulting in a reduction of U at load. For a voltage source with negative internal resistance, however, U rises for increasing load. The circuit in Fig. 2.25 has this behaviour. It follows from the circuit that

$$U_2 = V_1 = U_0 - I_1 R_1.$$

Since $I_1 = -I_2$, then

$$U_2 = U_0 + I_2 R_1.$$

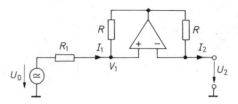

Fig. 2.25 Voltage source with negative output resistance.

$$\textit{Output voltage:} \qquad U_2 = U_0 + I_2 R_1$$

$$\textit{Output resistance:} \quad r_o = -\frac{\partial U_2}{\partial I_2} = -R_1$$

Here, the INIC is connected in such a way that the voltage source is stable at no-load.

Negative resistances can be connected in series and in parallel just as conventional resistors, and the same laws apply. For example, a voltage source with negative internal resistance can be used to compensate for the resistance of a long line so that at the end of the line, the voltage U_0 is obtained with zero internal resistance.

2.6   Gyrator

The gyrator is a converter circuit by which any impedance can be converted into its dual-transformed counterpart, e.g. a capacitance can

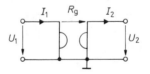

Fig. 2.26 Symbol of the gyrator

be changed into an inductance. The graphic symbol of a gyrator is represented in Fig. 2.26. The ideal transfer characteristics are

$$I_1 = 0 \cdot U_1 + \frac{1}{R_g} U_2, \quad I_2 = \frac{1}{R_g} U_1 + 0 \cdot U_2. \qquad (2.18)$$

Hence, the current at one port is proportional to the voltage at the other port. For this reason, the gyrator can be constructed from two voltage-controlled current sources having high input and output resistances, as is shown schematically in Fig. 2.27.

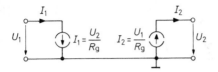

Fig. 2.27 Realization of a gyrator by two voltage-controlled current sources

Another method of realizing a gyrator is based on the combination of INICs [2.2] and is represented in Fig. 2.28. For the determination of the transfer characteristics, we apply KCL to the P-inputs and N-inputs of the amplifiers OA 1 and OA 2 and obtain for

$$\text{node } P_1: \quad \frac{V_3 - U_1}{R_g} - \frac{U_1}{R_g} + I_1 = 0,$$

$$\text{node } N_1: \quad \frac{V_3 - U_1}{R_g} + \frac{U_2 - U_1}{R_g} = 0,$$

$$\text{node } P_2: \quad \frac{V_4 - U_2}{R_g} + \frac{U_1 - U_2}{R_g} - I_2 = 0,$$

$$\text{node } N_2: \quad \frac{V_4 - U_2}{R_g} - \frac{U_2}{R_g} = 0,$$

By eliminating V_3 and V_4, the transfer characteristics become

$$I_1 = \frac{U_2}{R_g} \quad \text{and} \quad I_2 = \frac{U_1}{R_g},$$

which are the desired relationships as given in Eq. (2.18).

Some applications of the gyrator are described below. In the first example, a resistor R_2 is connected to the right-hand port. Since the arrows of I_2 and U_2 have the same direction, $I_2 = U_2/R_2$, following Ohm's law. Insertion of this relationship in the transfer characteristics gives

$$U_1 = I_2 R_g = \frac{U_2 R_g}{R_2} \quad \text{and} \quad I_1 = \frac{U_2}{R_g}.$$

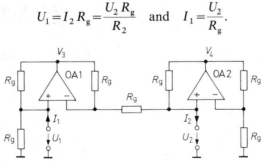

Fig. 2.28 Gyrator using two INICs

Port 1 therefore behaves as a resistance having the value

$$R_1 = \frac{U_1}{I_1} = \frac{R_g^2}{R_2},$$ (2.19)

and is thus proportional to the reciprocal of the load resistance connected to port 2.

The conversion of resistances is also valid for impedances and, according Eq. (2.19), gives

$$\underline{Z}_1 = \frac{R_g^2}{\underline{Z}_2}.$$ (2.20)

This relationship indicates an interesting application of the gyrator: if a capacitor of the value C_2 is connected to one side, the impedance measured on the other side is

$$\underline{Z} = R_g^2 \cdot j\omega\, C_2,$$

which is the impedance of an inductance

$$L_1 = R_g^2\, C_2.$$ (2.21)

The importance of the gyrator is due to the fact that it can be used to simulate large low-loss inductances. The appropriate circuit is depicted in Fig. 2.29. The two free terminals of the gyrator behave, according to Eq. (2.21), as if an inductance $L_1 = R_g^2\, C_2$ were connected between them. With $C_2 = 1\,\mu\text{F}$ and $R_g = 10\,\text{k}\Omega$, one obtains $L_1 = 100\,\text{H}$.

If a capacitor C_1 is connected in parallel to the inductance L_1, the result is a parallel resonant circuit which can be used to build "L"C-filters having high Q-factors.

The Q-factor of the parallel resonant circuit, for $C_1 = C_2$, is well suited to describe the deviation of a real gyrator from the ideal behaviour, and is also called the Q-factor of the gyrator. The losses of a real gyrator can be accounted for by two equivalent resistances R_1 connected in parallel to the two ports. For the circuit involving current sources shown in Fig. 2.27, these resistances are given by the parallel connection of the input resistance of one source with the output resistance of the other. For the circuit involving INICs shown in Fig. 2.28, the equivalent resistances are defined by the matching tolerance of

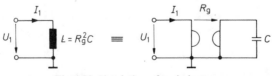

Fig. 2.29 Simulation of an inductance

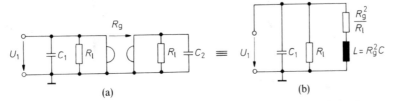

Fig. 2.30 (a) Simulation of a resonant circuit using a gyrator with losses. (b) Equivalent circuit of a resonant circuit having losses

the resistors. The equivalent circuit of a parallel-resonant circuit involving a real gyrator with losses, is shown in Fig. 2.30a. When applying the conversion equation (2.20) to the right-hand side, the transformed equivalent circuit in Fig. 2.30b is obtained. From this, the gyrator Q-factor can be determined as $Q = R_1/2R_g$.

This relationship, however, is only valid for low frequencies, as the Q-factor is very sensitive to phase displacements between current and voltage in the transfer characteristics (Eqs. (2.18)). According to [2.3], a first order approximation is

$$Q(\varphi) = \frac{1}{\dfrac{1}{Q_0} + \varphi_1 + \varphi_2},$$

where Q_0 is the low-frequency value of Q-factor. The terms φ_1 and φ_2 are the phase displacements between current I_1 and voltage U_2, and current I_2 and voltage U_1 respectively, at resonant frequency of the circuit. For lagging phase angles the Q-factor rises with increasing resonant frequency. For $|\varphi_1 + \varphi_2| \geq 1/Q_0$, the circuit becomes unstable, and an oscillation at the resonant frequency of the circuit occurs. For leading phase angles the Q-factor decreases with increasing resonant frequency.

Not only two-terminal, but also four-terminal networks can be converted using gyrators. For this purpose the four-terminal network to be converted is connected between two gyrators with identical gyration resistances R_g, as in Fig. 2.31. The dual-transformed counterpart of the middle two-port then appears between the two outer ports. For the deduction of the transfer characteristics, the product of the chain matrices is calculated. The four-terminal network to be trans-

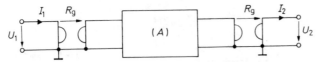

Fig. 2.31 Dual transformation of four-terminal networks

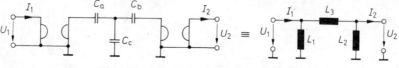

Fig. 2.32 Example for the dual transformation.

Transformation equations: $L_1 = R_g^2 C_a$
$L_2 = R_g^2 C_b$
$L_3 = R_g^2 C_c$

formed has the chain matrix

$$(A) = \begin{pmatrix} A_{11} & A_{12} \\ A_{21} & A_{22} \end{pmatrix}.$$

From Eq. (2.18), we obtain the following relationship for the gyrator

$$\begin{pmatrix} U_1 \\ I_1 \end{pmatrix} = \underbrace{\begin{pmatrix} 0 & R_g \\ 1/R_g & 0 \end{pmatrix}}_{(A_g)} \begin{pmatrix} U_2 \\ I_2 \end{pmatrix}. \tag{2.22}$$

The chain matrix $(\overline{A})$ of the resulting four-terminal network is then

$$(\overline{A}) = (A_g)(A)(A_g) = \begin{pmatrix} A_{22} & A_{21} \cdot R_g^2 \\ A_{12}/R_g^2 & A_{11} \end{pmatrix} \tag{2.23}$$

which is the matrix of the dual-transformed middle two-port.

As an example, Fig. 2.32 shows how a circuit of three inductors can be replaced by the dual circuit containing three capacitors.

If a capacitor is externally connected in parallel to each of the inductances L_1 and L_2, one obtains an inductance-coupled bandpass filter consisting exclusively of capacitors. When short-circuiting C_a and C_b, a floating inductance L_3 is obtained.

2.7 Circulator

A circulator is a circuit with three or more terminals, the graphic symbol for which is shown in Fig. 2.33. It has the characteristic that a signal applied to one of the terminals is relayed in the direction of the

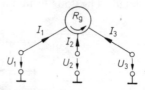

Fig. 2.33 Symbol for a circulator

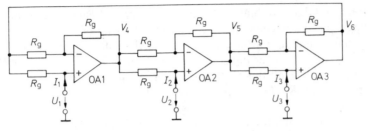

Fig. 2.34 One possible realization of a circulator

arrow. If a terminal is open, the signal passes unchanged, whereas at a short-circuited terminal the polarity of the signal voltage is inverted. If a resistor $R = R_g$ is connected between one terminal and ground, the signal voltage appears across this resistor and will, in this case, not be relayed to the next terminal.

A circuit having these properties is shown in Fig. 2.34 [2.4]. It can be seen that it consists of three identical stages, one of which is shown again in Fig. 2.35. The operation of the individual stage is examined below, where several cases can be distinguished: If terminal 1 is left open, $I_1 = 0$ and $V_P = U_i = V_N$. Hence, no current flows through the feedback resistor, and $U_o = U_i$.

If terminal 1 is short-circuited, $U_1 = 0$, and the circuit behaves as an inverting amplifier with unity gain. In this case, we obtain the output voltage $U_o = -U_i$.

If a resistor $R_1 = R_g$ is connected to terminal 1, the circuit operates as a subtracting amplifier for two equal input voltages U_i, and the voltage U_o is therefore zero.

If U_i is made zero and a voltage U_1 applied to terminal 1, the circuit behaves as a non-inverting amplifier having the gain 2, and we obtain the output voltage $U_o = 2U_1$.

From these characteristics, the operation of the circuit in Fig. 2.34 can be easily understood. Let us assume that the voltage U_1 is applied to terminal 1, that a resistor R_g is connected between terminal 2 and

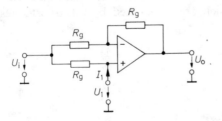

Fig. 2.35 One stage of a circulator

ground, and that terminal 3 is left open. We know already that the output voltage of OA 2 becomes zero. OA 3 has unity gain because of the open terminal 3 and its output voltage is therefore also zero. OA 1 hence operates as non-inverting amplifier with gain 2, so that its output voltage is $2U_1$. Half this voltage (U_1) appears at terminal 2 since this is terminated by R_g. Other special cases can be analyzed in an identical manner.

If a more general case is to be considered, the transfer characteristics of the circulator are used to determine the properties of the circuit. For this purpose we apply KCL to the P-inputs and N-inputs:

<table>
<tr><td align="center">P-inputs</td><td align="center">N-inputs</td></tr>
<tr>
<td>$$\frac{V_6 - U_1}{R_g} + I_1 = 0$$</td>
<td>$$\frac{V_6 - U_1}{R_g} + \frac{V_4 - U_1}{R_g} = 0$$</td>
</tr>
<tr>
<td>$$\frac{V_4 - U_2}{R_g} + I_2 = 0$$</td>
<td>$$\frac{V_4 - U_2}{R_g} + \frac{V_5 - U_2}{R_g} = 0$$</td>
</tr>
<tr>
<td>$$\frac{V_5 - U_3}{R_g} + I_3 = 0$$</td>
<td>$$\frac{V_5 - U_3}{R_g} + \frac{V_6 - U_3}{R_g} = 0$$</td>
</tr>
</table>

By elimination of V_4 to V_6, the transfer characteristics become

$$I_1 = \frac{1}{R_g}(U_2 - U_3),$$

$$I_2 = \frac{1}{R_g}(U_3 - U_1), \qquad (2.24)$$

$$I_3 = \frac{1}{R_g}(U_1 - U_2).$$

It is obvious from Eq. (2.24) that a circulator can also be realized by three voltage-controlled current sources with differential input, as is represented in Fig. 2.36. A current source suitable for this purpose is shown in Fig. 2.17.

Figure 2.37 shows the application of a circulator as an active echo suppressor for telephone circuits. It consists of a circulator having three ports all of which are terminated by the transfer resistance R_g. The

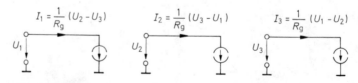

Fig. 2.36 Circulator composed of two voltage-controlled current sources

signal from the microphone is relayed to the exchange and does not reach the receiver. The signal from the exchange is transferred to the receiver but not to the microphone. The cross-talk attenuation is largely determined by the degree to which the terminating resistances are matched.

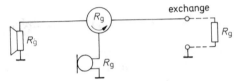

Fig. 2.37 Application of a circulator as echo suppressor in telephone circuits

3   Active filters

3.1   Basic theory of lowpass filters

The circuit of the simplest lowpass filter is shown in Fig. 3.1. The ratio of output voltage to input voltage can be expressed, for sinusoidal signals, as

$$\underline{A}(j\omega)=\frac{\underline{U}_o}{\underline{U}_i}=\frac{1}{1+j\omega RC},$$

and is called the frequency response of the circuit. Replacing $j\omega$ by $j\omega+\sigma=p$ gives the system function or generalized transfer function

$$A(p)=\frac{L\{U_o(t)\}}{L\{U_i(t)\}}=\frac{1}{1+pRC}.$$

This is the ratio of the Laplace transformed output and input voltage for signals of any time dependence. On the other hand, the transition from the transfer function $A(p)$ to the frequency response $\underline{A}(j\omega)$ for sinusoidal input signals, is made by setting σ to zero.

In order to present the problem in a more general form, it is useful to normalize the complex frequency variable p by defining

$$P=\frac{p}{\omega_c}.$$

Hence, for $\sigma=0$,

$$P=\frac{j\omega}{\omega_c}=j\frac{f}{f_c}=j\Omega.$$

The circuit in Fig. 3.1 has the cutoff frequency $f_c=1/2\pi RC$. Therefore, $P=pRC$ and

$$A(P)=\frac{1}{1+P}. \tag{3.1}$$

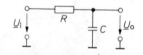

Fig. 3.1 Simplest passive lowpass filter

For the absolute value of the transfer function, i.e. for the amplitude ratio at sinusoidal input signals, we obtain

$$|\underline{A}(j\Omega)|^2 = \frac{1}{1 + \Omega^2}.$$

For $\Omega \gg 1$, i.e. for $f \gg f_c$, then $|\underline{A}| = 1/\Omega$; this corresponds to a reduction in gain of 20 dB per frequency decade.

If a steeper decrease in gain is required, n lowpass filters can be connected in series. The expression of the transfer function is then of the form

$$A(P) = \frac{1}{(1 + \alpha_1 P)(1 + \alpha_2 P) \dots (1 + \alpha_n P)}, \tag{3.2}$$

where the coefficients $\alpha_1, \alpha_2, \alpha_3 \dots$ are real and positive. For $\Omega \gg 1$, $|\underline{A}|$ is proportional to $1/\Omega^n$; the gain therefore slopes off at $n \cdot 20$ dB per decade. It can be seen that the transfer function possesses n real negative poles. This is characteristic for n-th order passive RC lowpass filters. If decoupled lowpass filters of identical cutoff frequencies are cascaded, then

$$\alpha_1 = \alpha_2 = \alpha_3 = \cdots = \alpha = \sqrt{\sqrt[n]{2} - 1},$$

this being the condition for which critical damping occurs. Each individual lowpass filter then has a cutoff frequency a factor $1/\alpha$ higher than that of the whole filter.

The transfer function of a lowpass filter has the general form

$$A(P) = \frac{A_0}{1 + c_1 P + c_2 P^2 + \cdots + c_n P^n}, \tag{3.3}$$

where $c_1, c_2 \dots c_n$ are positive and real. The order of the filter is equal to the highest power of P. It is advantageous for the realization of filters if the denominator polynomial is written in factored form. If complex poles are also permitted, a separation into linear factors as in Eq. (3.2) is no longer possible, and a product of quadratic expressions is obtained:

$$A(P) = \frac{A_0}{(1 + a_1 P + b_1 P^2)(1 + a_2 P + b_2 P^2) \dots}, \tag{3.4}$$

where a_i and b_i are positive and real. For odd orders n, the coefficient b_1 is zero.

There are several different theoretical aspects for which the frequency response can be optimized. Any such aspect leads to a different set of coefficients a_i and b_i. As will be seen, conjugate complex poles arise. They cannot be realized by passive RC elements, as a comparison with Eq. (3.2) shows. One way of implementing conjugate complex poles is the use of *LRC* networks. For high frequencies, the realization of the necessary inductances usually presents no difficulties, but in the low frequency range, large inductances are often needed. These are unwieldy and have poor electrical properties. However, the use of inductances at low frequencies can be avoided by the addition of active elements (e.g. operational amplifiers) to the *RC* network. Such circuits are called active filters.

Let us first compare the most important optimized frequency responses, the technical realizations of which are discussed in the following sections.

Butterworth lowpass filters have an amplitude-frequency response which is flat for as long as possible and drops sharply just before the cutoff frequency. Their step response shows a considerable overshoot which increases for higher-order filters.

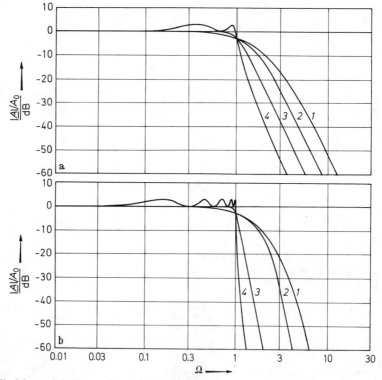

Fig. 3.2a and b Comparison of the amplitude-frequency response for different filter types. (a) Fourth order. (b) Tenth order.
Curve *1*: Lowpass filter with critical damping. Curve *2*: Bessel lowpass filter. Curve *3*: Butterworth lowpass filter. Curve *4*: Chebyshev lowpass filter with 3 dB ripple

Chebyshev lowpass filters have an even steeper drop in gain above the cutoff frequency. In the passband, however, the gain varies and has a ripple of constant amplitude. For a given order, the decrease above the cutoff frequency is steeper the larger the permitted ripple. The overshoot in the step response is even greater than for the Butterworth filters.

Bessel lowpass filters have the optimum square-wave response. The underlying condition is that the group delay is constant over the largest possible frequency range, i.e. that the phase shift in this frequency range is proportional to the frequency. The amplitude-frequency response of Bessel filters does not fall as sharply as that of Butterworth or Chebyshev filters.

Figure 3.2 shows the amplitude-frequency responses of the four described filter types for the orders 4 and 10. It can be seen that the Chebyshev lowpass filter has the most abrupt change-over from the passband to the stopband. This is advantageous but has the side effect of a ripple in the passband of the amplitude-frequency response. As this ripple is gradually reduced, the behaviour of the Chebyshev filter approaches that of a Butterworth filter [3.1]. Both kinds of filter show a considerable overshoot in the step response, as is seen in Fig. 3.3. Bessel filters on the other hand have only negligible overshoot. Despite their unfavourable amplitude-frequency response they will always be used where a good step response is important. A passive *RC* lowpass

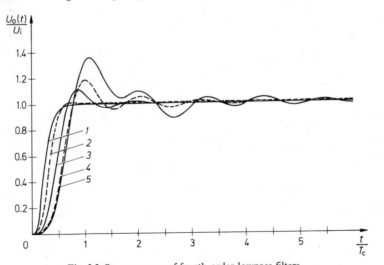

Fig. 3.3 Step response of fourth order lowpass filters.
Curve *1*: Lowpass filter with critical damping. Curve *2*: Bessel lowpass filter. Curve *3*: Butterworth lowpass filter. Curve *4*: Chebyshev lowpass filter with 0.5 dB ripple. Curve *5*: Chebyshev lowpass filter with 3 dB ripple. $T_c = 1/f_c$

filter shows no overshoot; however, the relatively small improvement over the Bessel filter involves a considerable deterioration of the amplitude-frequency response. In addition, the corners in the step response are much rounder than for the Bessel filter. The table in Fig. 3.4 compares the rise times, delay times and the overshoot. The rise time is the time in which the output signal rises from 10% to 90% of its final-state value. The delay time is that in which the output signal increases form 0 to 50% of the final-state value.

	order				
	2	4	6	8	10
critical damping					
normalized rise time t_r/T_c	0.344	0.342	0.341	0.341	0.340
normalized delay time t_d/T_c	0.172	0.254	0.316	0.367	0.412
overshoot %	0	0	0	0	0
Bessel					
normalized rise time t_r/T_c	0.344	0.352	0.350	0.347	0.345
normalized delay time t_d/T_c	0.195	0.329	0.428	0.505	0.574
overshoot %	0.43	0.84	0.64	0.34	0.06
Butterworth					
normalized rise time t_r/T_c	0.342	0.387	0.427	0.460	0.485
normalized delay time t_d/T_c	0.228	0.449	0.663	0.874	1.084
overshoot %	4.3	10.8	14.3	16.3	17.8
Chebyshev, 0.5 dB *ripple*					
normalized rise time t_r/T_c	0.338	0.421	0.487	0.540	0.584
normalized delay time t_d/T_c	0.251	0.556	0.875	1.196	1.518
overshoot %	10.7	18.1	21.2	22.9	24.1
Chebyshev, 1 dB *ripple*					
normalized rise time t_r/T_c	0.334	0.421	0.486	0.537	0.582
normalized delay time t_d/T_c	0.260	0.572	0.893	1.215	1.540
overshoot %	14.6	21.6	24.9	26.6	27.8
Chebyshev, 2 dB *ripple*					
normalized rise time t_r/T_c	0.326	0.414	0.491	0.529	0.570
normalized delay time t_d/T_c	0.267	0.584	0.912	1.231	1.555
overshoot %	21.2	28.9	32.0	33.5	34.7
Chebyshev, 3 dB *ripple*					
normalized rise time t_r/T_c	0.318	0.407	0.470	0.519	0.692
normalized delay time t_d/T_c	0.271	0.590	0.912	1.235	1.557
overshoot %	27.2	35.7	38.7	40.6	41.6

Fig. 3.4 Comparison of lowpass filters. Rise time and delay time are normalized to the reciprocal cutoff frequency $T_c = 1/f_c$

It can be seen that the rise time is not strongly dependent on the order or the type of the filter. Its value is approximately $1/3f_c$, as shown in [1]. On the other hand, the delay time and overshoot increase for rising order. The Bessel filters are an exception in that the overshoot decreases for orders higher than 4.

It will be seen later that a single circuit is sufficient to implement any of these filter types for a particular order; the values of resistances and capacitances determine the type of filter. In order that the circuit parameters can be defined, the frequency response of the individual filter types must be known for each order. We shall therefore discuss these in more detail in the following sections.

3.1.1 Butterworth lowpass filters

From Eq. (3.3), the absolute value of the gain of an n-th order lowpass filter has the general form

$$|\underline{A}|^2 = \frac{A_0^2}{1 + k_2 \Omega^2 + k_4 \Omega^4 + \cdots + k_{2n} \Omega^{2n}}. \tag{3.5}$$

Odd orders of Ω do not occur since the square of $|\underline{A}|$ must be an even function. Below the cutoff frequency of the Butterworth lowpass filter, the function $|\underline{A}|^2$ must be maximally flat. Since for this range $\Omega < 1$, this condition is best fulfilled if $|\underline{A}|^2$ is dependent only on the highest power of Ω. The reason for this is that for $\Omega < 1$, the lower orders of Ω contribute most to the denominator and therefore to the drop in gain. Hence,

$$|\underline{A}|^2 = \frac{A_0^2}{1 + k_{2n} \Omega^{2n}}.$$

The coefficient k_{2n} is defined by the "normalizing condition", namely, that the gain at $\Omega = 1$ is reduced by 3 dB. Thus

$$\frac{A_0^2}{2} = \frac{A_0^2}{1 + k_{2n}},$$

$$k_{2n} = 1.$$

Therefore, the square of the gain $|\underline{A}|^2$ of n-th order Butterworth lowpass filters is given by

$$|\underline{A}|^2 = \frac{A_0^2}{1 + \Omega^{2n}}. \tag{3.6}$$

n	
1	$1+P$
2	$1+\sqrt{2}P+P^2$
3	$1+2P+2P^2+P^3=(1+P)(1+P+P^2)$
4	$1+2.613P+3.414P^2+2.613P^3+P^4=(1+1.848P+P^2)(1+0.765P+P^2)$

Fig. 3.5 Butterworth polynomials

To build a Butterworth lowpass filter, a circuit must be designed in which the square of the gain has the form given above. However, the circuit analysis initially gives the complex gain $\underline{A}$, and not the gain square $|\underline{A}|^2$. It is therefore necessary to know the complex gain involved in Eq. (3.6). This is found by calculating the absolute value of Eq. (3.3) and by comparing the coefficients with those of Eq. (3.6). In this way, the desired coefficients $c_1 \ldots c_n$ can be defined. The denominators of Eq. (3.3) are then the Butterworth polynomials, the first four orders of which are shown in Fig. 3.5.

According to reference [3.2], it is possible to determine analytically the poles of the transfer function. By combining the conjugate complex poles, we immediately obtain the coefficients, a_i and b_i of the quadratic expressions in Eq. (3.4):

even order n:

$$a_i = 2 \cos \frac{(2i-1)\pi}{2n} \quad \text{for} \quad i=1 \ldots \frac{n}{2},$$

$$b_i = 1,$$

odd order n:

$$a_1 = 1,$$

$$b_1 = 0$$

and

$$a_i = 2 \cos \frac{(i-1)\pi}{n} \quad \text{for} \quad i=2 \ldots \frac{n+1}{2},$$

$$b_i = 1.$$

The coefficients of the Butterworth polynomials up to the order 10 are shown in Fig. 3.14.

It can be seen that the first-order Butterworth lowpass filter is a passive lowpass filter having the transfer function of Eq. (3.1). The higher Butterworth polynomials possess conjugate complex zeros. A comparison with Eq. (3.2) shows that such denominator polynomials cannot be realized by passive RC networks, as with these, all zeros are real. In such cases, the only choice is to use LRC circuits with all their disadvantages, or active RC filters. The frequency response of the gain is shown in Fig. 3.6.

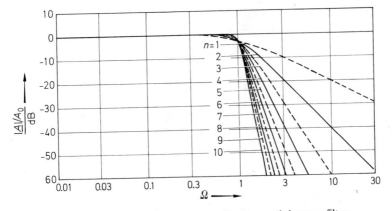

Fig. 3.6 Amplitude-frequency response of Butterworth lowpass filters

3.1.2   Chebyshev lowpass filters

At low frequencies, the gain of a Chebyshev lowpass filter has the value A_0, but varies below the cutoff frequency, having a predetermined ripple. Polynomials which have a constant ripple within a defined range (equal ripple) are the Chebyshev polynomials

$$T_n(x)=\begin{cases} \cos(n\cos^{-1}x) & \text{for } 0\leqq x\leqq 1 \\ \cosh(n\cosh^{-1}x) & \text{for } x>1, \end{cases}$$

the first four of which are shown in Fig. 3.7. For $0\leqq x\leqq 1$, $|T(x)|$ swings between 0 and 1; for $x>1$, $T(x)$ rises steadily. In order to obtain the

n	
1	$T_1(x)=x$
2	$T_2(x)=2x^2-1$
3	$T_3(x)=4x^3-3x$
4	$T_4(x)=8x^4-8x^2+1$

Fig. 3.7 Chebyshev polynomials

equation of a lowpass filter from the Chebyshev polynomials, one defines

$$|A|^2=\frac{kA_0^2}{1+\varepsilon^2 T_n^2(x)}. \tag{3.7}$$

The constant k is chosen so that for $x=0$ the square of the gain $|A|^2$ becomes A_0^2, i.e. $k=1$ for odd orders n and $k=1+\varepsilon^2$ for even n. The

	ripple			
	0.5 dB	1 dB	2 dB	3 dB
A_{max}/A_{min}	1.059	1.122	1.259	1.413
k	1.122	1.259	1.585	1.995
ε	0.349	0.509	0.765	0.998

Fig. 3.8 Comparison of some Chebyshev parameters

factor ε is a measure of the ripple. It follows from Eq. (3.7) that

$$\frac{A_{max}}{A_{min}} = \sqrt{1+\varepsilon^2}$$

and

$$\left.\begin{aligned} A_{max} &= A_0\sqrt{1+\varepsilon^2} \\ A_{min} &= A_0 \end{aligned}\right\} \quad \text{for even orders}$$

and

$$\left.\begin{aligned} A_{max} &= A_0 \\ A_{min} &= A_0/\sqrt{1+\varepsilon^2} \end{aligned}\right\} \quad \text{for odd orders.}$$

In Fig. 3.8, the appropriate values are listed for different ripples. In principle, the complex gain can be calculated from $|\underline{A}|^2$ and hence, the coefficients of the factored form can be determined. However, as shown in [3.3], it is possible to derive the poles of the transfer function directly from those of the Butterworth filters. By combining the conjugate complex poles, the coefficients a_i and b_i in Eq. (3.4) are determined as follows:

even order n:

$$\left.\begin{aligned} b_i' &= \frac{1}{\cosh^2\gamma - \cos^2\dfrac{(2i-1)\pi}{2n}} \\ a_i' &= 2b_i' \cdot \sinh\gamma \cdot \cos\frac{(2i-1)\pi}{2n} \end{aligned}\right\} \quad \text{for } i=1\ldots\frac{n}{2},$$

odd order n:

$$b_1' = 0,$$
$$a_1' = 1/\sinh\gamma,$$
$$\left.\begin{aligned} b_i' &= \frac{1}{\cosh^2\gamma - \cos^2\dfrac{(i-1)\pi}{n}} \\ a_i' &= 2b_i' \cdot \sinh\gamma \cdot \cos\frac{(i-1)\pi}{n} \end{aligned}\right\} \quad \text{for } i=2\ldots\frac{n+1}{2},$$

where $\gamma = \dfrac{1}{n}\sinh^{-1}\dfrac{1}{\varepsilon}$.

If the coefficients a_i' and b_i' found in this way replace a_i and b_i in Eq. (3.4), Chebyshev filters are obtained. However, P is then not normalized with respect to the 3 dB cutoff frequency ω_c, but rather to the frequency $\omega_{c\,min}$ at which the gain takes on the value A_{min} for the last time.

For an easy comparison of the different filter types, it is useful to normalize P to the 3 dB cutoff frequency. The variable P is replaced by αP and the normalizing constant α is determined such that the gain, for $P = j$, has the value $1/\sqrt{2}$. The quadratic expressions in the denominator of the complex gain are then

$$(1 + a_i' \alpha P + b_i' \alpha^2 P^2).$$

Hence, by comparing the coefficients with those of Eq. (3.4):

$$a_i = \alpha a_i' \quad \text{and} \quad b_i = \alpha^2 b_i'.$$

The coefficients a_i and b_i are shown in the table of Fig. 3.14 up to the tenth order, and for ripple values of 0.5, 1, 2 and 3 dB. The frequency response of the gain is shown in Fig. 3.9 for ripple values of 0.5 and

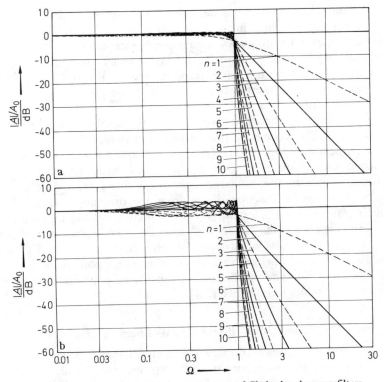

Fig. 3.9a and b Amplitude-frequency response of Chebyshev lowpass filters.
(a) 0.5 dB ripple. (b) 3 dB ripple

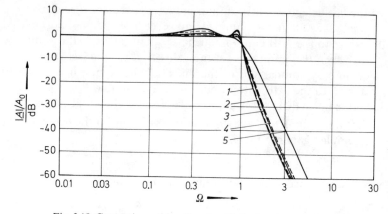

Fig. 3.10 Comparison of fourth order Chebyshev lowpass filters.
Ripple: Curve *1*: 3 dB. Curve *2*: 2 dB. Curve *3*: 1 dB. Curve *4*: 0.5 dB. Curve *5*: Fourth order Butterworth lowpass filter for comparison

3 dB. Figure 3.10 makes possible a direct comparison of fourth order Chebyshev filters having different amount of ripple. It can be seen that the differences in the frequency response in the stopband are very small, and that they become even smaller for higher orders. It is also obvious that even the Chebyshev filter response having the small ripple of 0.5 dB emerges from the passband much more steeply than that of the Butterworth filter.

The change-over from the passband to the stopband can be made even steeper. To accomplish this, zeros are introduced into the amplitude-frequency response above the cutoff frequency. One way of optimizing the design is to give the amplitude-frequency response a constant ripple in the stopband also. Such filters are called *Cauer* filters. The transfer function differs from the ordinary lowpass filter equation in that the numerator is a polynomial, instead of the constant A_0. For this reason, the "steepened" lowpass filters cannot be realized by the simple circuits of Section 3.4. However, in Section 3.11 we discuss a universal filter with which any numerator polynomial can be put into practice. The coefficients of the Cauer polynomials can be found in the tables of reference [3.4].

3.1.3   Bessel lowpass filters

As previously shown, Butterworth and Chebyshev lowpass filters have a considerable overshoot in their step response. An ideal square-wave response is achieved by filters having frequency-independent group delay, i.e. having a phase shift proportional to frequency. This behaviour is best approximated by Bessel filters, sometimes also called Thomson filters. The approximation consists of choosing the coefficients so that the group delay below the cutoff frequency $\Omega = 1$ is as little as possible dependent on Ω. This procedure is equivalent to a Butterworth approximation of the group delay, i.e. the realization of a maximally flat group delay.

From Eq. (3.4), with $P = j\Omega$, the gain of a second order lowpass filter is given by

$$\underline{A} = \frac{A_0}{1 + a_1 P + b_1 P^2} = \frac{A_0}{1 + j a_1 \Omega - b_1 \Omega^2}.$$

Therefore the phase

$$\varphi = -\tan^{-1} \frac{a_1 \Omega}{1 - b_1 \Omega^2}. \tag{3.8}$$

The group delay is defined as

$$t_{gr} = -\frac{d\varphi}{d\omega}.$$

To simplify further calculations, we introduce the normalized group delay

$$T_{gr} = \frac{t_{gr}}{T_c} = t_{gr} \cdot f_c = \frac{1}{2\pi} t_{gr} \cdot \omega_c, \tag{3.9a}$$

where T_c is the reciprocal of the cutoff frequency. We thus obtain

$$T_{gr} = -\frac{\omega_c}{2\pi} \cdot \frac{d\varphi}{d\omega} = -\frac{1}{2\pi} \cdot \frac{d\varphi}{d\omega} \tag{3.9b}$$

and with Eq. (3.8)

$$T_{gr} = \frac{1}{2\pi} \cdot \frac{a_1(1 + b_1 \Omega^2)}{1 + (a_1^2 - 2b_1)\Omega^2 + b_1^2 \Omega^4}. \tag{3.9c}$$

In order to find the Butterworth approximation of the group delay, we use the fact that, for $\Omega \ll 1$,

$$T_{gr} = \frac{a_1}{2\pi} \cdot \frac{1 + b_1 \Omega^2}{1 + (a_1^2 - 2b_1)\Omega^2} \qquad \text{for } \Omega \ll 1.$$

This expression becomes independent of Ω if the coefficients of Ω^2 in the numerator and denominator are identical. The condition for this is that

$$b_1 = a_1^2 - 2b_1$$

or

$$b_1 = \tfrac{1}{3} a_1^2. \tag{3.10}$$

A second relationship is derived from the normalizing condition, $|\underline{A}|^2 = \tfrac{1}{2}$ for $\Omega = 1$:

$$\frac{1}{2} = \frac{1}{(1 - b_1)^2 + a_1^2}.$$

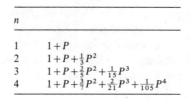

n	
1	$1+P$
2	$1+P+\frac{1}{3}P^2$
3	$1+P+\frac{2}{5}P^2+\frac{1}{15}P^3$
4	$1+P+\frac{3}{7}P^2+\frac{2}{21}P^3+\frac{1}{105}P^4$

Fig. 3.11 Bessel polynomials

Hence, with Eq. (3.10)

$$a_1 = 1.3617,$$
$$b_1 = 0.6180.$$

For higher filter orders, a corresponding calculation becomes very involved since a system of non-linear equations arises. Using a different concept [3.5], however, it is possible to define the coefficients c_i in Eq. (3.3) by a recursion formula

$$c'_1 = 1,$$
$$c'_i = \frac{2(n-i+1)}{i(2n-i+1)} c'_{i-1}.$$

The denominators of Eq. (3.3) obtained in such a way are the Bessel polynomials and are shown in Fig. 3.11 up to the fourth order. It must be considered that, in this representation, the frequency P is not normalized with respect to the 3 dB cutoff frequency, but to the reciprocal of the group delay for $\Omega = 0$. However, this is of little use for the design of lowpass filters. We have therefore recalculated the coefficients c_i for the 3 dB cutoff frequency, as in the previous section, and

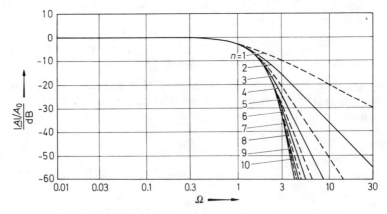

Fig. 3.12 Amplitude-frequency response of Bessel lowpass filters

in addition have broken down the denominator into quadratic expressions. The coefficients a_i and b_i of Eq. (3.4) so obtained, are listed in Fig. 3.14 for up to the tenth order. The frequency response of the gain is shown graphically in Fig. 3.12.

In order to demonstrate the amount of phase distortion of other filters in comparison with the Bessel filters, we have illustrated in Fig. 3.13 the frequency response of the phase shift and of the group delay for fourth order filters. These curves can best be calculated from the factored transfer function in Eq. (3.4) by adding together the phase shifts of each individual second order filter stage and by adding the individual group delays. For a filter of a given order, the following

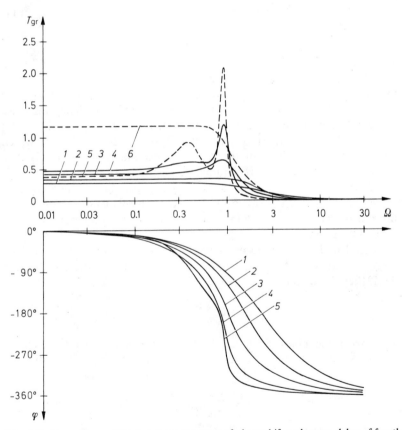

Fig. 3.13 Comparison of the frequency response of phase shift and group delay of fourth order filters.
Curve *1*: Lowpass filter with critical damping. Curve *2*: Bessel lowpass filter. Curve *3*: Butterworth lowpass filter. Curve *4*: Chebyshev lowpass filter with 0.5 dB ripple. Curve *5*: Chebyshev lowpass filter with 3 dB ripple. Curve *6*: Allpass filter for comparison

relationships are derived from Eqs. (3.8) and (3.9c):

$$\varphi = -\sum_i \tan^{-1} \frac{a_i \Omega}{1 - b_i \Omega^2}$$

and

$$T_{gr} = \frac{1}{2\pi} \sum_i \frac{a_i(1 + b_i \Omega^2)}{1 + (a_i^2 - 2b_i)\Omega^2 + b_i^2 \Omega^4}.$$

3.1.4 Summary of theory

We have seen that the transfer functions of all lowpass filters have the form

$$A(P) = \frac{A_0}{\prod_i (1 + a_i P + b_i P^2)}. \tag{3.11}$$

The order n of the filter is determined by the highest power of P in Eq. (3.11) when the denominator is multiplied out. It defines the slope of the asymptote of the amplitude-frequency response as having the value $-n \cdot 20\,\mathrm{dB/decade}$. The rest of the amplitude-frequency response curve for a particular order is determined by the type of filter. Of special interest are the Butterworth, Chebyshev and Bessel filters which all have different values for the coefficients a_i and b_i in Eq. (3.11). The values of the coefficients are summarized in Fig. 3.14 for up to the tenth order. In addition, the 3 dB cutoff frequency of each individual filter stage is indicated by the ratio f_{ci}/f_c. Although this value is not needed for the design, it is useful for checking the correct operation of the individual filter stages.

Also listed are the pole-pair quality factors Q_i of the individual filter stages. In analogy to the Q-factors of the bandpass filters in Section 3.6.1, they are defined as

$$Q_i = \frac{\sqrt{b_i}}{a_i}.$$

The larger the pole-pair Q-factor, the more likely the filter is to be unstable. Filters with real poles have pole-pair Q-factors of $Q \geq 0.5$.

n	i	a_i	b_i	f_{ci}/f_c	Q_i
critically damped filters					
1	1	1.0000	0.0000	1.000	—
2	1	1.2872	0.4142	1.000	0.50
3	1	0.5098	0.0000	1.961	—
	2	1.0197	0.2599	1.262	0.50
4	1	0.8700	0.1892	1.480	0.50
	2	0.8700	0.1892	1.480	0.50
5	1	0.3856	0.0000	2.593	—
	2	0.7712	0.1487	1.669	0.50
	3	0.7712	0.1487	1.669	0.50
6	1	0.6999	0.1225	1.839	0.50
	2	0.6999	0.1225	1.839	0.50
	3	0.6999	0.1225	1.839	0.50
7	1	0.3226	0.0000	3.100	—
	2	0.6453	0.1041	1.995	0.50
	3	0.6453	0.1041	1.995	0.50
	4	0.6453	0.1041	1.995	0.50
8	1	0.6017	0.0905	2.139	0.50
	2	0.6017	0.0905	2.139	0.50
	3	0.6017	0.0905	2.139	0.50
	4	0.6017	0.0905	2.139	0.50
9	1	0.2829	0.0000	3.534	—
	2	0.5659	0.0801	2.275	0.50
	3	0.5659	0.0801	2.275	0.50
	4	0.5659	0.0801	2.275	0.50
	5	0.5659	0.0801	2.275	0.50
10	1	0.5358	0.0718	2.402	0.50
	2	0.5358	0.0718	2.402	0.50
	3	0.5358	0.0718	2.402	0.50
	4	0.5358	0.0718	2.402	0.50
	5	0.5358	0.0718	2.402	0.50

Fig. 3.14 Coefficients of the individual filter types

n	i	a_i	b_i	f_{ci}/f_c	Q_i
Bessel filters					
1	1	1.0000	0.0000	1.000	–
2	1	1.3617	0.6180	1.000	0.58
3	1	0.7560	0.0000	1.323	–
	2	0.9996	0.4772	1.414	0.69
4	1	1.3397	0.4889	0.978	0.52
	2	0.7743	0.3890	1.797	0.81
5	1	0.6656	0.0000	1.502	–
	2	1.1402	0.4128	1.184	0.56
	3	0.6216	0.3245	2.138	0.92
6	1	1.2217	0.3887	1.063	0.51
	2	0.9686	0.3505	1.431	0.61
	3	0.5131	0.2756	2.447	1.02
7	1	0.5937	0.0000	1.684	–
	2	1.0944	0.3395	1.207	0.53
	3	0.8304	0.3011	1.695	0.66
	4	0.4332	0.2381	2.731	1.13
8	1	1.1112	0.3162	1.164	0.51
	2	0.9754	0.2979	1.381	0.56
	3	0.7202	0.2621	1.963	0.71
	4	0.3728	0.2087	2.992	1.23
9	1	0.5386	0.0000	1.857	–
	2	1.0244	0.2834	1.277	0.52
	3	0.8710	0.2636	1.574	0.59
	4	0.6320	0.2311	2.226	0.76
	5	0.3257	0.1854	3.237	1.32
10	1	1.0215	0.2650	1.264	0.50
	2	0.9393	0.2549	1.412	0.54
	3	0.7815	0.2351	1.780	0.62
	4	0.5604	0.2059	2.479	0.81
	5	0.2883	0.1665	3.466	1.42

Fig. 3.14 *continued*

n	i	a_i	b_i	f_{ci}/f_c	Q_i
Butterworth filters
1	1	1.0000	0.0000	1.000	–
2	1	1.4142	1.0000	1.0000	0.71
3	1	1.0000	0.0000	1.000	–
	2	1.0000	1.0000	1.272	1.00
4	1	1.8478	1.0000	0.719	0.54
	2	0.7654	1.0000	1.390	1.31
5	1	1.0000	0.0000	1.000	–
	2	1.6180	1.0000	0.859	0.62
	3	0.6180	1.0000	1.448	1.62
6	1	1.9319	1.0000	0.676	0.52
	2	1.4142	1.0000	1.000	0.71
	3	0.5176	1.0000	1.479	1.93
7	1	1.0000	0.0000	1.000	–
	2	1.8019	1.0000	0.745	0.55
	3	1.2470	1.0000	1.117	0.80
	4	0.4450	1.0000	1.499	2.25
8	1	1.9616	1.0000	0.661	0.51
	2	1.6629	1.0000	0.829	0.60
	3	1.1111	1.0000	1.206	0.90
	4	0.3902	1.0000	1.512	2.56
9	1	1.0000	0.0000	1.000	–
	2	1.8794	1.0000	0.703	0.53
	3	1.5321	1.0000	0.917	0.65
	4	1.0000	1.0000	1.272	1.00
	5	0.3473	1.0000	1.521	2.88
10	1	1.9754	1.0000	0.655	0.51
	2	1.7820	1.0000	0.756	0.56
	3	1.4142	1.0000	1.000	0.71
	4	0.9080	1.0000	1.322	1.10
	5	0.3129	1.0000	1.527	3.20

Fig. 3.14 *continued*

n	i	a_i	b_i	f_{ci}/f_c	Q_i

Chebyshev filters, 0.5 dB ripple

n	i	a_i	b_i	f_{ci}/f_c	Q_i
1	1	1.0000	0.0000	1.000	–
2	1	1.3614	1.3827	1.000	0.86
3	1	1.8636	0.0000	0.537	–
	2	0.6402	1.1931	1.335	1.71
4	1	2.6282	3.4341	0.538	0.71
	2	0.3648	1.1509	1.419	2.94
5	1	2.9235	0.0000	0.342	–
	2	1.3025	2.3534	0.881	1.18
	3	0.2290	1.0833	1.480	4.54
6	1	3.8645	6.9797	0.366	0.68
	2	0.7528	1.8573	1.078	1.81
	3	0.1589	1.0711	1.492	6.51
7	1	4.0211	0.0000	0.249	–
	2	1.8729	4.1795	0.645	1.09
	3	0.4861	1.5676	1.208	2.58
	4	0.1156	1.0443	1.517	8.84
8	1	5.1117	11.9607	0.276	0.68
	2	1.0639	2.9365	0.844	1.61
	3	0.3439	1.4206	1.284	3.47
	4	0.0885	1.0407	1.521	11.53
9	1	5.1318	0.0000	0.195	–
	2	2.4283	6.6307	0.506	1.06
	3	0.6839	2.2908	0.989	2.21
	4	0.2559	1.3133	1.344	4.48
	5	0.0695	1.0272	1.532	14.58
10	1	6.3648	18.3695	0.222	0.67
	2	1.3582	4.3453	0.689	1.53
	3	0.4822	1.9440	1.091	2.89
	4	0.1994	1.2520	1.381	5.61
	5	0.0563	1.0263	1.533	17.99

Fig. 3.14 *continued*

n	i	a_i	b_i	f_{ci}/f_c	Q_i
Chebyshev filters, 1 dB ripple					
1	1	1.0000	0.0000	1.000	–
2	1	1.3022	1.5515	1.000	0.96
3	1	2.2156	0.0000	0.451	–
	2	0.5442	1.2057	1.353	2.02
4	1	2.5904	4.1301	0.540	0.78
	2	0.3039	1.1697	1.417	3.56
5	1	3.5711	0.0000	0.280	–
	2	1.1280	2.4896	0.894	1.40
	3	0.1872	1.0814	1.486	5.56
6	1	3.8437	8.5529	0.366	0.76
	2	0.6292	1.9124	1.082	2.20
	3	0.1296	1.0766	1.493	8.00
7	1	4.9520	0.0000	0.202	–
	2	1.6338	4.4899	0.655	1.30
	3	0.3987	1.5834	1.213	3.16
	4	0.0937	1.0423	1.520	10.90
8	1	5.1019	14.7608	0.276	0.75
	2	0.8916	3.0426	0.849	1.96
	3	0.2806	1.4334	1.285	4.27
	4	0.0717	1.0432	1.520	14.24
9	1	6.3415	0.0000	0.158	–
	2	2.1252	7.1711	0.514	1.26
	3	0.5624	2.3278	0.994	2.71
	4	0.2076	1.3166	1.346	5.53
	5	0.0562	1.0258	1.533	18.03
10	1	6.3634	22.7468	0.221	0.75
	2	1.1399	4.5167	0.694	1.86
	3	0.3939	1.9665	1.093	3.56
	4	0.1616	1.2569	1.381	6.94
	5	0.0455	1.0277	1.532	22.26

Fig. 3.14 *continued*

n	i	a_i	b_i	f_{ci}/f_c	Q_i
Chebyshev filters, 2 dB ripple					
1	1	1.0000	0.0000	1.000	–
2	1	1.1813	1.7775	1.000	1.13
3	1	2.7994	0.0000	0.357	–
	2	0.4300	1.2036	1.378	2.55
4	1	2.4025	4.9862	0.550	0.93
	2	0.2374	1.1896	1.413	4.59
5	1	4.6345	0.0000	0.216	–
	2	0.9090	2.6036	0.908	1.78
	3	0.1434	1.0750	1.493	7.23
6	1	3.5880	10.4648	0.373	0.90
	2	0.4925	1.9622	1.085	2.84
	3	0.0995	1.0826	1.491	10.46
7	1	6.4760	0.0000	0.154	–
	2	1.3258	4.7649	0.665	1.65
	3	0.3067	1.5927	1.218	4.12
	4	0.0714	1.0384	1.523	14.28
8	1	4.7743	18.1510	0.282	0.89
	2	0.6991	3.1353	0.853	2.53
	3	0.2153	1.4449	1.285	5.58
	4	0.0547	1.0461	1.518	18.69
9	1	8.3198	0.0000	0.120	–
	2	1.7299	7.6580	0.522	1.60
	3	0.4337	2.3549	0.998	3.54
	4	0.1583	1.3174	1.349	7.25
	5	0.0427	1.0232	1.536	23.68
10	1	5.9618	28.0376	0.226	0.89
	2	0.8947	4.6644	0.697	2.41
	3	0.3023	1.9858	1.094	4.66
	4	0.1233	1.2614	1.380	9.11
	5	0.0347	1.0294	1.531	29.27

Fig. 3.14 *continued*

n	i	a_i	b_i	f_{ci}/f_c	Q_i
Chebyshev filters, 3 dB ripple					
1	1	1.0000	0.0000	1.000	–
2	1	1.0650	1.9305	1.000	1.30
3	1	3.3496	0.0000	0.299	–
	2	0.3559	1.1923	1.396	3.07
4	1	2.1853	5.5339	0.557	1.08
	2	0.1964	1.2009	1.410	5.58
5	1	5.6334	0.0000	0.178	–
	2	0.7620	2.6530	0.917	2.14
	3	0.1172	1.0686	1.500	8.82
6	1	3.2721	11.6773	0.379	1.04
	2	0.4077	1.9873	1.086	3.46
	3	0.0815	1.0861	1.489	12.78
7	1	7.9064	0.0000	0.126	–
	2	1.1159	4.8963	0.670	1.98
	3	0.2515	1.5944	1.222	5.02
	4	0.0582	1.0348	1.527	17.46
8	1	4.3583	20.2948	0.286	1.03
	2	0.5791	3.1808	0.855	3.08
	3	0.1765	1.4507	1.285	6.83
	4	0.0448	1.0478	1.517	22.87
9	1	10.1759	0.0000	0.098	–
	2	1.4585	7.8971	0.526	1.93
	3	0.3561	2.3651	1.001	4.32
	4	0.1294	1.3165	1.351	8.87
	5	0.0348	1.0210	1.537	29.00
10	1	5.4449	31.3788	0.230	1.03
	2	0.7414	4.7363	0.699	2.94
	3	0.2479	1.9952	1.094	5.70
	4	0.1008	1.2638	1.380	11.15
	5	0.0283	1.0304	1.530	35.85

Fig. 3.14 *continued*

3.2 Lowpass/highpass transformation

In the logarithmic presentation (Bode plot), the amplitude-frequency response of a lowpass filter is transformed into the analogous highpass filter response by drawing its mirror image about the cutoff frequency, i.e. by replacing Ω by $1/\Omega$ and P by $1/P$. The cutoff

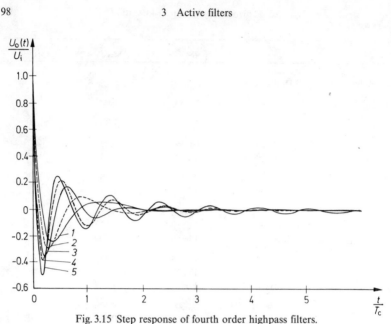

Fig. 3.15 Step response of fourth order highpass filters.
Curve 1: Highpass filter having critical damping. Curve 2: Bessel highpass filter. Curve 3:
Butterworth highpass filter. Curve 4: Chebyshev highpass filter with 0.5 dB ripple.
Curve 5: Chebyshev highpass filter with 3 dB ripple

frequency remains the same, and A_0 changes to A_∞. Equation (3.11) then becomes

$$A(P) = \frac{A_\infty}{\prod_i \left(1 + \frac{a_i}{P} + \frac{b_i}{P^2}\right)}. \qquad (3.12)$$

In the time domain, the performance cannot be transformed as the step response shows a basically different behaviour. This can be seen in Fig. 3.15, where an oscillation about the final-state value occurs even in highpass filters having critical damping. The analogy to the corresponding lowpass filters remains in as much as the transient oscillation decays more slowly, the higher the pole-pair Q-factors.

3.3   Realization of first order lowpass and highpass filters

According to Eq. (3.11), the transfer function of a first order lowpass filter has the general form

$$A(P) = \frac{A_0}{1 + a_1 P}. \qquad (3.13)$$

It can be implemented by the simple RC network in Fig. 3.1. From Section 3.1 it follows that for this circuit

$$A(P)=\frac{1}{1+pRC}=\frac{1}{1+\omega_c RCP}.$$

The low-frequency gain is defined by the value $A_0=1$, but the parameter a_1 can be chosen freely. Its value is found by comparing the coefficients

$$RC=\frac{a_1}{2\pi f_c}.$$

As can be seen from the table in Fig. 3.14, all filter types of the first order are identical and have the coefficient $a_1=1$. When higher order filters are built by cascade-connecting filter stages of lower orders, first order filter stages may be required for which $a_1 \neq 1$. The reason for this is that individual filter stages have, as a rule, a cutoff frequency different from that of the whole filter, namely $f_{c1}=f_c/a_1$.

The simple RC network in Fig. 3.1 has the disadvantage that its properties change when it is loaded. Therefore an impedance converter (buffer) is usually connected in series. If it is given the voltage gain A_0, the low-frequency gain can then be chosen freely. An appropriate circuit is presented in Fig. 3.16.

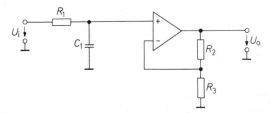

Fig. 3.16 First order lowpass filter with impedance converter.

Low-frequency voltage gain: $A_0=1+\dfrac{R_2}{R_3}$

In order to arrive at the corresponding highpass filter, the variable P in Eq. (3.13) must be replaced by $1/P$. In the circuit itself, the conversion is achieved by simply exchanging R_1 and C_1.

Somewhat simpler circuits for first order lowpass and highpass filters are obtained if the filtering network is included in the feedback loop of the operational amplifier. The corresponding lowpass filter is shown in Fig. 3.17, and its transfer function is

$$A(P)=-\frac{R_2/R_1}{1+\omega_c R_2 C_1 P}.$$

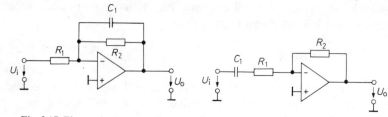

Fig. 3.17 First order lowpass filter
with inverting amplifier

Fig. 3.18 First order highpass filter
with inverting amplifier

For the actual design, one defines the cutoff frequency, the low-frequency gain A_0 which in this case is negative, and the capacitance C_1. From a comparison of the coefficients with those of Eq. (3.13) it follows that

$$R_2 = \frac{a_1}{2\pi f_c C_1} \quad \text{and} \quad R_1 = -\frac{R_2}{A_0}.$$

Figure 3.18 shows the corresponding highpass filter. Its transfer function is given by

$$A(P) = -\frac{R_2/R_1}{1 + \frac{1}{\omega_c R_1 C_1} \cdot \frac{1}{P}}.$$

By comparing coefficients with Eq. (3.12) it follows that

$$R_1 = \frac{1}{2\pi f_c a_1 C_1} \quad \text{and} \quad R_2 = -R_1 A_\infty.$$

The transfer functions given for the previous circuits apply only to the range of frequency for which the open-loop gain of the operational amplifier is large with respect to the absolute value of $\underline{A}$. This condition is difficult to fulfill for high frequencies as the magnitude of the open-loop gain falls at a rate of 6 dB/octave because of the necessary frequency compensation. For a standard operational amplifier at 10 kHz, $|\underline{A}_D|$ is only about 100.

On the other hand, this property can be used for the realization of lowpass filters at higher frequencies by giving the amplifier a purely resistive feedback [3.6]. The frequency response of the gain is then determined by the lowpass characteristic of the amplifier. A suitable amplifier circuit that leaves sufficient scope in its design, is shown in Fig. 3.19. Taking into account the finite complex open-loop gain $\underline{A}_D$, we obtain

$$\underline{U}_o = \alpha \frac{\underline{A}_D}{1 + k \underline{A}_D} \underline{U}_i. \tag{3.14}$$

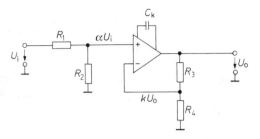

Fig. 3.19 First order lowpass filter with resistive feedback.

$$\alpha = \frac{R_2}{R_1 + R_2}; \quad k = \frac{R_4}{R_3 + R_4}$$

The frequency response of the open-loop gain of a frequency compensated operational amplifier can be described by a first order lowpass filter equation:

$$\underline{A}_D = \frac{A_D}{1 + j \dfrac{\omega}{\omega_{cA}}}, \tag{3.15}$$

where A_D is the low-frequency value and ω_{cA} the angular cutoff frequency of $\underline{A}_D$. Insertion in Eq. (3.14) gives

$$\underline{A} = \frac{\alpha/k}{1 + \dfrac{j\omega}{k A_D \omega_{cA}}},$$

with the approximation that $k A_D \gg 1$. The expression $A_D \cdot f_{cA} = f_T$ is the unity gain bandwidth of the operational amplifier. With $j\omega = p = \omega_c \cdot P$, we obtain the transfer function

$$A(P) = \frac{\alpha/k}{1 + \dfrac{f_c}{k f_T} P}. \tag{3.16}$$

Comparison of the coefficients with those of Eq. (3.13), gives the relationships

$$k = \frac{f_c}{a_1 f_T} \quad \text{and} \quad \alpha = A_0 k.$$

For the actual design of the circuit, certain additional conditions must be taken into account. The factors k and α are defined by the circuit as being ≤ 1. The output voltage swing decreases for higher frequencies due to the limited slew rate, and at frequency f_T is already very small. For this reason the cutoff frequency of the circuit, $f_{c1} = f_c/a_1$, should be chosen so as not to exceed approximately $0.1 f_T$. Hence, $k < 0.1$; but there is also a lower limit to the value of k: if k is chosen to be very small, one obtains a reduced loop gain $g=kA_D$ even at low frequency, and the d.c. voltage gain is thus poorly defined. With these two conditions, the specification is obtained, that $k\approx0.1$.

In order to be able to freely choose the cutoff frequency, f_T must be variable. This implies the use of an operational amplifier having external frequency compensation facilities, as for example the types µA 748 or LM 301. For these amplifiers, f_T can be approximated by

$$f_T=\frac{1\,\text{MHz}\cdot30\,\text{pF}}{C_k},$$

where C_k is the capacitance required for the frequency compensation. For reasons of stability, C_k must be chosen larger than $k\cdot30\,\text{pF}$, i.e. for our example at least 3 pF. The maximum attainable cutoff frequency is therefore

$$f_{c\,1\,\text{max}}=\frac{f_{c\,\text{max}}}{a_1}=k\cdot f_{T\,\text{max}}\approx0.1\cdot10\,\text{MHz}=1\,\text{MHz}.$$

3.4   Realization of second order lowpass and highpass filters

According to Eq. (3.11), the transfer function of a second order lowpass filter has the general form

$$A(P)=\frac{A_0}{1+a_1P+b_1P^2}. \tag{3.17}$$

As can be seen from the table in Fig. 3.14, the optimized transfer functions of second and higher orders have conjugate complex poles. Such transfer functions cannot be implemented by passive RC networks, as discussed in Section 3.1. One possibility of realizing these circuits is to use inductances and this is demonstrated in the following example.

3.4.1   *LRC* filters

The transfer function of the circuit in Fig. 3.20 is expressed by

$$A(P)=\frac{1}{1+\omega_c RCP+\omega_c^2 LCP^2}.$$

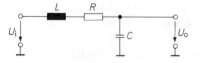

Fig. 3.20 Passive second order lowpass filter

Comparison of the coefficients with those of Eq. (3.17) yields

$$R = \frac{a_1}{2\pi f_c C} \quad \text{and} \quad L = \frac{b_1}{4\pi^2 f_c^2 C}.$$

From Fig. 3.14, the coefficients of a second order Butterworth lowpass filter are $a_1 = 1.414$ and $b_1 = 1.000$. For a given cutoff frequency of $f_c = 10\,\text{Hz}$ and a capacitance of $C = 10\,\mu\text{F}$, the remaining design parameters are $R = 2.25\,\text{k}\Omega$ and $L = 25.3\,\text{H}$. Obviously, such a filter is extremely hard to build because of the size of the inductance. However, this can be avoided by simulating the inductance with an active RC circuit. The gyrator in Fig. 2.32 is useful for this purpose, although its application involves a considerable number of components.

The desired transfer functions can be put into practice much more simply without inductance simulation by connecting suitable RC networks around operational amplifiers.

3.4.2 Filter with multiple negative feedback

The transfer function of the active lowpass filter in Fig. 3.21 is given by

$$A(P) = -\frac{R_2/R_1}{1 + \omega_c C_1 \left(R_2 + R_3 + \frac{R_2 R_3}{R_1}\right) P + \omega_c^2 C_1 C_2 R_2 R_3 P^2}.$$

By comparing the coefficients with those of Eq. (3.17), we obtain the relationships

$$A_0 = -R_2/R_1,$$

$$a_1 = \omega_c C_1 \left(R_2 + R_3 + \frac{R_2 R_3}{R_1}\right),$$

$$b_1 = \omega_c^2 C_1 C_2 R_2 R_3.$$

For the actual specification, the values of the resistors R_1 and R_3, for example, can be predetermined; the parameters R_2, C_1 and C_2 can

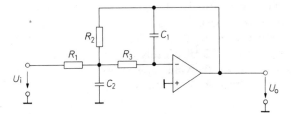

Fig. 3.21 Active second order lowpass filter with multiple negative feedback

then be calculated from the above equations. Such a determination is possible for all positive values of a_1 and b_1, so that any desired filter type can be realized. The gain at zero frequency, A_0, is negative. At low frequencies, therefore, the filter inverts the signal.

In order to attain the desired frequency response, the circuit elements must not have too large a tolerance. This requirement is easy to fulfill for resistors as they can be obtained off-the-shelf with one percent tolerance, in the E 96 standard series. The situation is different with capacitors, which, as a rule, are available off-the-shelf only in the E 6 series. It is therefore advantageous in filter design to predetermine the capacitors and calculate the values for the resistors. We therefore solve the design equations for the resistances and arrive at

$$R_2 = \frac{a_1 C_2 - \sqrt{a_1^2 C_2^2 - 4 C_1 C_2 b_1 (1 - A_0)}}{4 \pi f_c C_1 C_2},$$

$$R_1 = \frac{R_2}{-A_0},$$

$$R_3 = \frac{b_1}{4 \pi^2 f_c^2 C_1 C_2 R_2}.$$

In order that the value for R_2 is real, the condition

$$\frac{C_2}{C_1} \geqq \frac{4 b_1 (1 - A_0)}{a_1^2}$$

must be fulfilled. The most favourable design is obtained if the ratio C_2/C_1 is chosen not much larger than is prescribed by this condition. The data of the filter are relatively insensitive to the tolerances of the components, and therefore the circuit is particularly suited to the realization of filters having high Q-factors.

3.4.3 Filter with single positive feedback

Active filters can also be designed using amplifiers with positive feedback. However, the gain must be fixed at a precise value by an internal negative feedback ("controlled source"). The voltage divider R_3, $(\alpha - 1) R_3$ in Fig. 3.22 effects this negative feedback and defines the internal gain as having the value α. The positive feedback is caused by capacitor C_2. The transfer function is given by

$$A(P) = \frac{\alpha}{1 + \omega_c [C_1 (R_1 + R_2) + (1 - \alpha) R_1 C_2] P + \omega_c^2 R_1 R_2 C_1 C_2 P^2}.$$

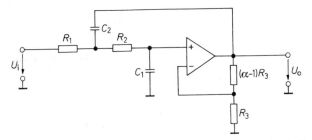

Fig. 3.22 Active second order lowpass filter with single positive feedback

The design can be considerably simplified if it is restricted to certain special cases. One possible case is to define the internal gain as $\alpha = 1$. Then $(\alpha - 1)R_3 = 0$, and both resistors R_3 can be omitted. Such operational amplifiers with unity feedback factor are available as integrated voltage followers (e.g. type LM 310). A simple impedance converter or buffer, for instance a Darlington pair, is often sufficient. In this way, filters in the MHz-range can also be realized. For the special case when $\alpha = 1$, the transfer function is given by

$$A(P) = \frac{1}{1 + \omega_c C_1 (R_1 + R_2) P + \omega_c^2 R_1 R_2 C_1 C_2 P^2}.$$

Defining C_1 and C_2 and comparing the coefficients with those of Eq. (3.17) gives

$$A_0 = 1,$$

$$R_{1/2} = \frac{a_1 C_2 \mp \sqrt{a_1^2 C_2^2 - 4 b_1 C_1 C_2}}{4 \pi f_c C_1 C_2}.$$

In order to arrive at values that are real, the condition

$$\frac{C_2}{C_1} \geq \frac{4 b_1}{a_1^2}$$

must be fulfilled. As for the filter with multiple negative feedback, the most favourable design is obtained if the ratio C_2/C_1 is chosen not much larger than is prescribed by this condition.

A further interesting special case occurs when identical resistors and capacitors are employed in the circuit, i.e. when $R_1 = R_2 = R$ and $C_1 = C_2 = C$. To enable implementation of the different filter types, the internal gain must, in this case, be variable. The transfer function then becomes

$$A(P) = \frac{\alpha}{1 + \omega_c R C (3 - \alpha) P + (\omega_c R C)^2 P^2}.$$

	critical	Bessel	Butterworth	3 dB-Chebyshev	undamped
α	1.000	1.268	1.586	2.234	3.000

Fig. 3.23 Internal gain for single positive feedback

By comparing the coefficients with those in Eq. (3.17) we arrive at the relationships

$$RC = \frac{\sqrt{b_1}}{2\pi f_c},$$

$$\alpha = A_0 = 3 - \frac{a_1}{\sqrt{b_1}} = 3 - \frac{1}{Q_1}.$$

As can be seen, the internal gain α is dependent only on the pole-pair Q-factor and not on the cutoff frequency f_c. The value of α therefore determines the type of filter. Insertion of the coefficients given in Fig. 3.14 for a second order filter results in the values for α given in Fig. 3.23. For $\alpha = 3$, the circuit produces a self-contained oscillation at the frequency $f = 1/2\pi RC$. It can be seen that the adjustment of the internal gain becomes more difficult as one approaches the value $\alpha = 3$. For the Chebyshev filter in particular, a very precise adjustment is therefore necessary. This is a slight disadvantage when compared with the previous filters. A considerable advantage, however, is due to the fact that the type of filter is solely determined by α and is not dependent on R and C. Hence, the cutoff frequency of this filter circuit can be changed particularly easily, for example by a dual-gang potentiometer for the two identical resistors R_1 and R_2 in Fig. 3.22.

If the resistors and capacitors are interchanged, one arrives at the *highpass filter* in Fig. 3.24, the transfer function of which is

$$A(P) = \frac{\alpha}{1 + \frac{R_2(C_1 + C_2) + R_1 C_2(1 - \alpha)}{R_1 R_2 C_1 C_2 \omega_c} \cdot \frac{1}{P} + \frac{1}{R_1 R_2 C_1 C_2 \omega_c^2} \cdot \frac{1}{P^2}}.$$

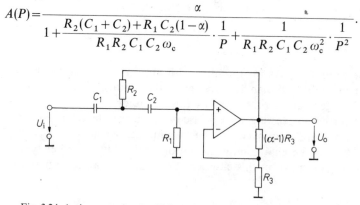

Fig. 3.24 Active second order highpass filter with single positive feedback

To simplify the design process, we choose the special case where $\alpha = 1$ and $C_1 = C_2 = C$. Comparing the coefficients with those of Eq. (3.12) gives

$$A_\infty = 1,$$

$$R_1 = \frac{1}{\pi f_c C a_1},$$

$$R_2 = \frac{a_1}{4 \pi f_c C b_1}.$$

3.4.4   Lowpass filters with resistive negative feedback

In Section 3.3, a method is shown for the design of first order lowpass filters for higher frequencies. It employs the frequency response of the open-loop gain of an operational amplifier by giving the amplifier a purely resistive feedback. The same method can also be used for second order lowpass filters having complex poles by giving two amplifiers a common feedback loop, as is represented in Fig. 3.25.

The gain-bandwidth product f_T of the two amplifiers is assumed to be the same. With Eq. (3.15) we obtain the transfer function

$$A(P) = - \frac{R_3/R_1}{1 + \dfrac{R_3 f_c}{\alpha R_2 f_T} P + \left[1 + \dfrac{R_3 (R_1 + R_2)}{R_1 R_2} \right] \dfrac{f_c^2}{\alpha f_T^2} P^2}. \qquad (3.18)$$

Comparing the coefficients with those of Eq. (3.17) gives

$$A_0 = - R_3/R_1,$$

$$a_1 = \frac{R_3 f_c}{\alpha R_2 f_T},$$

$$b_1 = \left[1 + \frac{R_3 (R_1 + R_2)}{R_1 R_2} \right] \frac{f_c^2}{\alpha f_T^2}.$$

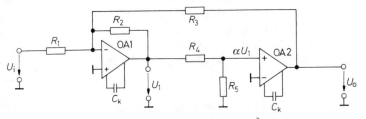

Fig. 3.25 Active second order lowpass filter with resistive feedback. $\alpha = R_5/(R_4 + R_5)$. The output U_1 has bandpass behaviour. Its operation is described in Section 3.7.4

Therefore we arrive at the design equations

$$\alpha = \frac{(f_c/f_T)^2(1-A_0)}{b_1 - \alpha_1(f_c/f_T)} \quad \text{where} \quad A_0 < 0, \tag{3.19}$$

$$R_2 = -\frac{A_0 f_c}{\alpha a_1 f_T} R_1, \tag{3.20a}$$

$$R_3 = -A_0 R_1. \tag{3.20b}$$

For the actual design of the circuit, the factor (f_c/f_T) is predetermined as approximately 0.1 to attain a sufficient large-signal bandwidth. In order that the required cutoff frequency is obtained, f_T must be fixed to the appropriate value using the two capacitors C_k, as described in Section 3.3. The Eq. (3.19) then enables the calculation of the divider ratio α which should be within the range $0.01...0.1$. If this is not possible, either (f_c/f_T) or A_0 must be modified. When R_1 has been defined, the resistances R_2 and R_3 may be calculated from the Eq. (3.20a) and (3.20b).

The design process is illustrated by the following example: a Butterworth filter having a cutoff frequency of 100 kHz and a d.c. voltage gain of $A_0 = -2$ is required. We choose $(f_c/f_T) = 0.1$, that is $f_T = 1$ MHz. From Fig. 3.14, we take $a_1 = 1.4142$ and $b_1 = 1$. From Eq. (3.19), it then follows that $\alpha = 0.035$. We choose R_1 to be 1 kΩ and therefore obtain from Eq. (3.20a), $R_2 = 4.04$ kΩ and from Eq. (3.20b), $R_3 = 2$ kΩ.

3.5 Realization of lowpass and highpass filters of higher orders

In cases where the filter characteristic is not steep enough, filters of higher orders must be employed. For this purpose, first and second order filters are cascade-connected, thereby multiplying the frequency responses of the individual filters. It would however be wrong to cascade two second order Butterworth filters, for example, to obtain a fourth order Butterworth filter. The resulting filter would have a different cutoff frequency and also a different filter characteristic. The coefficients of the individual filters must therefore be chosen so that the product of the frequency responses gives the desired optimized filter type.

To simplify the determination of the circuit parameters of the individual filters, we have factored the polynomials of the different filter types. The coefficients a_i and b_i of the individual filter stages are given in Fig. 3.14. Each factor can be implemented by one of the first or second order filters described previously, where the coefficients a_1 and b_1 are replaced by a_i and b_i. For the calculation of the circuit parameters from the given formulae, the desired cutoff frequency of the *resulting total filter* must be inserted. As a rule, the individual filter stages possess cutoff frequencies different from that of the whole filter, as can be seen in Fig. 3.14.

The sequence in which the individual filter stages are cascaded is, in principle, not significant, as the resulting frequency response remains the same. In practice, however, there are several design considerations for the best sequence of the filter stages. One such aspect is the permissible voltage swing for which it is useful to arrange the filter stages according to their cutoff frequencies. The one having the lowest cutoff frequency is at the input; the first stage may otherwise be saturated while the output of the second stage is still below the maximum permissible voltage swing. The reason for this is that the filter stages having the higher cutoff frequencies also invariably have the higher pole-pair Q, and therefore show a rise in gain in the vicinity of their cutoff frequency. This can be seen in Fig. 3.26 where the amplitude-frequency responses of a 0.5 dB Chebyshev lowpass filter of tenth order and of its five individual stages are shown. It is obvious that the permissible output voltage swing is highest if the filter stages having low cutoff frequencies are at the input of the filter cascade.

Another aspect which may have to be considered for a suitable arrangement of the filter stages is noise. In this case, just the reverse sequence is most favourable as then the filters with the lower cutoff

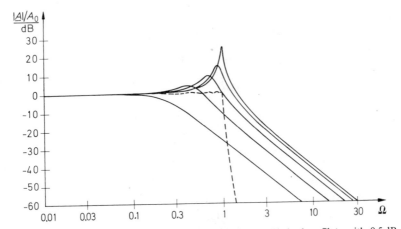

Fig. 3.26 Amplitude-frequency responses of a tenth order Chebyshev filter with 0.5 dB ripple and of its five individual filter stages

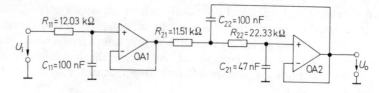

Fig. 3.27 Third order Bessel lowpass filter having a cutoff frequency $f_c = 100\,\text{Hz}$

frequencies at the end of the filter chain reduce the noise due to the input stages.

The design process is demonstrated for a third order Bessel lowpass filter. It is to be constructed from the first order lowpass filter in Fig. 3.16 and from the second order lowpass filter of Fig. 3.22 for which the special case of $\alpha = 1$ (described in Section 3.4.3) is chosen. The low-frequency gain is defined as unity. To achieve this, the impedance converter in the first order filter stage must have the gain $\alpha = 1$. The resulting circuit is represented in Fig. 3.27.

The desired cutoff frequency is $f_c = 100\,\text{Hz}$. For the calculation of the first filter stage, we predetermine $C_{11} = 100\,\text{nF}$ and obtain, according to Section 3.3, with the coefficients from Fig. 3.14:

$$R_{11} = \frac{a_1}{2\pi f_c C_{11}} = \frac{0.7560}{2\pi \cdot 100\,\text{Hz} \cdot 100\,\text{nF}} = 12.03\,\text{k}\Omega.$$

For the second filter stage, we set $C_{22} = 100\,\text{nF}$ and obtain, according to Section 3.4.3, the condition for C_{21}:

$$C_{21} \leqq C_{22}\frac{a_2^2}{4b_2} = 100\,\text{nF} \cdot \frac{(0.9996)^2}{4 \cdot 0.4772},$$
$$C_{21} \leqq 52.3\,\text{nF}.$$

We choose the nearest standard value $C_{21} = 47\,\text{nF}$ and arrive at

$$R_{21/22} = \frac{a_2 C_{22} \mp \sqrt{a_2^2 C_{22}^2 - 4b_2 C_{21} C_{22}}}{4\pi f_c C_{21} C_{22}},$$
$$R_{21} = 11.51\,\text{k}\Omega,$$
$$R_{22} = 22.33\,\text{k}\Omega.$$

For third order filters, it is possible to omit the first operational amplifier. The simple lowpass filter of Fig. 3.1 is then connected in front of the second order filter. Due to the mutual loading of the filters, a different method of calculation must be employed which is considera-

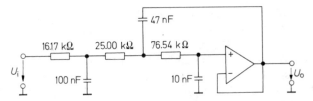

Fig. 3.28 Simplified third order Bessel lowpass filter having a cutoff frequency
$f_c = 100\,\text{Hz}$

bly more difficult than in the decoupled case. Fig. 3.28 shows such a circuit having the same characteristics as that in Fig. 3.27.

3.6   Lowpass/bandpass transformation

In Section 3.2 it is shown how, by a transformation of the frequency variable, a given lowpass frequency response can be converted to the corresponding highpass frequency response. With a very similar transformation, the frequency response of a bandpass filter can be created, i.e. by replacing the frequency variable P in the lowpass transfer function by the expression

$$\frac{1}{\Delta\Omega}\left(P+\frac{1}{P}\right). \tag{3.21}$$

By means of this transformation, the amplitude response of the lowpass filter in the range $0\leq\Omega\leq1$ is converted into the pass range of a bandpass filter between the centre frequency $\Omega=1$ and the upper cutoff frequency Ω_{max}. On a logarithmic frequency scale, it also appears as a mirror image below the centre frequency. The lower cutoff frequency is then $\Omega_{min}=1/\Omega_{max}$ [3.7]. Figure 3.29 illustrates this process.

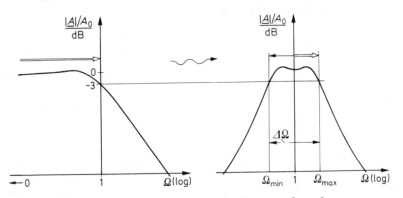

Fig. 3.29 Illustration of the lowpass/bandpass transformation

The normalized bandwidth $\Delta\Omega = \Omega_{max} - \Omega_{min}$ can be chosen freely. The described transformation results in the bandpass filter having the same gain at Ω_{min} and Ω_{max} as the corresponding lowpass filter at $\Omega = 1$. If the lowpass filter, as in the table Fig. 3.14, is normalized with respect to the 3 dB cutoff frequency, $\Delta\Omega$ represents the normalized 3 dB bandwidth of the bandpass filter. Since $\Delta\Omega = \Omega_{max} - \Omega_{min}$ and $\Omega_{max} \cdot \Omega_{min} = 1$, we obtain for the normalized 3 dB cutoff frequencies

$$\Omega_{max/min} = \tfrac{1}{2}\sqrt{(\Delta\Omega)^2 + 4} \pm \tfrac{1}{2}\Delta\Omega.$$

3.6.1   Second order bandpass filters

The simplest bandpass filter results from the application of the transformation equation (3.21) to a first order lowpass filter where

$$A(P) = \frac{A_0}{1 + P}.$$

Therefore the transfer function of the second order bandpass filter is

$$A(P) = \frac{A_0}{1 + \dfrac{1}{\Delta\Omega}\left(P + \dfrac{1}{P}\right)} = \frac{A_0 \Delta\Omega P}{1 + \Delta\Omega P + P^2}. \tag{3.22}$$

The interesting parameters of bandpass filters are the gain A_r at the resonant frequency f_r, and the quality factor Q. It follows directly from the given transformation characteristic that $A_r = A_0$. This can be easily verified by making $\Omega = 1$, i.e. $P = j$ in Eq. (3.22). Since A_r is real, the phase shift at resonant frequency is zero.

As for a resonant circuit, the Q-factor is defined as the ratio of the resonant frequency f_r to the bandwidth B. Therefore,

$$Q = \frac{f_r}{B} = \frac{f_r}{f_{max} - f_{min}} = \frac{1}{\Omega_{max} - \Omega_{min}} = \frac{1}{\Delta\Omega}. \tag{3.23}$$

Inserting this in Eq. (3.22) gives

$$\boxed{A(P) = \frac{(A_r/Q)\,P}{1 + \dfrac{1}{Q}P + P^2}.} \tag{3.24}$$

This equation is the transfer function of a second order bandpass filter and enables the direct identification of all parameters of interest.

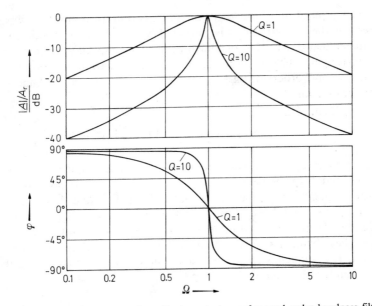

Fig. 3.30 Frequency response of amplitude and phase of second order bandpass filters having the quality factors $Q=1$ and $Q=10$

With $P=j\Omega$, we obtain from Eq. (3.24) the frequency response of the amplitude and the phase

$$|\underline{A}| = \frac{(A_r/Q)\,\Omega}{\sqrt{1+\Omega^2\left(\dfrac{1}{Q^2}-2\right)+\Omega^4}}, \qquad (3.24\,\text{a})$$

$$\varphi = \tan^{-1}\frac{Q(1-\Omega^2)}{\Omega}. \qquad (3.24\,\text{b})$$

These two functions are shown in Fig. 3.30 for the Q-factors 1 and 10.

3.6.2   Fourth order bandpass filters

The amplitude-frequency response of second order bandpass filters becomes more peaked the larger the Q-factor chosen. However, there are many applications where the curve must be as flat as possible in the region of the resonant frequency but must also have a steep change-over to the stopband. This optimization problem may be solved by the application of the lowpass/bandpass transformation to higher order lowpass filters. It is then possible to choose freely not only the bandwith $\Delta\Omega$, but also the most suitable filter type.

The application of the lowpass/bandpass transformation to lowpass filters of the second order is particularly important. It results in a mathematical description of fourth order bandpass filters. Such filters are investigated below. By inserting the transformation equation (3.21) in the second order lowpass equation (3.17), we obtain the bandpass transfer function

$$A(P) = \frac{P^2 A_0 (\Delta\Omega)^2/b_1}{1 + \frac{a_1}{b_1}\Delta\Omega P + \left[2 + \frac{(\Delta\Omega)^2}{b_1}\right]P^2 + \frac{a_1}{b_1}\Delta\Omega P^3 + P^4}. \tag{3.25}$$

It can be seen that the asymptotes of the amplitude-frequency response at low and high frequencies have the slope $\pm 12\,\text{dB/octave}$. At the centre frequency $\Omega = 1$, the gain is real and has the value $A_m = A_0$.

In Fig. 3.31 we have plotted the frequency response of the amplitude and the phase of a Butterworth bandpass filter, and of a 0.5 dB Chebyshev bandpass filter, both having a normalized bandwidth $\Delta\Omega = 1$. The frequency response of a second order bandpass having the same bandwidth is shown for comparison.

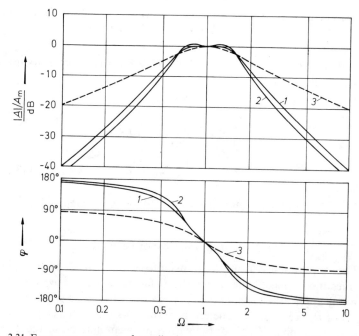

Fig. 3.31 Frequency response of amplitude and phase for bandpass filters having a bandwidth $\Delta\Omega = 1$.
Curve 1: Fourth order Butterworth bandpass filter. Curve 2: Fourth order Chebyshev bandpass filter with 0.5 dB ripple. Curve 3: Second order bandpass for comparison

As for the lowpass filters, we shall simplify the design process by splitting the denominator into quadratic factors. For reasons of symmetry we can choose a more simple formulation and write

$$A(P) = \frac{P^2 A_m (\Delta\Omega)^2 / b_1}{\left[1 + \dfrac{\alpha P}{Q_i} + (\alpha P)^2\right]\left[1 + \dfrac{1}{Q_i}\left(\dfrac{P}{\alpha}\right) + \left(\dfrac{P}{\alpha}\right)^2\right]}. \tag{3.26}$$

By multiplying out and comparing the coefficients with those in Eq. (3.25), we arrive at the equation for α:

$$\alpha^2 + \left[\frac{\alpha \Delta\Omega a_1}{b_1(1+\alpha^2)}\right]^2 + \frac{1}{\alpha^2} - 2 - \frac{(\Delta\Omega)^2}{b_1} = 0. \tag{3.27}$$

For any particular application it can be easily solved numerically with the help of a pocket calculator. Having determined α, the pole-pair quality factor Q_i of an individual filter stage is

$$Q_i = \frac{(1+\alpha^2) b_1}{\alpha \Delta\Omega a_1}. \tag{3.28}$$

There are two possible ways of realizing the filter, depending on how the numerator is factored. Splitting-up into a constant factor and a factor containing P^2 yields the cascade connection of a highpass and a lowpass filter. This design is of advantage for the realization of large bandwidths $\Delta\Omega$.

For smaller bandwidths, $\Delta\Omega \lesssim 1$, it is more suitable to use a cascade connection of two second order bandpass filters which are not tuned to quite the same centre frequency. This method is called "staggered tuning". For the design of the individual bandpass filter stages, we split the numerator of Eq. (3.26) into two factors containing P and obtain

$$A(P) = \frac{(A_r/Q_i)(\alpha P)}{1 + \dfrac{\alpha P}{Q_i} + (\alpha P)^2} \cdot \frac{(A_r/Q_i)(P/\alpha)}{1 + \dfrac{1}{Q_i}\left(\dfrac{P}{\alpha}\right) + \left(\dfrac{P}{\alpha}\right)^2}. \tag{3.29}$$

Comparing the coefficients with those of Eq. (3.26) and Eq. (3.24), we obtain the parameters of the two individual bandpass filters:

	f_r	Q	A_r
1st filter stage	f_m/α	Q_i	$Q_i \Delta\Omega \sqrt{A_m/b_1}$
2nd filter stage	$f_m \cdot \alpha$	Q_i	$Q_i \Delta\Omega \sqrt{A_m/b_1}$

$$(3.30)$$

where f_m is the centre frequency of the resulting bandpass filter and A_m is the gain at this frequency. The factors α and Q_i are given by Eqs. (3.27) and (3.28).

The determination of the parameters of the individual filter stages is demonstrated by an example. A Butterworth bandpass is required with a centre frequency of 1 kHz and a bandwidth of 100 Hz. The gain at the centre frequency is required to be $A_m = 1$. To begin with, we take the coefficients of a second order Butterworth lowpass filter from the table in Fig. 3.14: $a_1 = 1.4142$ and $b_1 = 1.000$. As $\Delta\Omega = 0.1$, Eq. (3.27) gives $\alpha = 1.0360$. Equation (3.28) yields $Q_i = 14.15$, and from Eq. (3.30) $A_r = 1.415$, $f_{r1} = 965$ Hz and $f_{r2} = 1.036$ kHz.

3.7   Realization of second order bandpass filters

The cascade connection of a highpass and a lowpass filter of first order as in Fig. 3.32, gives a bandpass filter with the transfer function

$$A(p) = \frac{1}{1 + \dfrac{1}{\alpha\, pRC}} \cdot \frac{1}{1 + \dfrac{pRC}{\alpha}} = \frac{\alpha\, pRC}{1 + \dfrac{1+\alpha^2}{\alpha}\, pRC + (pRC)^2}.$$

With the resonant frequency $\omega_r = 1/RC$, the normalized form is

$$A(P) = \frac{\alpha P}{1 + \dfrac{1+\alpha^2}{\alpha}\, P + P^2}.$$

Comparison of the coefficients with those of Eq. (3.24) gives the Q-factor

$$Q = \frac{\alpha}{1+\alpha^2}.$$

For $\alpha = 1$, Q is at maximum, $Q_{max} = \frac{1}{2}$, which is the highest Q-factor that can be attained by a cascade connection of first order filters. For higher Q-factors, the denominator of Eq. (3.24) must have complex zeros, but such a transfer function can only be implemented by LRC circuits or by special active RC circuits which are discussed below.

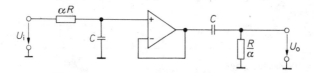

Fig. 3.32 Bandpass filter consisting of a first order highpass and lowpass filter

3.7.1 *LRC* bandpass filter

A common method of designing selective filters having high quality factors is the use of resonant circuits. Figure 3.33 shows such a circuit, the transfer function of which is

$$A(p) = \frac{pRC}{1 + pRC + p^2 LC}.$$

Since the resonant frequency $\omega_r = 1/\sqrt{LC}$, it follows that

$$A(P) = \frac{R\sqrt{\dfrac{C}{L}}\,P}{1 + R\sqrt{\dfrac{C}{L}}\,P + P^2}.$$

Comparison of the coefficients with those of Eq. (3.24) gives

$$Q = \frac{1}{R}\sqrt{\frac{L}{C}} \quad \text{and} \quad A_r = 1.$$

For high frequencies the necessary inductances can be easily realized and have very little loss. For the low-frequency range the inductances become unwieldy and have a poor electrical performance. If, for example, a filter having the resonant frequency $f_r = 10\,\text{Hz}$ is to be built with the circuit in Fig. 3.33, the inductance $L = 25.3\,\text{H}$ is needed when a

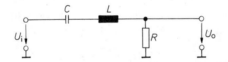

Fig. 3.33 *LRC* bandpass filter

capacitance of $10\,\mu\text{F}$ is chosen. As has been already shown in Section 3.4.1 for the lowpass and highpass filters, such inductances can be simulated, for instance by means of gyrators. In most cases, the desired transfer function of Eq. (3.24) can be put into practice much more easily by inserting suitable RC networks in the feedback loop of an operational amplifier.

3.7.2 Bandpass filter with multiple negative feedback

The principle of multiple negative feedback can also be applied to bandpass filters. The appropriate circuit is shown in Fig. 3.34. Its

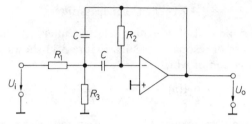

Fig. 3.34 Bandpass filter with multiple negative feedback.

Resonant frequency: $\qquad f_r = \dfrac{1}{2\pi C}\sqrt{\dfrac{R_1 + R_3}{R_1 R_2 R_3}}$

Gain at f_r: $\qquad\qquad -A_r = \dfrac{R_2}{2R_1}$

Q-factor: $\qquad\qquad\quad Q = \pi R_2 C f_r$

Bandwidth: $\qquad\qquad B = \dfrac{1}{\pi R_2 C}$

transfer function is

$$A(P) = \frac{-\dfrac{R_2 R_3}{R_1 + R_3}\, C\omega_r P}{1 + \dfrac{2R_1 R_3}{R_1 + R_3}\, C\omega_r P + \dfrac{R_1 R_2 R_3}{R_1 + R_3}\, C^2 \omega_r^2 P^2}.$$

As can be seen by comparison with Eq. (3.24), the coefficient of P^2 must be unity. Therefore, the resonant frequency is given by

$$f_r = \frac{1}{2\pi C}\sqrt{\frac{R_1 + R_3}{R_1 R_2 R_3}}. \tag{3.31}$$

Inserting this relationship in the transfer function and comparing the remaining coefficients with those of Eq. (3.24) gives the other parameters

$$-A_r = \frac{R_2}{2R_1}, \tag{3.32}$$

$$Q = \frac{1}{2}\sqrt{\frac{R_2(R_1 + R_3)}{R_1 R_3}} = \pi R_2 C f_r. \tag{3.33}$$

It can be seen that the gain, the Q-factor and the resonant frequency can be freely chosen.

From Eq. (3.33), we obtain the bandwidth of the filter

$$B = \frac{f_r}{Q} = \frac{1}{\pi R_2 C},$$

which is independent of R_1 and R_3. On the other hand, it can be seen from Eq. (3.32) that A_r is not dependent on R_3. It is therefore possible to vary the resonant frequency by R_3 without influencing the bandwidth B and the gain A_r.

If resistor R_3 is omitted, the filter remains functional, but the Q-factor becomes dependent on A_r. This is because, with $R_3 \to \infty$, Eq. (3.33) becomes

$$-A_r = 2Q^2.$$

In order that the loop gain of the circuit is very much larger than unity, the open-loop gain of the operational amplifier must be large compared to $2Q^2$. With the resistor R_3, high Q-factors can be attained even for a low gain A_r. As can be seen in Fig. 3.34, the low gain is due only to the fact that the input signal is attenuated by the voltage divider R_1, R_3. Therefore, the open-loop gain of the operational amplifier must in this case also be large compared to $2Q^2$. This requirement is particularly strict as it must be met around the resonant frequency; it determines the choice of the operational amplifier, especially for applications at higher frequency.

The determination of the circuit parameters is shown by the following example. A selective filter is required, having the resonant frequency $f_r = 10\,\text{Hz}$ and the quality factor $Q = 100$. Therefore, the cutoff frequencies lie at approximately 9.95 Hz and 10.05 Hz. The gain at resonant frequency is required to be $-A_r = 10$. One of the parameters can be freely chosen, e.g. $C = 1\,\mu\text{F}$, and the remainder must be calculated. To begin with, from Eq. (3.33)

$$R_2 = \frac{Q}{\pi f_r C} = 3.18\,\text{M}\Omega.$$

Hence, from Eq. (3.32)

$$R_1 = \frac{R_2}{-2A_r} = 159\,\text{k}\Omega.$$

The resistance R_3 is given by Eq. (3.31) as

$$R_3 = \frac{-A_r R_1}{2Q^2 + A_r} = 79.5\,\Omega.$$

The open-loop gain of the operational amplifier must, at resonant frequency, still be large compared with $2Q^2 = 20 \cdot 10^3$.

One advantage of the circuit is that it has no tendency to oscillate at resonant frequency, even if the circuit elements do not quite match their theoretical values. Naturally, this is true only if the operational amplifier is correctly frequency compensated; otherwise high-frequency oscillations occur.

3.7.3   Bandpass filter with single positive feedback

The application of single positive feedback results in the bandpass circuit in Fig. 3.35. The negative feedback via the resistors R_1 and $(k-1)R_1$ gives the internal gain the value k. The transfer function is given by

$$A(P) = \frac{kRC\omega_r P}{1 + RC\omega_r(3-k)P + R^2 C^2 \omega_r^2 P^2}.$$

Comparison of the coefficients with those of Eq. (3.24) yields the equations given for the determination of the circuit parameters. A disadvantage is that Q and A_r cannot be chosen independently of one another. The advantage, however, is that the Q-factor may be altered by a variation of k without at the same time changing the resonant frequency.

For $k=3$, the gain is infinite, and an undamped oscillation occurs. The adjustment of the internal gain therefore becomes more critical the closer it approaches the value 3.

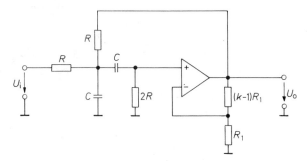

Fig. 3.35   Bandpass filter with single positive feedback.

$$\text{Resonant frequency:} \quad f_r = \frac{1}{2\pi RC}$$

$$\text{Gain at } f_r: \qquad A_r = \frac{k}{3-k}$$

$$\text{Q-factor:} \qquad Q = \frac{1}{3-k}$$

3.7.4   Bandpass filter with resistive negative feedback

In Section 3.4.4, we show that it is of advantage at higher frequencies to use the operational amplifier itself as the element determining the frequency response. Using this method, not only lowpass but also bandpass filters can be built. For this purpose, the lowpass filter circuit of Fig. 3.25 may be used again, but with the voltage U_1 as output voltage. For the calculation of the bandpass transfer function we begin with the lowpass transfer function of Eq. (3.18) and insert the relationship between the U_o and U_1, which is given by Eq. (3.15)

$$\frac{U_o}{U_1} = \frac{\alpha A_D}{1 + j \dfrac{\omega}{\omega_{cA}}} .$$

By introducing this into Eq. (3.18), we obtain the transfer function for the voltage U_1

$$A_{BP}(P) = - \frac{\dfrac{R_3 f_r}{\alpha R_1 f_T} P}{1 + \dfrac{R_3 f_r}{\alpha R_2 f_T} P + \left[1 + \dfrac{R_3(R_1 + R_2)}{R_1 R_2} \right] \dfrac{f_r^2}{\alpha f_T^2} P^2} .$$

Comparison of the coefficients with those in Eq. (3.24) gives the conditions

$$A_r = -\frac{R_2}{R_1}, \tag{3.34}$$

$$Q = \frac{\alpha R_2 f_T}{R_3 f_r}, \tag{3.35}$$

$$\left[1 + \frac{R_3(R_1 + R_2)}{R_1 R_2} \right] \frac{f_r^2}{\alpha f_T^2} = 1. \tag{3.36}$$

The actual design of the circuit must follow the same principles as in Section 3.4.4, that is f_r/f_T should be approximately $0.1 \ldots 0.2$ and α about $0.01 \ldots 0.1$. Initially, the resistor R_1 is defined and R_2 calculated from Eq. (3.34). The ratio f_r/f_T is then chosen, and from Eqs. (3.35) and (3.36) the values for α and R_3 are determined as

$$\alpha = \frac{Q(f_r/f_T)^2}{Q - \dfrac{f_r}{f_T}(1 - A_r)} \quad \text{and} \quad R_3 = \frac{\alpha R_2 f_T}{Q f_r} .$$

Should this value for α be unfavourable, the ratio f_r/f_T or the gain A_r must be suitably modified.

   An actual example should clarify the design process. A bandpass filter is required having $f_r = 100\,\text{kHz}$, $Q = 3$ and $A_r = -5$. We choose $R_1 = 1.5\,\text{k}\Omega$ and obtain $R_2 = 7.5\,\text{k}\Omega$. We then define f_r/f_T as 0.2, and therefore $f_T = 500\,\text{kHz}$. Hence, we arrive at $\alpha = 0.067$ and $R_3 = 833\,\Omega$.

3.8   Lowpass/band-rejection filter transformation

For a selective rejection of a particular frequency, a filter is needed, the gain of which is zero at the resonant frequency and rises to a constant value at higher and lower frequencies. Such filters are called *rejection filters* or notch filters. To characterize the selectivity, the rejection quality factor is defined as $Q = f_r/B$, where B is the 3 dB bandwidth. The larger the Q-factor of the filter, the more steeply the gain falls in the vicinity of the resonant frequency f_r.

As for the bandpass filter, the amplitude-frequency response of the band-rejection filter can be found from the frequency response of a lowpass filter by using a suitable frequency transformation. To accomplish this, the variable P is replaced by the expression

$$\frac{\Delta\Omega}{P + \dfrac{1}{P}}, \tag{3.37}$$

where $\Delta\Omega = 1/Q$ is the normalized 3 dB bandwidth. By means of this transformation, the amplitude of the lowpass filter in the range $0 \leq \Omega \leq 1$ is converted to the pass range of the band-rejection filter between $0 \leq \Omega \leq \Omega_{c\,1}$. In addition, it appears on a logarithmic scale as a mirror image about the resonant frequency. At the resonant frequency $\Omega = 1$, the transfer function is zero. As for the bandpass filter, the order of the filter is doubled by the transformation. It is of particular interest to apply the transformation to a first order lowpass filter. This results in a notch filter of the second order, having a transfer function

$$A(P) = \frac{A_0(1 + P^2)}{1 + \Delta\Omega P + P^2} = \frac{A_0(1 + P^2)}{1 + \dfrac{1}{Q}P + P^2}. \tag{3.38}$$

From this, we obtain the relationships for the amplitude-frequency response and the phase-frequency response

$$|\underline{A}| = \frac{A_0|(1 - \Omega^2)|}{\sqrt{1 + \Omega^2\left(\dfrac{1}{Q^2} - 2\right) + \Omega^4}},$$

$$\varphi = \tan^{-1}\frac{\Omega}{Q(\Omega^2 - 1)}.$$

The corresponding curves are shown in Fig. 3.36 for the rejection quality factors 1 and 10.

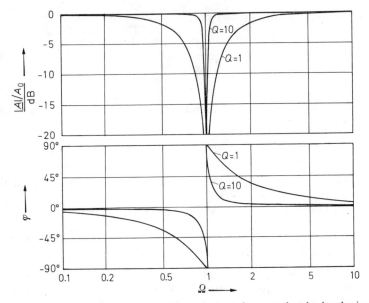

Fig. 3.36 Frequency response of amplitude and phase for second order band-rejection filters having the quality factors $Q=1$ and $Q=10$

The denominator of Eq. (3.38) is identical to that of Eq. (3.24) for bandpass filters. It follows from Eq. (3.24) that a maximum Q-factor of only $Q=\frac{1}{2}$ can be attained with passive RC circuits; for higher Q-factors, LRC networks or special active RC circuits must be employed.

3.9 Realization of second order rejection filters

3.9.1 LRC rejection filter

A well known method for the construction of rejection filters is the application of series-tuned oscillating circuits as in Fig. 3.37. At resonant frequency, the arrangement represents a short-circuit and the output voltages is zero. The transfer function of the circuit is

$$A(p)=\frac{1+p^2 LC}{1+p\,RC+p^2 LC}.$$

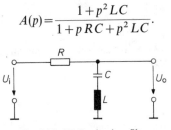

Fig. 3.37 LRC rejection filter

Hence, the resonant frequency $\omega_r = 1/\sqrt{LC}$, and we obtain the normalized form

$$A(P) = \frac{1+P^2}{1+R\sqrt{\dfrac{C}{L}}\,P+P^2}.$$

The rejection quality factor is found by comparing the coefficients with those of Eq. (3.24),

$$Q = \frac{1}{R}\sqrt{\frac{L}{C}}.$$

This holds for lossless coils only, because only for such coils can the output voltage fall to zero. In addition, the same considerations for the use of inductances apply here as for the bandpass filters.

3.9.2   Active parallel-T *RC* rejection filter

It can be shown [1] that the parallel-T filter represents a passive *RC* rejection filter. Its rejection *Q*-factor is $Q = 0.25$, which can be raised by incorporating the parallel-T filter in the feedback loop of an amplifier. One practicable version is shown in Fig. 3.38.

For high and low frequencies, the parallel-T filter transfers the input signal unchanged. The output voltage of the impedance converter is then $k\underline{U_i}$. At the resonant frequency, the output voltage is zero. In this case, the parallel-T filter behaves as if the resistor $R/2$ were

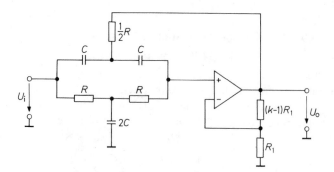

Fig. 3.38 Active parallel-T rejection filter.

Resonant frequency: $f_r = \dfrac{1}{2\pi RC}$

Gain: $A_r = k$

Rejection Q-factor: $Q = \dfrac{1}{2(2-k)}$

connected to ground. Therefore, the resonant frequency $f_r = 1/2\pi RC$, remains unchanged. The transfer function of the whole circuit is

$$A(P) = \frac{k(1+P^2)}{1+2(2-k)P+P^2}$$

from which the filter data given below Fig. 3.38 can be directly deduced. If the voltage follower has unity gain, $Q = 0.5$. For an increase in gain, Q rises towards infinity as k approaches 2.

A precondition for the correct operation of the circuit is the precise adjustment of the resonant frequency and gain of the parallel-T filter. This is difficult to achieve for higher Q-factors since the variation of one resistance always influences both parameters simultaneously. The active Wien-Robinson rejection filter is more favourable in this respect.

3.9.3   Active Wien-Robinson rejection filter

As is shown in [1], the Wien-Robinson bridge also behaves as a notch filter. However, its Q-factor is not very much higher than that of the parallel-T filter. As for the parallel-T filter, the Q-factor can be raised to any desired value by incorporating the filter in the feedback loop of an operational amplifier. The corresponding circuit is shown in

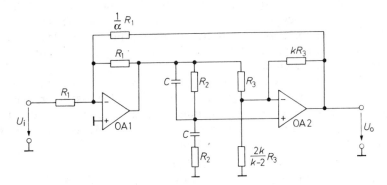

Fig. 3.39 Active Wien-Robinson rejection filter.

Resonant frequency: $f_r = \dfrac{1}{2\pi R_2 C}$

Gain: $A_r = \dfrac{k}{2+\alpha k}$

Rejection Q-factor: $Q = \dfrac{2+\alpha k}{6}$

Fig. 3.39, and its transfer function is

$$A(P) = -\frac{\dfrac{k}{2+\alpha k}(1+P^2)}{1+\dfrac{6}{2+\alpha k}P+P^2},$$

from which the filter data given below Fig. 3.39 can be directly determined. For the actual design of the circuit, the parameters f_r, A_0, Q and C are defined and the remaining parameters are then

$$R_2 = \frac{1}{2\pi f_r C}, \qquad \alpha = \frac{3Q-1}{3A_0 Q} \quad \text{and} \quad k = 6A_0 Q.$$

In order to tune the filter to the resonant frequency, the capacitors C are changed in steps and the two resistors R_2 can be varied smoothly using potentiometers. If the resonant frequency is not fully suppressed due to a slight mismatch of the bridge components, the final adjustment can be made by a slight variation of the resistor $2kR_3/(k-2)$.

3.10   Allpass filters

3.10.1   Underlying principles

The filters discussed so far are circuits, the gain and phase shift of which are frequency dependent. In this section, we examine circuits for which the gain remains constant but the phase shift is dependent on the frequency. These are called allpass filters, and they are used for phase correction and signal delay.

Initially, we show how the frequency response of an allpass filter can be derived from the frequency response of a lowpass filter. To do this, the constant factor A_0 in the numerator of Eq. (3.11) is replaced by the conjugate complex denominator and, in this way, constant unity gain and phase shift doubling is obtained:

$$A(P) = \frac{\prod_i (1 - a_i P + b_i P^2)}{\prod_i (1 + a_i P + b_i P^2)} = \frac{\prod_i \sqrt{(1 - b_i \Omega^2)^2 + a_i^2 \Omega^2}\; e^{-j\alpha}}{\prod_i \sqrt{(1 - b_i \Omega^2)^2 + a_i^2 \Omega^2}\; e^{+j\alpha}} \tag{3.39}$$

$$= 1 \cdot e^{-2j\alpha} = e^{j\varphi},$$

where

$$\varphi = -2\alpha = -2\sum_i \tan^{-1} \frac{a_i \Omega}{1 - b_i \Omega^2}. \tag{3.40}$$

The use of allpass filters for signal delay is of particular interest. Constant gain is one prerequisite for an undistorted signal transfer, a condition always fulfilled by allpass filters. The second prerequisite is that the group delay of the circuit is constant for all frequencies considered. The filters which best fulfill this condition, are the Bessel lowpass filters for which the group delay is Butterworth approximated. Therefore, in order to obtain a "Butterworth allpass filter", the Bessel coefficients must be inserted in Eq. (3.39).

It is reasonable, however, to re-normalize the frequency responses so obtained, as the 3 dB cutoff frequency of the lowpass filters is in this

n	i	a_i	b_i	f_i/f_c	Q_i	$T_{gr\,0}$
1	1	0.6436	0.0000	1.554	–	0.2049
2	1	1.6278	0.8832	1.064	0.58	0.5181
3	1	1.1415	0.0000	0.876	–	0.8437
	2	1.5092	1.0877	0.959	0.69	
4	1	2.3370	1.4878	0.820	0.52	1.1738
	2	1.3506	1.1837	0.919	0.81	
5	1	1.2974	0.0000	0.771	–	1.5060
	2	2.2224	1.5685	0.798	0.56	
	3	1.2116	1.2330	0.901	0.92	
6	1	2.6117	1.7763	0.750	0.51	1.8395
	2	2.0706	1.6015	0.790	0.61	
	3	1.0967	1.2596	0.891	1.02	
7	1	1.3735	0.0000	0.728	–	2.1737
	2	2.5320	1.8169	0.742	0.53	
	3	1.9211	1.6116	0.788	0.66	
	4	1.0023	1.2743	0.886	1.13	
8	1	2.7541	1.9420	0.718	0.51	2.5084
	2	2.4174	1.8300	0.739	0.56	
	3	1.7850	1.6101	0.788	0.71	
	4	0.9239	1.2822	0.883	1.23	
9	1	1.4186	0.0000	0.705	–	2.8434
	2	2.6979	1.9659	0.713	0.52	
	3	2.2940	1.8282	0.740	0.59	
	4	1.6644	1.6027	0.790	0.76	
	5	0.8579	1.2862	0.882	1.32	
10	1	2.8406	2.0490	0.699	0.50	3.1786
	2	2.6120	1.9714	0.712	0.54	
	3	2.1733	1.8184	0.742	0.62	
	4	1.5583	1.5923	0.792	0.81	
	5	0.8018	1.2877	0.881	1.42	

Fig. 3.40 Allpass filter coefficients for a maximally flat group delay

case meaningless. For this reason, we recalculate the coefficients a_i and b_i so that the group delay at $\Omega=1$ reduced to $1/\sqrt{2}$ of is low-frequency value. The coefficients obtained in this way are shown in Fig. 3.40 for filters of up to the tenth order.

The group delay is the time interval by which the signal is delayed in the allpass filter. According to the definition, Eq. (3.9b), it can be determined from Eq. (3.40):

$$T_{gr} = \frac{t_{gr}}{T_c} = t_{gr} \cdot f_c = -\frac{1}{2\pi} \cdot \frac{d\varphi}{d\Omega}$$

$$= \frac{1}{\pi} \sum_i \frac{a_i(1+b_i\Omega^2)}{1+(a_i^2-2b_i)\Omega^2+b_i^2\Omega^4}, \qquad (3.41)$$

and therefore at low frequencies has the value

$$T_{gr\,0} = \frac{1}{\pi}\sum_i a_i,$$

which is given for each order in Fig. 3.40. In addition, the pole-pair quality factor $Q_i = \sqrt{b_i}/a_i$ is given. As it is not influenced by the renormalization, it has the same values as for the Bessel filters.

To enable a check to be made of the correct functioning of the individual filter stages, we have also shown the ratio f_i/f_c in Fig. 3.40. Here, f_i is the frequency at which the phase of the particular filter stage approaches the value $-180°$ if of the second order, or $-90°$ for a first order filter stage. This frequency is considerably easier to measure than the cutoff frequency of the group delay.

The frequency response of the group delay is shown in Fig. 3.41 for allpass filters of the first to the tenth order.

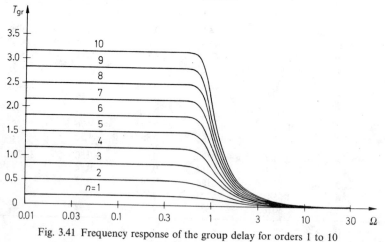

Fig. 3.41 Frequency response of the group delay for orders 1 to 10

The following example shows the steps in the design of an allpass filter. A signal having a frequency spectrum of 0 to 1 kHz is to be delayed by $t_{gr\,0} = 2\,\text{ms}$. In order that the phase distortion is not too large, the cutoff frequency of the allpass filter must be $f_c \geqq 1\,\text{kHz}$. From Eq. (3.9a), it is therefore necessary that

$$T_{gr\,0} \geqq 2\,\text{ms} \cdot 1\,\text{kHz} = 2.00.$$

From Fig. 3.40, it can be seen that a filter of at least seventh order is needed for which $T_{gr\,0} = 2.1737$. To make the group delay exactly 2 ms, the cutoff frequency must be chosen according to Eq. (3.9a), as

$$f_c = \frac{T_{gr\,0}}{t_{gr\,0}} = \frac{2.1737}{2\,\text{ms}} = 1.087\,\text{kHz}.$$

3.10.2 Realization of first order allpass filters

The circuit in Fig. 3.42 has the gain $+1$ at low frequencies and the gain -1 at high frequencies, i.e. the phase shift changes from zero to $-180°$. The circuit is an allpass filter if the magnitude of the gain is also unity for the middle frequency range. To examine this, we determine the transfer function

$$A(P) = \frac{1 - pRC}{1 + pRC} = \frac{1 - RC\,\omega_c P}{1 + RC\,\omega_c P}.$$

The absolute value of the gain is indeed constant and unity. Comparing the coefficients with those of Eq. (3.39) gives

$$RC = \frac{a_1}{2\pi f_c}.$$

With Eq. (3.41), the low-frequency value of the group delay is therefore

$$t_{gr\,0} = 2RC.$$

The first order allpass filter in Fig. 3.42 is very well suited for use as a phase shifter over a wide range of phase delays. By variation of the

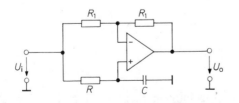

Fig. 3.42 First order allpass filter

resistor R, the phase delay can be adjusted to values between 0 and $-180°$ without influencing the amplitude. The phase shift is

$$\varphi = -2\tan^{-1}(\omega RC).$$

3.10.3　Realization of second order allpass filters

The second order allpass filter transfer function can be implemented, for example, by subtracting the output voltage of a bandpass filter from its input voltage. The transfer function of this circuit is then

$$A(P') = 1 - \frac{\dfrac{A_r}{Q}P'}{1+\dfrac{1}{Q}P'+P'^2} = \frac{1+\dfrac{1-A_r}{Q}P'+P'^2}{1+\dfrac{1}{Q}P'+P'^2}.$$

It can be seen that for $A_r = 2$, the transfer function of an allpass filter is obtained. It is not normalized to the cutoff frequency of the allpass, but to the resonant frequency of the bandpass filter. For a correct normalization, we take

$$\omega_c = \beta\omega_r$$

and obtain

$$P' = \frac{p}{\omega_r} = \frac{\beta p}{\omega_c} = \beta P.$$

Hence, the transfer function

$$A(P) = \frac{1-\dfrac{\beta}{Q}P+\beta^2 P^2}{1+\dfrac{\beta}{Q}P+\beta^2 P^2}.$$

Comparing the coefficients with those of Eq. (3.39) yields

$$a_1 = \frac{\beta}{Q} \quad \text{and} \quad b_1 = \beta^2.$$

The data of the required bandpass filter are therefore

$$A_r = 2,$$
$$f_r = f_c/\sqrt{b_1},$$
$$Q = \sqrt{b_1}/a_1 = Q_1.$$

As an example, the implementation using the bandpass of Fig. 3.34 is given. As the Q-factors are relatively small, the resistor R_3 may be

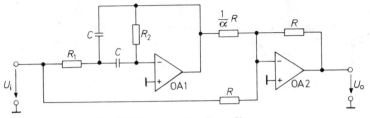

Fig. 3.43 Second order allpass filter

omitted and the gain adjusted by the resistor R/α in Fig. 3.43 instead. The transfer function is

$$A(P) = -\frac{1 + (2R_1 - \alpha R_2) C\omega_c P + R_1 R_2 C^2 \omega_c^2 P^2}{1 + 2R_1 C\omega_c P + R_1 R_2 C^2 \omega_c^2 P^2}.$$

Comparing the coefficients with those of Eq. (3.39) gives the relationships

$$R_1 = \frac{a_1}{4\pi f_c C}, \quad R_2 = \frac{b_1}{\pi f_c C a_1} \quad \text{and} \quad \alpha = \frac{a_1^2}{b_1} = \frac{1}{Q_1^2}.$$

From the transfer function, a further application of the circuit in Fig. 3.43 can be deduced. If

$$2R_1 - \alpha R_2 = 0,$$

a rejection filter circuit is obtained.

3.11 Adjustable universal filter

As shown previously, the transfer function of a second order filter element has the general form

$$A(P) = \frac{d_0 + d_1 P + d_2 P^2}{c_0 + c_1 P + c_2 P^2}. \tag{3.42}$$

The filter families described so far can be deduced from Eq. (3.42) by assigning special values to the coefficients of the numerator:

lowpass filter:	$d_1 = d_2 = 0$;
highpass filter:	$d_0 = d_1 = 0$;
bandpass filter:	$d_0 = d_2 = 0$;
band-rejection filter:	$d_1 = 0, \quad d_0 = d_2$;
allpass filter:	$d_0 = c_0, \quad d_1 = -c_1, \quad d_2 = c_2$.

The numerator coefficients may have either sign, whereas the coefficients of the denominator must always be positive for reasons of stability. The pole-pair Q-factor is defined by the denominator coefficients

$$Q_i = \frac{\sqrt{c_0 c_2}}{c_1}. \tag{3.43}$$

In the previous sections we have shown special and, if possible, simple circuits for each filter type. Sometimes, however, a single circuit is required to realize all the described kinds of filters, as well as the more general types of Eq. (3.42) with any numerator coefficients. This problem can be solved by using the circuit in Fig. 3.44. Moreover, the circuit has the advantage that the individual coefficients can be set independently of one another, as each coefficient is determined by only one circuit element. The transfer function of the circuit is

$$A(P) = \frac{k_0 - k_1 \omega_0 \tau P + k_2 \omega_0^2 \tau^2 P^2}{l_0 + l_1 \omega_0 \tau P + l_2 \omega_0^2 \tau^2 P^2}, \tag{3.44}$$

where ω_0 is the normalizing frequency and $\tau = RC$ is the time constant of the two integrators. The coefficients k_i and l_i are resistance ratios and therefore always positive. If the sign of a numerator coefficient is to be changed, the input voltage of the filter must be inverted by an additional amplifier to which the corresponding resistor is connected.

To realize a higher-order filter, the number of integrators can be raised accordingly. However, it is usually easier to split the filter into second order stages and cascade them.

The design process for the circuit is illustrated by the following example: a second order allpass filter is required, the group delay curve of which is maximally flat and has, at low frequencies, the value of 1 ms.

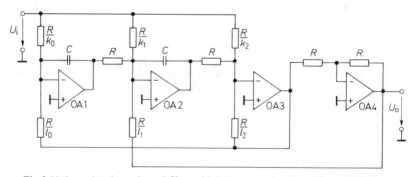

Fig. 3.44 Second order universal filter with independently adjustable coefficients

From the table in Fig. 3.40 we take the values $a_1 = 1.6278$, $b_1 = 0.8832$ and $T_{gr\,0} = 0.5181$. Equation (3.9a) gives the cutoff frequency

$$f_c = \frac{T_{gr\,0}}{t_{gr\,0}} = \frac{0.5181}{1\,\text{ms}} = 518.1\,\text{Hz}.$$

We choose $\tau = 1\,\text{ms}$ and, by comparing the coefficients of Eqs. (3.44) and (3.39), and using $\omega_0 = 2\pi f_c = 3.26\,\text{kHz}$, obtain the values

$$l_0 = k_0 = 1$$

$$l_1 = k_1 = \frac{a_1}{\omega_0 \tau} = 0.500,$$

$$l_2 = k_2 = \frac{b_1}{(\omega_0 \tau)^2} = 0.0833.$$

Such a low value of l_2 is difficult to obtain in practice. However, it increases more rapidly than the other coefficients when τ is reduced. We therefore choose $\tau = 0.3\,\text{ms}$ and obtain

$$l_0 = k_0 = 1, \quad l_1 = k_1 = 1.67 \quad \text{and} \quad l_2 = k_2 = 0.926.$$

For some applications of a bandpass filter, it is desirable that the resonant frequency, the Q-factor and the gain at resonant frequency can be set independently. As a comparison of Eq. (3.44) and Eq. (3.24) shows, the two coefficients l_1 and k_1 should be simultaneously adjustable to enable setting of the Q-factor without affecting the gain. Figure 3.45 shows a circuit in which such a dependence is eliminated.

The circuit is interesting in that it acts simultaneously as a bandpass filter, as a rejection filter, as a lowpass and as a highpass filter, depending on which output is used. For the determination of the filter

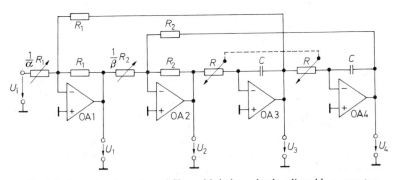

Fig. 3.45 Second order universal filter with independently adjustable parameters

data, we take from the circuit the following relationships:

$$\underline{U}_1 = -\alpha \underline{U}_i - \underline{U}_3,$$

$$\underline{U}_2 = -\beta \underline{U}_1 - \underline{U}_4,$$

$$\underline{U}_3 = -\frac{\underline{U}_2}{pRC},$$

$$\underline{U}_4 = -\frac{\underline{U}_3}{pRC}.$$

By eliminating three of the four output voltages, we obtain the gain with respect to the individual outputs:

$$\frac{\underline{U}_1}{\underline{U}_i} = -\frac{\alpha(1+R^2C^2\omega_0^2 P^2)}{1+\beta RC\omega_0 P+R^2C^2\omega_0^2 P^2} \quad \text{(band-rejection filter)},$$

$$\frac{\underline{U}_2}{\underline{U}_i} = +\frac{\alpha\beta R^2 C^2 \omega_c^2 P^2}{1+\beta RC\omega_c P+R^2C^2\omega_c^2 P^2} \quad \text{(second order highpass filter)},$$

$$\frac{\underline{U}_3}{\underline{U}_i} = -\frac{\alpha\beta RC\omega_0 P}{1+\beta RC\omega_0 P+R^2C^2\omega_0^2 P^2} \quad \text{(bandpass filter)},$$

$$\frac{\underline{U}_4}{\underline{U}_i} = +\frac{\alpha\beta}{1+\beta RC\omega_c P+R^2C^2\omega_c^2 P^2} \quad \text{(second order lowpass filter)}.$$

By comparing the coefficients with those in Eqs. (3.38), (3.12), (3.24) and (3.11), the filter data may be determined as

$$\left.\begin{array}{l}\text{bandpass filter:}\\ \text{rejection filter:}\end{array}\right\}\quad \begin{array}{l} f_r = 1/2\pi RC \\ A_r = A_0 = \alpha \\ Q = 1/\beta \end{array}$$

lowpass filter:

$$f_c = \sqrt{b_i}/2\pi RC$$
$$\beta = a_i/\sqrt{b_i} = 1/Q_i$$
$$\alpha = A_0/\beta$$

highpass filter

$$f_c = 1/2\pi RC\sqrt{b_i}$$
$$\beta = a_i/\sqrt{b_i} = 1/Q_i$$
$$\alpha = A_\infty/\beta$$

When the circuit is used as a bandpass filter or as a band-rejection filter, it can be seen that the resonant frequency, the gain and the Q-factor can each be changed without influencing the others. This is because the resonant frequency is determined solely by the product RC. As these values (R and C) do not appear in the equations for A and Q, a variation of the resonant frequency is possible without affecting A and Q. These two parameters can be adjusted independently of one another by the potentiometers R_1/α and R_2/β.

It follows from the design equations for highpass and lowpass filters that β determines the filter type, RC the cutoff frequency and α the gain. For a given type of filter ($\beta=$ const.), the cutoff frequency and the gain can be varied smoothly and independently of one another.

The coefficient β is equal to the reciprocal of the pole-pair quality factor Q_i which is given in the table of Fig. 3.14. It is identical to the Q-factor at the bandpass filter output. Hence, the rather abstract definition of the pole-pair Q-factor $Q_i=\sqrt{b_i/a_i}$ can now be interpreted as the quality factor of a corresponding bandpass filter having the same denominator polynomial.

The resistors R have rather large values for low filter frequencies. It may in this case be advantageous to replace them by fixed resistors connected to voltage dividers. The voltage divider can then be realized by low resistance potentiometers. This method may also be used for the resistors R_1 and R_2.

If a filter parameter is to be voltage controlled, the voltage divider can be replaced by an analog multiplier, where the control voltage is connected to the second input, as represented in Fig. 3.46. The effective resistance is then

$$R_x = R_0 \cdot \frac{E}{U_{\text{control}}},$$

where U_{control} is the controlling voltage. If two such circuits are inserted instead of the two frequency determining resistors R, the resonant frequency of the bandpass filter becomes

$$f_r = \frac{1}{2\pi R_0 C} \cdot \frac{U_{\text{control}}}{E},$$

which is proportional to the control voltage.

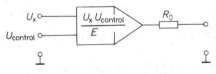

Fig. 3.46 Multiplier as a controlled resistor

4 Broadband amplifiers

For the design of amplifier circuits, in which the upper cutoff frequency must be higher than 100 kHz, certain special aspects must be considered. These are discussed below. There are two main effects which influence the value of the upper cutoff frequency:

(1) the frequency dependence of the current gain which is given by the internal structure of the transistor;
(2) stray capacitances which, together with the external resistances, form lowpass filters.

4.1 Frequency dependence of the current gain

The frequency response of the current gain $\underline{\beta} = \underline{I}_C/\underline{I}_B$ of a bipolar transistor can be described with good accuracy by a first order lowpass filter characteristic, i.e.

$$\underline{\beta} = \frac{\beta}{1 + j\dfrac{f}{f_\beta}}, \qquad (4.1)$$

where β is the current gain at low frequencies and f_β is the 3 dB cutoff frequency.

Instead of the 3 dB cutoff frequency, the gain-bandwidth product f_T is often given. This is the frequency at which the magnitude of $\underline{\beta}$ becomes unity. From Eq. (4.1) with $\beta \gg 1$, it follows that

$$\boxed{f_T = \beta f_\beta}. \qquad (4.2)$$

Hence the name gain-bandwidth product, also known as the transit frequency.

The best way to study the influence of the frequency response of the current gain on the frequency response of the voltage gain of a circuit is to use the Johnson-Giacoletto model depicted in Fig. 4.1. The frequency dependence of the current gain is represented by the "diffusion capacitance", C_D, of the forward-biased base-emitter diode. The additional depletion layer capacitance $C_{CB'}$ is ignored for the moment. The relationship between C_D and f_T is, according to ref. [4.1]

$$C_D = \frac{I_C}{2\pi f_T U_T} = \frac{g_{mi}}{2\pi f_T} = \frac{\beta}{2\pi f_T r_{BE}}. \qquad (4.3)$$

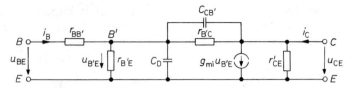

Fig. 4.1 Giacoletto equivalent circuit for the common-emitter connection. Relation of the equivalent parameters to the low-frequency values of the y-parameters:

Internal base-emitter resistance: $r_{B'E} \approx r_{BE}$

Base spreading resistance: $r_{BB'} \approx \frac{1}{10} r_{BE}$

Internal transconductance: $g_{mi} \approx g_m$

Internal collector-base resistance: $r_{B'C} \approx 1/g_{mr}$

Internal collector-emitter resistance: $r'_{CE} \approx r_{CE}$

In a first approximation, the gain-bandwidth product f_T is independent of the mean value of the emitter current. The capacitance C_D must therefore be proportional to I_C, see Eq. (4.3).

If a transistor in common-emitter connection is operated from a high-impedance signal voltage source, i.e. with an impressed base current I_B, the cutoff frequency is determined by the lowpass filter consisting of $r_{B'E}$ and C_D, according to

$$\underline{I}_C = g_{mi}\underline{U}_{B'E} = g_{mi}\frac{r_{B'E}}{1+j\omega r_{B'E}C_D}\underline{I}_B = \frac{\beta}{1+j\dfrac{f}{f_T/\beta}}\underline{I}_B. \qquad (4.4)$$

The cutoff frequency is therefore f_T/β and thus equal to f_β, as required by the definition.

If the common-emitter circuit is controlled by a low-impedance voltage source, the cutoff frequency of the circuit is determined by the time constant

$$\tau = (r_{BB'} \| r_{B'E})\, C_D \approx r_{BB'} C_D.$$

Therefore the frequency response of the transconductance

$$\underline{g}_m = \frac{\underline{I}_C}{\underline{U}_{BE}} = \frac{g_m}{1+j\omega r_{BB'}C_D} = \frac{g_m}{1+j\dfrac{f}{f_g}}, \qquad (4.5)$$

where g_m is the low-frequency value of the transconductance and $f_g = 1/2\pi r_{BB'} C_D$, the cutoff frequency of the transconductance. The latter is larger than the β-cutoff frequency, f_β, by a factor of $r_{B'E}/r_{BB'} \approx 10$.

If the transistor is operated as a common-base amplifier with voltage control, the same result is obtained, as the control voltage is applied across the same two terminals.

The conditions change if the emitter current is impressed. Since the collector current is practically the same as the emitter current for $|\beta| \gg 1$, the gain begins to decrease only in the region of the transit frequency f_T. The relationship between collector and emitter current is given with $i_E = i_C + i_B$ and $i_B = i_C/\beta$ as

$$\alpha = \frac{i_C}{i_E} = \frac{\beta}{1+\beta}.$$

In complex notation, using Eq. (4.1),

$$\underline{\alpha} = \frac{\underline{\beta}}{1+\underline{\beta}} = \frac{\alpha}{1+j\dfrac{\alpha f}{\beta f_\beta}}.$$

Hence, we obtain the α-cutoff frequency

$$\boxed{f_\alpha = \frac{\beta f_\beta}{\alpha} \approx f_T}.$$

When operating the transistor as an emitter follower, a cutoff frequency of the voltage gain is obtained which lies between the values f_g and f_T, depending on the load resistance.

To summarize,

$$\boxed{f_\beta \ll f_g \ll f_\alpha \approx f_T}.$$

4.2   Influence of transistor and stray capacitances

In every circuit, there are a number of unavoidable transistor and stray capacitances which, together with the circuit's resistances, make lowpass filters, as shown in Fig. 4.2. The most important spurious capacitances are:

C_1: stray capacitance, in particular that of the input lead,
C_2: emitter-base capacitance,
C_3: collector-base capacitance,
C_4: collector-emitter capacitance.

It can be seen that the circuit contains two lowpass filters. The capacitances C_3 and C_4 make, together with the parallel resistance R_C, a lowpass filter at the output. They reduce the dynamic collector impedance at high frequencies and thereby reduce the voltage gain. At the input, the capacitances C_1, C_2 and C_3 form a lowpass filter together with R_g. The effective input capacitance of the circuit is

$$C_i = C_1 + C_2 + |A| C_3,$$

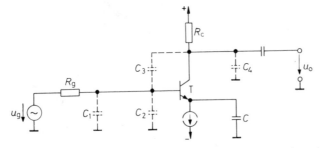

Fig. 4.2 Effect of transistor and stray capacitances on the common-emitter circuit

where A is the voltage gain of the circuit. The increase of the collector-base capacitance is known as the Miller effect and is caused by the fact that a voltage appears across the capacitance C_3 which is $(|A|+1)$ times that at the input. If $|A|\gg1$, the term $|A|\,C_3$ outweighs the remaining transistor capacitances, and we obtain approximately

$$C_i \approx |A|\,C_3.$$

For this reason, the input lowpass filter has a large effect and the bandwidth of a common-emitter circuit is therefore relatively small.

The conditions are more favourable for the common-base amplifier. As can be seen from Fig. 4.3, the effective input capacitance in this application is

$$C_i = C_1 + C_2 - AC_4 \quad \text{with } A > 0.$$

Here, the input capacitance is reduced instead of increased. The low input resistance, however, is a disadvantage.

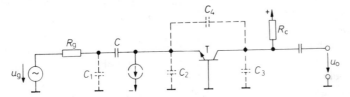

Fig. 4.3 Effect of transistor and stray capacitances on the common-base circuit

4.3 Cascode amplifier

The disadvantage of the low input resistance of the common-base circuit can be overcome by connecting two transistors in series in the form of a "cascode circuit", as shown in Fig. 4.4. Here, the input transistor T_1 operates in common-emitter connection, and the output transistor T_2 is operated in common-base connection with current control. Since T_2 has

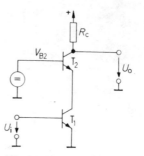

Fig. 4.4 Cascode amplifier.

Voltage gain:	$\underline{A} = -g_m R_C$
Input resistance:	$r_i = r_{BE1}$
Output resistance:	$r_o = R_C$

at its emitter terminal the low input resistance $1/g_m$, the voltage gain of the input stage is

$$A_1 = -g_m \cdot \frac{1}{g_m} = -1.$$

In this way, the Miller effect is eliminated. As practically the same collector current flows through the two transistors, the voltage gain of the whole circuit becomes

$$\underline{A} = -g_m \cdot R_C,$$

as for the normal common-emitter circuit. The transconductance cutoff frequency of the circuit is not adversely influenced by transistor T_2. This is because T_2 is in common-base connection and is current controlled, and therefore the high frequency $f_\alpha \approx f_T \gg f_g$ is the determining cutoff frequency of this stage.

The base potential V_{B2} of T_2 determines the collector potential of T_1. It is given such a value that the collector-emitter voltages of T_1 and T_2 do not fall below a few volts; this is to keep the voltage-dependent collector-base capacitances as small as possible.

4.4　Differential amplifier as a broadband amplifier

A further possibility of increasing the low input resistance of the common-base circuit is to connect an emitter follower to its input. This results in the asymmetric differential amplifier shown in Fig. 4.5. Since the transistor T_1 is operated with a constant collector potential, the Miller effect does not occur. The transistor T_2 is working in common-base connection with voltage control; the cutoff frequency of this stage is

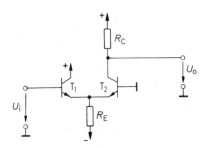

Fig. 4.5 Asymmetric differential amplifier.

Voltage gain: $\underline{A} = \frac{1}{2}\underline{g}_m R_C$

Input resistance: $r_i = 2r_{BE}$

Output resistance: $r_o \approx R_C$

therefore the transconductance cutoff frequency f_g. As the cutoff frequency of the emitter follower lies above this frequency, f_g is also the transconductance cutoff frequency of the entire arrangement and is the same as for the cascode amplifier. However, the absolute value of the total transconductance is not the same as that of the cascode amplifier. For its calculation we use the fact that an emitter follower with low-impedance control has the output resistance $r_{o\,1} = 1/g_{m\,1}$, and that the common-base circuit has the input resistance $r_{i\,2} = 1/g_{m\,2}$. Both transistors are operated with the same collector quiescent current and therefore have the same transconductance g_m. Therefore,

$$r_{o\,1} = r_{i\,2}.$$

Thus, exactly half of the input alternating voltage appears at the emitter of T_2, and we obtain the total transconductance as

$$\underline{g}_{m\,tot} = \frac{I_{C\,2}}{\underline{U}_i} = \frac{I_{C\,2}}{-2\underline{U}_{BE\,2}} = -\frac{1}{2}\underline{g}_m$$

and the voltage gain

$$\underline{A} = \frac{1}{2}\underline{g}_m R_C.$$

The voltage gain is therefore half that of the cascode amplifier.

The advantage of the differential amplifier over the cascode circuit is that the base-emitter voltages of the two transistors compensate one another.

The good high-frequency properties of the differential amplifier are only attained if, as in Fig. 4.5, the collector of the input transistor and the base of the output transistor have constant potential. In order to obtain a symmetrical broadband differential amplifier, some extensions are needed, and they are discussed in the following section.

4.5   Symmetrical broadband amplifiers

4.5.1   Differential amplifier using cascode circuits

In Fig. 4.6, a broadband differential amplifier is shown having a symmetrical input and output. In order to avoid the Miller effect, the two transistors of the basic differential amplifier circuit are each replaced by a cascode amplifier.

In broadband amplifiers, negative feedback across several stages is usually associated with considerable problems regarding stability. To attain a defined gain, however, each stage can be given its own feedback. This is the purpose of the two resistors R_E, each effecting series feedback. They reduce the transconductance $\underline{g}_m$ of the input transistors to the value

$$\underline{g}'_m = \frac{1}{R_E + 1/\underline{g}_m}. \tag{4.6}$$

It can be seen that the larger the value of R_E in comparison with $|1/\underline{g}_m|$, the more the transconductance is determined by the feedback resistance. In addition, the cutoff frequency of the transconductance is increased: by inserting Eq. (4.5) in (4.6), we obtain the frequency response of the reduced transconductance as

$$\underline{g}'_m = \frac{g'_m}{1 + j\dfrac{f}{f_g(g_m/g'_m)}}. \tag{4.7}$$

The transconductance cutoff frequency is therefore increased to

$$f'_g = f_g(1 + g_m R_E) = f_g \frac{g_m}{g'_m}. \tag{4.8}$$

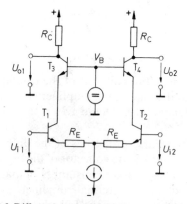

Fig. 4.6 Differential amplifier using cascode circuits

The design process is shown by an example: a bandwidth of $B = 100\,\text{MHz}$ is required. It is useful to choose the cutoff frequency f_c of the lowpass filter at the output and the transconductance cutoff frequency f_g', as being roughly the same. For a cascade connection of n lowpass filters all having the same cutoff frequency f_c, the approximation holds that

$$f_{cn} \approx \frac{1}{\sqrt{n}} \cdot f_c. \tag{4.9}$$

Hence, for this example, it is necessary that

$$f_g' \approx f_c \approx 100\,\text{MHz} \cdot \sqrt{2} \approx 150\,\text{MHz}.$$

The transistor and stray capacitances are assumed to have a total value of $6\,\text{pF}$. Thus, for the collector resistance

$$R_C = \frac{1}{2\pi f_c C_S} \approx 180\,\Omega.$$

In view of this low resistance value, a large transconductance, i.e. a large collector current, is required to attain sufficient voltage gain. The upper limit is given by the permissible power dissipation of the transistor and by the reduction of the gain-bandwidth product at higher collector currents. We choose $I_C = 10\,\text{mA}$ and obtain $1/g_m = U_T/I_C \approx 3\,\Omega$. In order to attain effective feedback, we make $R_E \gg 1/g_m$. With $R_E = 15\,\Omega$,

$$g_m' = \frac{1}{3\,\Omega + 15\,\Omega} = \frac{1}{18\,\Omega} = 56\,\frac{\text{mA}}{\text{V}},$$

and therefore a low-frequency voltage gain

$$A_D = \frac{u_{o1}}{u_{i1} - u_{i2}} = -\tfrac{1}{2} g_m' R_C = -5.$$

It is now clear that with a broadband amplifier stage involving feedback, only a relatively low voltage gain can be achieved. It is also obvious that FETs cannot be used to attain sufficient voltage gain as their transconductance is too small. If a high input resistance is required, FET source followers may be connected to the input transistors.

From Eq. (4.8), we can determine the necessary transconductance cutoff frequency of the input transistors:

$$f_g = \frac{g_m'}{g_m} f_g' = \frac{3\,\Omega}{18\,\Omega} \cdot 150\,\text{MHz} = 25\,\text{MHz}.$$

The gain-bandwidth product must therefore be over $250\,\text{MHz}$. Such values can be attained even with audio transistors but these are unsuitable because of their large capacitances.

There are several ways of increasing the bandwidth by incorporating pre-emphasis in the circuit. For instance, the current feedback can be made ineffective at high frequencies by connecting a capacitor between the emitter terminals of T_1 and T_2. For a lower cutoff frequency of 100 MHz, we obtain, in our example, the value of 53 pF.

A further possibility is to increase the impedance of the collector resistors in the vicinity of the cutoff frequency by series-connecting an inductance. In our example, it would have the approximate value of 0.3 μH. This method is known as 'peaking'.

4.5.2 Differential amplifier using inverting amplifiers

A broadband differential amplifier which operates in a similar way to that described above, is represented in Fig. 4.7. The input stage is identical. To obtain a large bandwidth, the collector potentials of the input transistors must again remain constant. For this reason, the transistors T_3 and T_4 of the adjoining differential amplifier have individual feedback by means of the resistors R_C. Their base terminals therefore represent summing points for which the changes in voltage remain small [4.2]. Hence, the alternating voltage at the output is

$$\underline{U}_{o1} = R_C \underline{I}_{C1} = \tfrac{1}{2} g'_m R_C (\underline{U}_{i1} - \underline{U}_{i2}).$$

For this circuit, the feedback resistors R_C determine the voltage gain. The resistors R_1 serve to adjust the collector quiescent potential

$$U_{o1} = V^+ - (I_{C1} + I_{C3}) R_1.$$

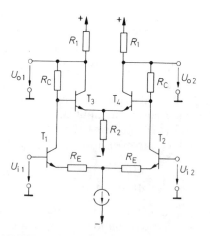

Fig. 4.7 Differential amplifier using inverting amplifiers

These two circuits are particularly suited for application as d.c. amplifiers in wide-band oscilloscopes. When transistors having a gain-bandwidth product of several GHz are used in suitably low-impedance circuits, bandwidths of over 500 MHz can be achieved [4.3].

4.5.3 Differential amplifier using complementary cascode circuits

With a single broadband amplifier stage, a voltage gain of much more than 10 is not possible and it is therefore necessary to cascade a number of amplifiers. When d.c. coupling is involved, a problem arises in that the quiescent output potentials of the circuits described are higher than the quiescent input potentials. Therefore, the quiescent potentials increase from stage to stage, and the number of stages is limited.

This disadvantage can be avoided if pnp-transistors are used in the output stage of the cascode differential amplifier of Fig. 4.6, as represented in Fig. 4.8. It is then possible to make the input and output quiescent potentials zero.

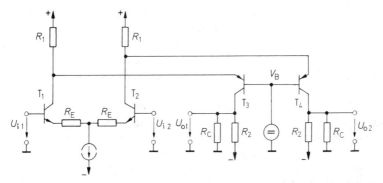

Fig. 4.8 Differential amplifier using complementary cascode circuits

Section 4.5.1 shows that the cutoff frequency of the circuit is determined by the transconductance cutoff frequency f_g' of the input transistors in feedback, as the output stage has the very much higher cutoff frequency $f_\alpha \approx f_T$. Therefore, the fact that the gain-bandwidth product for pnp-transistors is usually lower than that for npn-transistors, is of no significance for the circuit of Fig. 4.8.

The base potential V_B determines the collector potential of the input differential amplifier, since $V_{C1} = V_B + 0.7\,\mathrm{V}$. Hence, the current through the resistors R_1 is constant and has the magnitude

$$I = I_{C1} + I_{C3} = \frac{V^+ - V_{C1}}{R_1} = \frac{V^+ - V_B - 0.7\,\mathrm{V}}{R_1}.$$

If the collector current I_{C1} increases, I_{C3} decreases by the same amount. For the a.c. collector currents therefore

$$\underline{I}_{C3} = -\underline{I}_{C1}.$$

Apart from the polarity, this is the same relationship as for the normal cascode amplifier.

The resistors R_2 are given values such that the output quiescent potential is zero for the chosen collector current. The resistances obtained in this way are usually larger than can be tolerated for a given bandwidth. Therefore, the resistors R_C are inserted, the values of which can be chosen freely as no direct voltage appears across them. Hence, for the voltage gain

$$\underline{A} = \frac{\underline{U}_{o1}}{\underline{U}_{i1} - \underline{U}_{i2}} = -\tfrac{1}{2}\underline{g}'_m(R_C \| R_2).$$

4.5.4 Push-pull differential amplifier

At large-signal operation, the broadband amplifiers described have different responses for the leading and the trailing edge of a pulse. This is due to the fact that the transistor current can usually be increased faster than it can be decreased. To obtain equal slew rates for both edges, the principle of push-pull operation is applied. Such an amplifier consists of transistors which are controlled in opposition so that, at leading as well as at trailing edges, an increase in current occurs in one half of the circuit, and a simultaneous decrease in the other half.

To achieve this, the circuit in Fig. 4.8 can be extended symmetrically by complementary transistors, as represented in Fig. 4.9.

For zero input, the current through the transistors T_3 and T'_3 is the same, and the output quiescent potential is nil. If a positive input voltage difference $U_D = U_{i1} - U_{i2}$ is applied, the collector current of T_3 increases by the amount $U_D \cdot g'_m$ whereas I'_{C3} decreases by the same amount. The difference between the two currents flows through resistor R_C, and the voltage gain becomes

$$\underline{A}_D = \frac{\underline{U}_o}{\underline{U}_D} = \underline{g}'_m R_C,$$

where

$$\underline{g}'_m = \frac{1}{R_E + 1/\underline{g}_m},$$

which is the reduced transconductance of the input transistors.

As for the previous circuit, no d.c. voltage appears across the resistor R_C, the value of which can therefore be chosen freely for the bandwidth required.

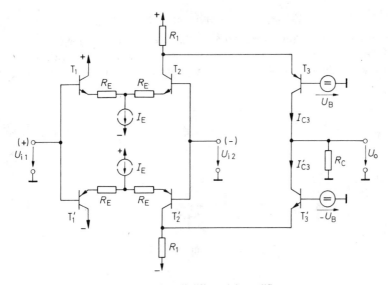

Fig. 4.9 Push-pull differential amplifier

If an output voltage opposite in polarity to that of U_o is required, a second output stage may be connected to the transistors T_1 and T'_1.

4.6 Broadband voltage follower

An emitter follower as in Fig. 4.10 is, in principle, well suited for use as a broadband voltage follower since its cutoff frequency is higher than the transconductance cutoff frequency. However, for large signals, an ex-

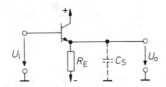

Fig. 4.10 Emitter follower with stray capacitance

tremely unsymmetrical transient behaviour is incurred. This is because the stray capacitance C_S can be charged relatively quickly since the conducting transistor has an output resistance of only $1/g_m$, whereas the discharge can take place only via resistor R_E, the transistor being reverse biased for trailing edges. This difficulty can again be overcome by the use of push-pull circuits.

4.6.1   Push-pull source follower

Push-pull circuits can be particularly easily realized with junction FETs, as no auxiliary voltages are required to define the operating point. In the circuit in Fig. 4.11, the desired quiescent current is adjusted by

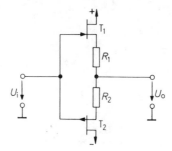

Fig. 4.11 Push-pull source follower

means of the series feedback resistors R_1 and R_2. In the first step of the design, the quiescent current I_{D0} is defined. In order that the output quiescent potential is zero, the resistor R_1 must have the voltage $|U_{GS1}(I_{D0})|$ across it. Because of the parabolic transfer characteristic [1],

$$U_{GS} = U_p(1 - \sqrt{I_D/I_{DS}}).$$

Hence

$$R_1 = \frac{|U_{p1}|}{I_{D0}}\left(1 - \sqrt{\frac{I_{D0}}{I_{DS1}}}\right) \quad \text{and} \quad R_2 = \frac{U_{p2}}{I_{D0}}\left(1 - \sqrt{\frac{I_{D0}}{I_{DS2}}}\right).$$

The output resistance of the circuit is

$$r_o = \left(\frac{1}{g_{fs1}} + R_1\right) \left\| \left(\frac{1}{g_{fs2}} + R_2\right).\right.$$

FETs having sufficiently large g_{fs} allow values of down to $50\,\Omega$.

4.6.2   Push-pull emitter follower

If output currents of more than approx. $10\,\text{mA}$ are required, bipolar transistors must be used. A suitable circuit is shown in Fig. 4.12. In order that a quiescent current flows through the two output transistors T_3 and T_4, the voltage between the two base terminals must be about $1.4\,\text{V}$. This voltage is generated by means of the two emitter followers T_1 and T_2, which also act as impedance converters. The quiescent current is stabilized by the series feedback effect of the resistors R_2. Values between $3\,\Omega$ and $30\,\Omega$ are usually chosen.

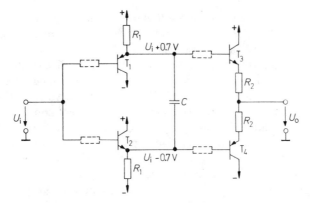

Fig. 4.12 Push-pull emitter follower

The resistors R_1 provide the emitter currents for the input circuit and the base currents for the output stage. They must be of such a low resistance that, even for the maximum input signal, their current is larger than the base current needed by the appropriate output transistor. At large-signal operation it is better to replace them by constant current sources.

The capacitor C improves the slew rate: The emitter follower T_1 produces steep falling edges, T_2 steep rising edges. The two sides of the output stage are coupled by the capacitor C so that the output always responds to the steeper edge.

The base resistors, shown by a broken line, are inserted to prevent highfrequency oscillations of the individual transistors. They are chosen as small as possible so as not to affect the output resistance and cutoff frequency. Practicable values lie between $20\,\Omega$ and $200\,\Omega$.

The design of the circuit is described in more detail in Chapter 5.2 in connection with its use as a power amplifier.

4.7   Broadband operational amplifiers

Figure 4.9 shows a differential amplifier having a good large-signal transient response and zero output quiescent potential. If it is to be used as an operational amplifier, the voltage gain must be high and the output resistance low.

A high voltage gain can be attained by omitting the "load" resistance R_C. The high internal resistance r_i of the circuit at these terminals then determines the voltage gain. In this way, the bandwidth is reduced by the same factor by which the gain is increased. However, the gain-bandwidth product characteristic for the amplifier performance, remains constant.

In order to attain the required low output resistance, an impedance converter must be added to the circuit. For example, the push-pull emitter follower of Fig. 4.12 can be used for this purpose.

In order that an external feedback loop may be applied to the resulting operational amplifier, a sufficient phase margin must be ensured, i.e. the magnitude of the loop gain must be less than unity before the phase shift has reached $-180°$. In principle, an RC element connected to the output of the amplifier in Fig. 4.9 can be used for this purpose. However, this results in a relatively poor slew rate for the output voltage. It is considerably better to adjust the required transient response by varying the series-feedback resistors R_E.

The low-frequency properties of the circuit are obviously worse than one would expect for integrated operational amplifiers. The series-feedback resistors R_E necessary for stable operation cause a low d.c. voltage gain and a high offset voltage. As the input transistors must carry a relatively high collector current because of the required bandwidth, the input bias currents are correspondingly high.

These disadvantages may be overcome if the broadband amplifier OA 2 is combined with a d.c. amplifier OA 1, as in Fig. 4.13. The broadband amplifier then determines the high-frequency characteristics and the d.c. amplifier the low-frequency characteristics of the circuit. The only disadvantage is that a non-inverting input is no longer available.

At low frequencies the circuit has the gain

$$A = -A_{D1} A_{D2}.$$

Since the voltage appearing at the input of OA 2 is already amplified by the very large factor A_{D1}, the offset voltage of OA 2 is no longer of any importance. The input bias current of OA 2 flows through the resistor R_2 and so bypasses the input. The circuit has therefore the low input bias current of the d.c. amplifier.

At high frequencies the output voltage of OA 1 is zero. In this case the total gain is given by

$$\underline{A} = -\underline{A}_{D2}.$$

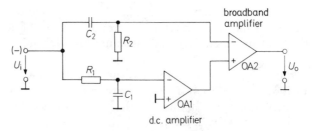

Fig. 4.13 D.C. stabilized broadband amplifier

The lowpass filter $R_1 C_1$ serves solely to keep high frequencies from reaching the d.c. amplifier and therefore to avoid uncontrollable effects.

Figure 4.14 shows an example for the frequency response of the individual gains and the total gain [4.4].

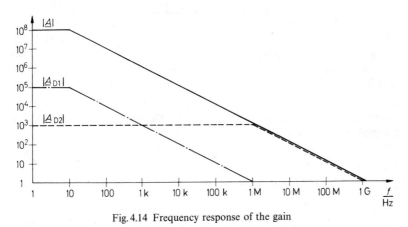

Fig. 4.14 Frequency response of the gain

5 Power amplifiers

Power amplifiers are designed to provide large output powers and therefore, the voltage gain plays only a minor role. Normally, the voltage gain of a power output stage is near unity and the power gain is thus mainly due to the current gain of the circuit. Output voltage and current must be able to assume positive and negative values. Power amplifiers for which the output current can flow in only one direction are discussed in Chapter 6, and are called power supplies.

5.1 Emitter follower as a power amplifier

The operation of the emitter follower is described elsewhere [1]. Here, we define some of the parameters which are of particular interest for its application as a power amplifier. Firstly, we calculate the load resistance for which the circuit delivers maximum power without distortion. If the output is negative, R_L carries some of the current flowing through R_E. The limit for control of the output voltage is reached when the current through

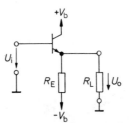

Fig. 5.1 Emitter follower as a power amplifier.

Voltage gain:	$A \approx 1$
Current gain if load is matched to internal resistance:	$A_i = \frac{1}{2}\beta$
Load resistance if matched:	$R_L = R_E$
Output power for matched load and full sinusoidal output swing:	$P_{L\,max} = \dfrac{V_b^2}{8R_E}$
Efficiency:	$\eta = \dfrac{P_{L\,max}}{P_{ges}} = 6.25\%$
Max. dissipation of the transistor:	$P_T = \dfrac{V_b^2}{R_E} = 8P_{L\,max}$

the transistor becomes zero. This is the case for the output voltage

$$U_{o\,min} = -\frac{V_b R_L}{R_E + R_L}.$$

If the output voltage is to be controlled sinusoidally around $0\,V$, its amplitude must not exceed the value

$$\hat{U}_{o\,max} = \frac{V_b R_L}{R_E + R_L}.$$

The power delivered to R_L is then

$$P_L = \frac{1}{2}\frac{\hat{U}_{o\,max}^2}{R_L} = \frac{V_b^2 R_L}{2(R_E + R_L)^2}.$$

With $\dfrac{dP_L}{dR_L} = 0$, it follows that for $R_L = R_E$ the maximum output power

$$P_{L\,max} = \frac{V_b^2}{8R_E}$$

is attained. This result is surprising in that one would normally expect maximum power output when the load resistance equals the output resistance r_o of the voltage source. However, this is so only for constant open-circuit voltage. Here, the open-circuit voltage is not constant as it must be reduced when R_L is small.

In the next step we determine the power consumption within the circuit for any output voltage amplitude and any load resistance. At sinusoidal voltage, the power

$$P_L = \frac{1}{2}\frac{\hat{U}_o^2}{R_L}$$

is supplied to the load resistance R_L. The power dissipation of the transistor is

$$P_T = \frac{1}{T}\int_0^T (V_b - U_o(t))\left(\frac{U_o(t)}{R_L} + \frac{U_o(t) + V_b}{R_E}\right)\,dt.$$

For $U_o(t) = \hat{U}_o \sin \omega t$ it follows that

$$P_T = \frac{V_b^2}{R_E} - \frac{1}{2}\hat{U}_o^2\left(\frac{1}{R_L} + \frac{1}{R_E}\right).$$

The power dissipation of the transistor is largest for zero input signal. Similarly, the power in R_E is given as

$$P_E = \frac{V_b^2}{R_E} + \frac{1}{2}\frac{\hat{U}_o^2}{R_E}.$$

The circuit therefore draws from the power supplies the total power of

$$P_{\text{tot}} = P_{\text{L}} + P_{\text{T}} + P_{\text{E}} = 2\frac{V_{\text{b}}^2}{R_{\text{E}}}.$$

This is a surprising result since it shows that the total power of the circuit is independent of the drive voltage and of the load, and since it remains constant as long as the circuit is not overdriven. The efficiency η is defined as the ratio of the maximum attainable output power to the power consumption at full voltage swing. With the results for $P_{\text{L max}}$ and P_{tot}, then $\eta = 1/16 = 6.25\%$. The following characteristics are typical for this circuit:

(1) The current through the transistor is never zero.
(2) The total power consumption is independent of the input drive voltage and of the load.

These are the characteristics of *class-A operation*.

5.2   Complementary emitter followers

The output power of the emitter follower in Fig. 5.1 is limited in that the resistor R_{E} restricts the maximum output current. A considerably higher output power and a better efficiency can be attained if R_{E} is replaced by a second emitter follower, as in Fig. 5.2.

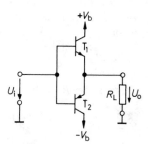

Fig. 5.2 Complementary class-B emitter follower.

Voltage gain:	$A \approx 1$
Current gain:	$A_{\text{i}} = \beta$
Output power at full sinusoidal output swing:	$P_{\text{L}} = \dfrac{V_{\text{b}}^2}{2R_{\text{L}}}$
Efficiency at full sinusoidal output swing:	$\eta = \dfrac{P_{\text{L}}}{P_{\text{tot}}} = 78.5\%$
Max. dissipation of one transistor:	$P_{\text{T}1} = P_{\text{T}2} = \dfrac{V_{\text{b}}^2}{\pi^2 R_{\text{L}}} = 0.2P_{\text{L}}$

5.2.1 Complementary class-B emitter follower

For positive input voltages, T_1 operates as an emitter follower and T_2 is reverse biased; vice versa for negative drive. The transistors thus carry the current alternately, each for half a period. Such a mode of operation is known as *push-pull class-B operation*. For $U_i = 0$, both transistors are turned off and therefore no quiescent current flows in the circuit. The current taken from the positive and negative power supply, respectively, is thus the same as the output current. The circuit therefore has a considerably better efficiency than the normal emitter follower. A further advantage is that, at any load, the output can be driven between $\pm V_b$ as the transistors do not limit the output current. The difference between the input and output voltage is determined by the base-emitter voltage of the current-carrying transistor. It changes only very little with load, therefore $U_i \approx U_o$, independent of the load current. The output power is inversely proportional to the resistance R_L and has no extremum, and therefore no matching is required between the load resistance and any internal circuit resistance. The maximum power output is determined instead by the permissible peak currents and the maximum power dissipation of the transistors, which, for full sinusoidal drive, is given by

$$P_L = \frac{V_b^2}{2R_L}.$$

We now determine the power dissipation P_{T1} of transistor T_1. As the circuit is symmetrical, it is identical to that of T_2.

$$P_{T1} = \frac{1}{T} \int_0^{T/2} (V_b - U_o(t)) \frac{U_o(t)}{R_L} \, dt.$$

For $U_o(t) = \hat{U}_o \sin \omega t$, this is

$$P_{T1} = \frac{1}{R_L} \left(\frac{\hat{U}_o V_b}{\pi} - \frac{\hat{U}_o^2}{4} \right).$$

For $\hat{U}_o = 0$, the power dissipation in the transistors is zero, as expected. For $\hat{U}_o = V_b$ it rises to

$$\frac{V_b^2}{R_L} \cdot \frac{4 - \pi}{4\pi} \approx 0.0685 \frac{V_b^2}{R_L}.$$

Hence, the efficiency of the circuit

$$\eta = \frac{P_L}{P_{ges}} = \frac{P_L}{2P_{T1} + P_L} = \frac{0.5}{2 \cdot 0.0685 + 0.5} = 78.5\,\%.$$

The power dissipation of the transistors reaches its maximum not at full output voltage swing, but at

$$\hat{U}_o = \frac{2}{\pi} V_b.$$

This follows from the extremum condition

$$\frac{dP_{T1}}{d\hat{U}_o} = 0.$$

In this case the power dissipation for each transistor is

$$P_{T\,max} = \frac{1}{\pi^2} \frac{V_b^2}{R_L} = 0.1 \frac{V_b^2}{R_L}.$$

The curves of output power, power dissipation and total power as functions of the relative output voltage swing $\hat{U}_o/V_b$ are given in Fig. 5.3.

As described above, only one transistor carries current at any time. However, this is so only for input frequencies which are small in comparison with the gain-bandwidth product of the transistors. Some time is needed for the transistor to change from the ON to the OFF mode. If the period of the input voltage is smaller than this time, both transistors may carry current simultaneously. Very high currents then flow through both transistors from $+V_b$ to $-V_b$ and will destroy them instantaneously. Oscillations at this critical frequency can occur in amplifiers with feedback or if the emitter followers feed leading loads. For the protection of the transistors, a current limiter should therefore be incorporated.

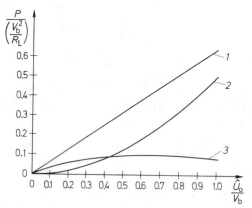

Fig. 5.3 Power and loss curves for the complementary emitter follower.
Curve 1: Total power taken from the supply. Curve 2: Output power. Curve 3: Power dissipation of each transistor

5.2.2   Complementary class-AB emitter followers

Figure 5.4 represents the transfer characteristic $U_o = U_o(U_i)$ for push-pull class-B operation, as described for the previous circuit. Near zero voltage, the current in the forward-biased transistor becomes very small and the transistor impedance rises. The output voltage at load therefore no longer changes linearly with the input voltage, and this is seen as a bend in the characteristic, near the origin. This gives rise to distortion of the output voltage, known as *crossover distortion*. If a small quiescent current flows through the transistors, their impedance near the origin is reduced and the transfer characteristic in Fig. 5.5 is obtained. The transfer characteristic of each individual emitter follower is shown by a broken line. It can be seen that the crossover distortion is considerably reduced. If the quiescent current is made as large as the maximum output current, the resulting mode of operation would be called a push-pull class-A operation, in analogy to Section 5.1. However, the crossover distortion is already greatly reduced if only a fraction of the maximum output current is permitted to flow as quiescent current. This mode is called push-pull class-AB operation and its crossover distortion is so small that it can be easily reduced to tolerable values by means of feedback.

Additional distortion may arise if positive and negative voltages are amplified at different gains. This is the case if the complementary emitter followers are driven from a high-impedance source and the transistors have different current transfer ratios. If strong feedback is undesirable, the transistors must be selected for identical current transfer ratios.

Figure 5.6 shows the basic circuit for the realization of class-AB operation. To obtain a small quiescent current, a direct voltage of about 1.4 V is applied between the base terminals of T_1 and T_2. If the two voltages, U_1 and U_2 are the same, the quiescent potential of the

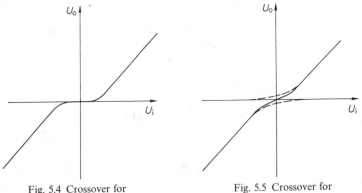

Fig. 5.4 Crossover for
push-pull class-B operation

Fig. 5.5 Crossover for
push-pull class-AB operation

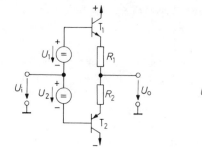

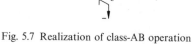

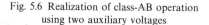

Fig. 5.6 Realization of class-AB operation   Fig. 5.7 Realization of class-AB operation
          using two auxiliary voltages   using a single auxiliary voltage

output approximately equals that of the input. The bias voltage can
also be supplied by a single voltage source, $U_3 = U_1 + U_2$, as represented
in Fig. 5.7. In this case, the potential difference between output and
input is about 0.7 V.

The main problem of class-AB operation is keeping the required
quiescent current constant over a wide temperature range. As the
transistors get warmer, the quiescent current increases which may
itself further increase the temperature and finally lead to destruction of
the transistors. This effect is known as positive thermal feedback. The
increase of the quiescent current can be avoided if the voltages U_1 and U_2
are each reduced by 2 mV for every degree of temperature rise. For this
purpose, diodes or thermistors can be used mounted on the heat sinks of
the power transistors.

However, temperature compensation is never quite perfect as the
temperature difference between junction and case is usually considerable.
Therefore, additional stabilization is required, provided by the resistors
R_1 and R_2 which cause series feedback. This is more effective the larger the
resistances chosen. As the resistors are connected in series with the load,
they reduce the available output power and must therefore be small
compared to the load resistance. This dilemma can be avoided by using
Darlington circuits, as is shown in Section 5.4.

5.2.3   Generation of the bias voltage

One way of providing a bias voltage is shown in Fig. 5.8. The voltage of
$U_1 = U_2 \approx 0.7$ V across each of the diodes D_1 and D_2 just allows a small
quiescent current to flow through the transistors T_1 and T_2. To attain a
higher input resistance, the diodes can be replaced by emitter followers,
this resulting in the circuit of Fig. 5.9.

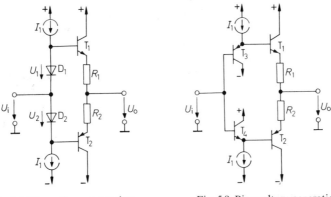

Fig. 5.8 Bias voltage generation
by diodes

Fig. 5.9 Bias voltage generation
by transistors

Figure 5.10 represents a driver arrangement which allows the adjustment of the bias voltage and its temperature coefficient over a wide range. Feedback is applied to the transistor T_3 by means of the voltage divider R_5, R_6. For negligible base current, its collector-emitter voltage has the value

$$U_{CE} = U_{BE} \left(1 + \frac{R_5}{R_6}\right).$$

To obtain the desired temperature coefficient, R_5 is, in practice, a resistor network containing an NTC resistor mounted on the heat sink of the output transistors. In this way, the quiescent current can to a large extent be made temperature-independent even though the temperature of the case is lower than that of the output transistor junctions.

In circuits using diodes for bias voltage generation, no current can flow from the input into the base of the output transistors. The base

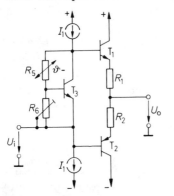

Fig. 5.10 Generation of the bias voltage
having an adjustable temperature coefficient

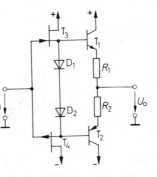

Fig. 5.11 Bias voltage generation
by FETs

current for the output transistors must therefore be supplied from the constant-current sources. The constant current I_1 must be larger than the maximum base current of T_1 and T_2 so that the diodes D_1 and D_2 (or the transistors T_3 and T_4) do not block before maximum permissible output voltage swing is reached. It would for this reason be disadvantageous to replace the constant-current sources by resistors, as this would cause the current to decrease with rising output voltage.

The most favourable driver circuit would be one supplying a larger base current for increasing output voltage, and such a circuit is represented in Fig. 5.11. The FETs T_3 and T_4 operate as source followers. The difference in their source voltages settles, due to series feedback, to a value of about 1.4 V; FETs having at $|U_{GS}| \approx 0.7$ V a drain current of a few mA are most suitable.

5.3   Electronic current limiter

Due to their low output resistance, power amplifiers can be easily overloaded and thereby destroyed. It is therefore advisable to limit the output current to a defined maximum value by an additional control circuit. Figure 5.12 shows a particularly simple solution. The limit cuts in when the diode D_3 or D_4 becomes conducting, as the voltage across R_1 or R_2 can then no longer increase. The maximum output current is therefore

$$I_{o\,max}^{+} = \frac{U_{D3} - U_{BE1}}{R_1}, \qquad I_{o\,max}^{-} = -\frac{U_{D4} - |U_{BE2}|}{R_2}.$$

It is obvious that the forward voltage U_{D3} and U_{D4} of the diodes must be higher than $U_{BE} \approx 0.7$ V. This can be achieved, for example, by series-connecting several silicon diodes. However, the use of light-emitting diodes (LEDs) effects a sharper limit. Red LEDs are highly suitable as they have a forward voltage of about 1.6 V.

Another way of limiting the output current is shown in Fig. 5.13. If the voltage across either R_1 or R_2 exceeds the value of about 0.6 V, either transistor T_3 or T_4 becomes conducting. A further rise in the base current of T_1 or T_2 is therefore prevented. By this method, the output current is limited to the maximum value of

$$I_{o\,max}^{+} \approx \frac{0.6\,\mathrm{V}}{R_1} \quad \text{or} \quad I_{o\,max}^{-} \approx \frac{0.6\,\mathrm{V}}{R_2}.$$

This has the advantage that the current limit is no longer influenced by the widely varying base-emitter voltage of the power transistors, but by the base-emitter voltage of the limiting transistors only. The resistors R_3 and R_4 serve to protect these transistors against high base current surges.

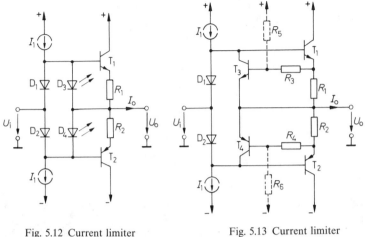

Fig. 5.12 Current limiter using diodes

Fig. 5.13 Current limiter using transistors

In the case of short circuit, the current $I_{o\,max}$ flows for one half period through T_1, and for the other half through T_2, while the output voltage is zero. The power dissipation in the output transistors is therefore

$$P_{T1}=P_{T2}\approx\tfrac{1}{2}V_b I_{o\,max}.$$

According to Section 5.2, this value is five times that of the dissipation at normal operation.

In many applications, power amplifiers have a defined resistive load R_L. The amplifier can supply the maximum power without distortion if

$$I_{o\,max}=\frac{U_{o\,max}}{R_L}.$$

For smaller output voltages, the load takes a smaller current $I_o=U_o/R_L$. The current limit can be therefore brought down for decreasing output voltage, and the power dissipation at short-circuit thereby reduced. This is the purpose of the two resistors R_5 and R_6, shown in Fig. 5.13 by a broken line. Their operation is described below for the example of a positive output voltage.

At large output voltages, $U_o\approx V_b$ and no current will flow through the resistor R_5. It remains ineffective and, as before, the current limit is at $I_{o\,max}\approx 0.6\,V/R_1$. For smaller output voltages, a positive base-emitter biasing voltage occurs at T_3, due to the voltage divider R_5, R_3. This causes T_3 to become conducting at smaller output currents. With the approximation $R_5\gg R_3\gg R_1$, we obtain for the current limit

$$I_{o\,max}\approx\frac{0.6\,V}{R_1}-\frac{R_3}{R_1 R_5}(V_b-U_o). \tag{5.1}$$

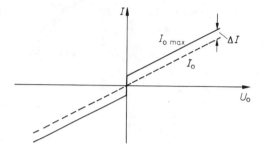

Fig. 5.14 Current limit and output current characteristics

Figure 5.14 shows schematically the current limit as a function of the output voltage. The circuit can be designed so that for every output voltage, the current limit has a value higher by ΔI than that of the output current for rated load R_L. This is the case when the two characteristics in Fig. 5.14 are running parallel. According to Eq. (5.1), the ratio

$$\frac{R_1 R_5}{R_3} = R_L$$

must then be chosen. Correspondingly, for negative output voltages

$$\frac{R_2 R_6}{R_4} = R_L.$$

5.4   Complementary emitter followers using Darlington circuits

With the circuits described so far, output currents of up to a few hundred milli-amperes can be obtained. For higher output currents, transistors having higher current transfer ratios must be used. They can be made up of two or more individual transistors if they are operated in a Darlington, or even in a complementary Darlington connection. Such circuits and their parameters are discussed in literature, e.g. in [1]. The basic circuit of a Darlington power amplifier is shown in Fig. 5.15, where the transistor pairs T_1, T_1' and T_2, T_2' are Darlington connected.

For the realization of push-pull class-AB, the adjustment of the quiescent current presents problems as four temperature-dependent base-emitter voltages must now be compensated. These difficulties can be avoided by allowing the quiescent current to flow only through the driver transistors T_1 and T_2. The output transistors then become conducting only for larger output currents. To achieve this, the bias voltage U_1 is selected so that a voltage of approx. 0.4 V appears across each of the

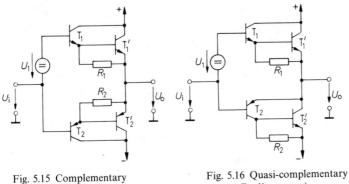

Fig. 5.15 Complementary
Darlington pairs

Fig. 5.16 Quasi-complementary
Darlington pairs

resistors R_1 and R_2; thus $U_1 \approx 2(0.4\,\text{V} + 0.7\,\text{V}) = 2.2\,\text{V}$. Then, for zero input, the output transistors have virtually no current, even at higher junction temperatures.

At higher output currents, the base-emitter voltages of the output transistors rise to about 0.8 V. This limits the current through R_1 and R_2 to double the quiescent value, and therefore most of the emitter current of the driver transistors is available as base current for the output transistors.

The resistors R_1 and R_2 also discharge the base of the output transistors. The lower their resistance, the faster the output transistors can be turned off. This is particularly important when the input voltage changes polarity as one transistor can become conducting before the other one is turned off. In this way, a large current can flow through both output transistors, and the resulting *second breakdown* will destroy them immediately. This effect determines the attainable large-signal band-width.

Sometimes it is preferable to use output power transistors of the same type. In such cases the Darlington circuit T_2, T_2' in Fig. 5.15 is replaced by the complementary Darlington connection [1]. The resulting circuit, shown in Fig. 5.16, is known as a *quasi-complementary* power amplifier. To arrive at the same quiescent current conditions as for the previous circuit, a voltage of about 0.4 V is impressed across resistor R_1. The voltage U_1 must then be $0.4\,\text{V} + 2 \cdot 0.7\,\text{V} = 1.8\,\text{V}$. The quiescent current flows via T_2 and R_2 to the negative supply. If $R_1 = R_2$, a bias voltage of 0.4 V is also obtained for T_2'. The purpose of the resistors R_1 and R_2 is the same as for the previous circuit, that is to discharge the base of the output current transistors.

The entire arrangement is available as a monolithic integrated circuit. Type TDA 1420 of SGS can supply a maximum output current of 3 A. The permissible power dissipation is 30 W for a case temperature of 60 °C.

To limit the current, the measures described in Section 5.3 can be applied. For this purpose, a current measuring resistor must be inserted in the emitter line of each Darlington connection. However, the method of Fig. 5.12 is not very favourable if applied to Darlington circuits as two base-emitter voltages are added to the voltage of the measuring resistor. The current measurement is therefore rather inaccurate.

5.5 Rating a power output stage

To illustrate the design process for a power output stage in more detail, we use the circuit of Fig. 5.17 and determine the rating of its components for an output power of 50 W. The circuit is based on the power amplifier of Fig. 5.15.

The amplifier is required to supply a load, $R_L = 5\,\Omega$, with a power of 50 W at sinusoidal output. The amplitude of the output voltage is then $\hat{U}_o = 22.4\,\mathrm{V}$, and that of the current $\hat{I}_o = 4.48\,\mathrm{A}$. For the determination of the supply voltage, we calculate the minimum voltage across T_1', T_1, T_3 and R_3. For the base-emitter voltages of T_1 and T_1' at I_{max}, we must allow about 1.8 V in all. If for D_1 a red LED is used, having a forward voltage of 1.6 V, the voltage across R_3 is 1 V. The collector-emitter voltage of T_3 should not fall below 0.8 V at full input drive. The output stage is intended for operation from an unregulated supply, the voltage of which may fall by about 3 V at full load. We therefore obtain the no-load supply voltage

$$V_b = 22.4\,\mathrm{V} + 1.8\,\mathrm{V} + 1\,\mathrm{V} + 0.8\,\mathrm{V} + 3\,\mathrm{V} = 29\,\mathrm{V}.$$

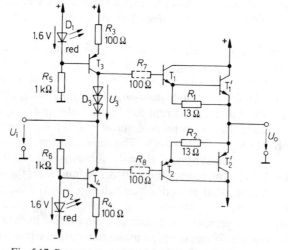

Fig. 5.17 Power output stage for 50 W sinusoidal output

For reasons of symmetry the negative supply voltage must have the same value. The maximum ratings of the transistors T'_1 and T'_2 can now be determined. The maximum collector current is 4.48 A. To be on the safe side, we choose $I_{C\,max} = 10$ A. The maximum collector-emitter voltage occurs at full output swing and is $V_b + \hat{U}_o = 51.4$ V. We choose the reverse collector-emitter voltage as $U_{CER} = 60$V. With the relationship

$$P_T = 0.1 \frac{V_b^2}{R_L},$$

of Section 5.2.1, we obtain $P_{T\,1'} = P_{T\,2'} = 17$ W. It can be shown that the relationship between the power dissipation and the thermal resistance is

$$P_{\vartheta j} = \frac{\vartheta_j - \vartheta_A}{R_{\theta A} + R_{\theta C}}.$$

The maximum junction temperature ϑ_j is normally 175 °C for silicon transistors. The ambient temperature ϑ_A within the housing of the amplifier should not exceed 55 °C. The thermal resistance between heat sink and air is assumed to be $R_{\theta A} = 4$ K/W. Hence, the value of the thermal resistance between junction and the transistor case must not be larger than

$$17\,\text{W} = \frac{(175 - 55)\,\text{K}}{\dfrac{4\,\text{K}}{\text{W}} + R_{\theta C}},$$

$$R_{\theta C} = \frac{3.1\,\text{K}}{\text{W}}.$$

Instead of the thermal resistance $R_{\theta C}$, the maximum power dissipation P_{25} at 25 °C case temperature is often given in the specifications. It can be calculated from

$$P_{25} = \frac{\vartheta_j - 25\,°\text{C}}{R_{\theta C}} = \frac{150\,\text{K}}{\dfrac{3.1\,\text{K}}{\text{W}}} = 48\,\text{W}.$$

The transistors selected in this way are assumed to have a current transfer ratio of 30 at maximum output current. We can therefore determine the data of the driver transistors T_1 and T_2. Their maximum collector current is

$$\frac{4.48\,\text{A}}{30} = 149\,\text{mA},$$

although this value applies to low frequencies only. For frequencies above $f_c \approx 20$ kHz, the current transfer ratio of audio power transistors falls markedly. When the current rises steeply, the driver transistor must momentarily supply the largest proportion of the output current. To obtain the largest possible bandwidth we choose $I_{C\,max} = 1\,A$. Transistors within this range of collector currents, having gain-bandwidth products in the region of 50 MHz, are still reasonably priced.

We have shown in Section 5.4 that it is useful to allow the quiescent current to flow only through the driver transistors, and to have a voltage of about 400 mV across the resistors R_1 and R_2. This is the purpose of the three silicon diodes D_3 which have a total forward voltage of about 2.2 V. A green light-emitting diode could also be used. We choose a quiescent current of approx. 30 mA to keep the crossover distortion reasonably small. Thus

$$R_1 = R_2 = \frac{400\,mV}{30\,mA} = 13\,\Omega.$$

The power dissipation of the driver transistors is, at zero input voltage, $30\,mA \cdot 29\,V = 0.9\,W$, and at maximum input is still 0.75 W. A small power transistor in a TO-5 case with cooling fins is obviously sufficient. A value of 100 is usual for the current transfer ratio of such a transistor. The maximum base current is then

$$I_{B\,max} = \frac{1}{100}\left(\frac{4.48\,A}{30} + \frac{0.8\,V}{13\,\Omega}\right) \approx 2\,mA.$$

The current through the constant current sources T_3 and T_4 must be large compared to this value, and we choose approx. 10 mA.

Emitter followers are prone to unwanted oscillations in the region of the transit frequency of the output transistors. These oscillations can be damped by additionally loading the output by a series RC element (approx. $1\,\Omega$; $0.22\,\mu F$). However, this also reduces the efficiency at higher frequencies. Another way of damping which may also be used in addition to that above, involves series resistors in the base lead of the driver transistors. If $R_7 = R_8 = 100\,\Omega$, and are inserted as shown in Fig. 5.17, the voltage across these resistors remains below 0.2 V. The attainable output voltage swing is therefore not significantly reduced.

5.6   Driver circuits with voltage gain

The power amplifiers described have a certain amount of crossover distortion in the region of zero output voltage, and this can be eliminated by feedback. The output stage is connected to a pre-amplifier stage, and negative feedback is applied across both stages. The voltage gain of the power amplifier is close to unity. In many cases therefore, the output voltage of the pre-amplifier is insufficient to attain full output swing of the power stage. The driver circuit must then be given a voltage gain. Figure 5.18 shows one possibility, where the output of the pre-amplifier T_5, T_6 is

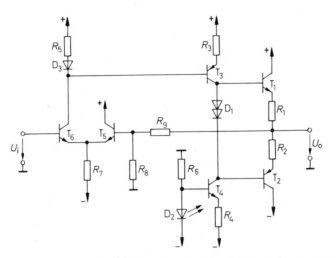

Fig. 5.18 Simple driver circuit, based on the principle of integrated power amplifiers (e.g. type TDA 2002 A from SGS)

connected to the base of T_3. At this node, an input of the power stage is available at which only a small voltage swing is required for full drive of the output stage, but this input has a high quiescent potential. In this circuit, the transistor T_3 operates as a controlled current source. Together with diode D_3 and resistor R_5, it constitutes a low-drift current mirror, as described in [1].

The whole arrangement is in feedback, effected by the resistors R_8 and R_9, and therefore has a voltage gain $A = 1 + R_9/R_8$. The offset voltage drift is determined to a large extent by the matching of the characteristics of the differential amplifier and the current mirror. It can be improved by replacing the differential amplifier by an operational amplifier, as in Fig. 5.19.

Once again, the current sources T_3 and T_4 provide the voltage gain. They are controlled at the emitter so that the largest possible bandwidth is attained. Together with the transistors T_5 and T_6, each forms a complementary cascode circuit as described for broadband amplifiers in Chapter 4.5.3.

In order not to limit the bandwidth by the relatively slow operational amplifier, the signal path is split into a low-frequency and a high-frequency branch, according to the principle of Fig. 4.13. The circuit has therefore the low drift of an operational amplifier and, at the same time, optimum broadband properties. The complete circuit behaves as an inverting operational amplifier, having a feedback via the resistors R_{15} and R_{16}. Its gain is then $A = -R_{16}/R_{15}$.

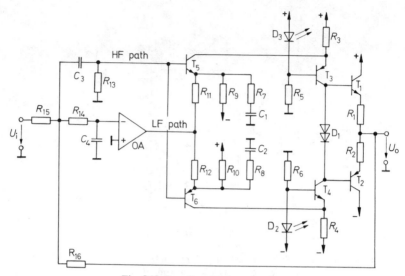

Fig. 5.19 Broadband driver circuit

To begin the dimensioning of the circuit, the collector currents of the transistors T_3 to T_6 are defined. We choose 10 mA. A current of 20 mA must then flow through the resistors R_3 and R_4. If the diodes D_2 and D_3 are red LEDs with a forward voltage of 1.6 V, a voltage of 1 V is measured across each of the resistors R_3 and R_4. Hence

$$R_3 = R_4 = \frac{1\,V}{20\,mA} = 50\,\Omega.$$

The output quiescent potential of the operational amplifier is determined by the offset voltage of the power output stage and is close to zero. Hence, with no input drive, the current through the resistors R_{11} and R_{12} is virtually nil. The collector currents of T_5 and T_6 must thus flow through the resistors R_9 and R_{10}. With supply potentials of ± 15 V, it follows that

$$R_9 = R_{10} \approx \frac{15\,V}{10\,mA} = 1.5\,k\Omega.$$

To attain full swing of the current sources T_3 and T_4, the collector currents of T_5 and T_6 must be controlled between zero and 20 mA. These values should be reached for full output swing of the operational amplifier. Thus, for the resistors R_{11} and R_{12},

$$R_{11} = R_{12} \approx \frac{10\,V}{10\,mA} = 1\,k\Omega.$$

It is suitable to give the lowpass filter R_{14}, C_4 at the operational amplifier input an upper cutoff frequency of about 10 kHz. In this way, uncontrollable reactions of the operational amplifier to higher-frequency input signals are avoided. The lower cutoff frequency of the highpass filter C_3, R_{13} in the high-frequency path must have a lower value, e.g. 1 kHz.

The total gain of the circuit can be set by the resistors R_{15} and R_{16} to values between 1 and 10. A higher gain is not to be recommended as the loop gain in the high-frequency path then becomes too low. The open-loop gain of the high-frequency path can be varied by means of the resistors R_7 and R_8. They are adjusted so as to obtain the desired transient response for the whole circuit. For the operational amplifier, the internal standard frequency compensation is sufficient. To avoid oscillations in the VHF range, it may be necessary to insert resistors in the base leads of the transistors.

5.7 Boosting the output current of integrated operational amplifiers

The output current of integrated operational amplifiers is normally limited, the maximum being about 20 mA. There are many applications for which about ten times this current is needed but where the number of additional components must be kept to a minimum. In such cases the described power output stages may be employed. For low signal frequencies, the number of components can be reduced by the use of push-pull class-B emitter followers. However, owing to the finite slew rate of the operational amplifier, noticeable crossover distortion occurs even with feedback. It can be reduced considerably by inserting a resistance R_1 as in Fig. 5.20, which by-passes the emitter followers in the region of zero voltage. In this case, the required slew rate of the amplifier is reduced from infinity to a value which is $1 + R_1/R_L$ times that of the rate of change of the output voltage.

In this way, a large-signal bandwidth of about 1 kHz is attained if a load resistance of $R_L = 5\,\Omega$ and a standard operational amplifier (e.g. type 741) are used. Operational amplifiers employing FETs usually possess a considerably higher slew rate, the type LF 356 for instance having the value of 12 V/µs. If such amplifiers are operated in the described circuit, a large-signal bandwidth of 20 kHz can be achieved.

The arrangement in Fig. 5.21 has the same properties as the previous circuit. Here however, control of the output transistors is effected by the supply terminals of the operational amplifier. This, together with the output transistors of the operational amplifier, results in two complementary Darlington connections.

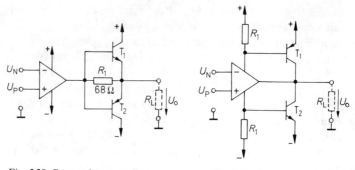

Fig. 5.20 Current booster with complementary emitter followers

Fig. 5.21 Current booster with complementary common-emitter circuits

At small output currents, the two output transistors T_1 and T_2 are turned off. The operational amplifier then supplies the whole output current. At larger output currents the transistors T_1 and T_2 become alternately conducting and supply the largest proportion of the output current. The contribution of the operational amplifier remains limited to approximately the value $0.6\,\text{V}/R_1$.

The circuit has an advantage over the previous one in that the quiescent current of the operational amplifier causes bias of the base-emitter junctions of the power output transistors. The values of the resistors are such that the bias is about 400 mV. This considerably reduces the range of crossover without the need for a quiescent current in the output transistors, the stabilization of which would require additional measures.

6 Power supplies

Every electronic circuit requires a power supply; this must provide one or more d.c. voltages. For larger power requirements, batteries are not economical. The d.c. voltage is therefore provided from the mains supply voltage by transformation and subsequent rectification. The d.c. voltage so obtained usually has considerable ripple, and changes depending on the variations of the mains supply voltage and the load. Therefore, a voltage regulator is often connected to the rectifier to keep the d.c. output voltage constant and counteract these variations. The following two sections describe ways of providing the unregulated d.c. voltage; regulator circuits will be dealt with later.

6.1 Properties of mains transformers

The internal resistance R_i of the mains transformer plays an important part in the design of rectifier circuits. It can be calculated from the rating of the secondary winding (U_n, I_n), and from the loss factor f_1 which is defined as the ratio of the no-load to the rated voltage

$$f_1 = \frac{U_0}{U_n}. \tag{6.1}$$

Hence, the relationship for the internal resistance

$$R_i = \frac{U_0 - U_n}{I_n} = \frac{U_n(f_1 - 1)}{I_n}. \tag{6.2}$$

We define a rated load $R_n = U_n/I_n$ and obtain from Eq. (6.2)

$$R_i = R_n(f_1 - 1). \tag{6.3}$$

The data of mains transformers normally used are shown in the table of Fig. 6.1. The values are based on the mains supply voltage $U_{p\,r.m.s.} = 220\,V$ at $50\,Hz$, a maximum flux density of $\hat{B} = 1.2\,T$ and a temperature above ambient of $\vartheta_{AA} = 40\,K$. More details can be obtained from [6.1] and [6.2].

core type	rated power	loss factor	number of primary turns	primary wire gauge	normalized number of secondary turns	normalized secondary wire gauge
	P_n [W]	f_1	w_1	d_1 [mm]	w_2/U_2 [1/V]	$d_2/\sqrt{I_2}$ [mm/$\sqrt{A}$]
M 42	4	1.31	4716	0.09	28.00	0.61
M 55	15	1.20	2671	0.18	14.62	0.62
M 65	33	1.14	1677	0.26	8.68	0.64
M 74	55	1.11	1235	0.34	6.24	0.65
M 85a	80	1.09	978	0.42	4.83	0.66
M 85b	105	1.06	655	0.48	3.17	0.67
M 102a	135	1.07	763	0.56	3.72	0.69
M 102b	195	1.05	513	0.69	2.45	0.71

Fig. 6.1 Typical data of M-core mains transformers for a primary voltage of $U_{p\,r.m.s.} = 220\,V$ at $50\,Hz$

6.2   Transformer rectifiers

6.2.1   Half-wave rectifier

The easiest way to rectify an a.c. voltage is to charge a capacitor via a diode, as in Fig. 6.2. If the output is not loaded, the capacitor C is charged during the positive half cycle to the peak value $U_{o\,0} = \sqrt{2}U_{0\,r.m.s.} - U_D$, where U_D is the forward voltage of the diode. The peak reverse voltage of the diode occurs when the transformer voltage is at its negative maximum, and therefore has the value of $2\sqrt{2}\,U_{0\,r.m.s.}$.

When a load resistance R_L is connected to the d.c. output, it discharges the capacitor C for as long as the diode is reverse-biased. Only when the no-load voltage of the transformer exceeds that of the output by the amount U_D, is the capacitor recharged. The voltage reached by recharging depends on the internal resistance R_i of the transformer. Figure 6.3 shows the shape of the output voltage at steady state. Owing to the unfavourable ratio of recharge/discharge time, the output voltage is considerably reduced even for a small load, and for this reason the circuit is unsuitable for use in power supplies.

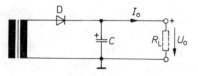

Fig. 6.2 Half-wave rectifier

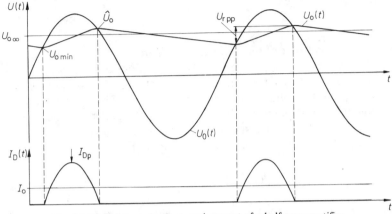

Fig. 6.3 Waveforms of voltage and current of a half-wave rectifier

6.2.2 Full-wave bridge rectifier

The ratio of recharge/discharge time can be greatly improved by charging the capacitor C during the positive *and* the negative half-cycle. This is achieved by the full-wave bridge rectifier in Fig. 6.4.

During the recharge period, the diodes connect whichever terminal of the transformer is negative to ground, and the positive terminal to the output. The repetitive peak reverse voltage is identical to the no-

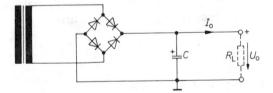

Fig. 6.4 Bridge rectifier.

No-load output voltage: $\qquad U_{o\,0} = \sqrt{2}\,U_{0\,\text{r.m.s}} - 2U_D$

On-load output voltage (infinitely large capacitor C): $\quad U_{o\,\infty} = U_{o\,0}\left(1 - \sqrt{\dfrac{R_i}{2R_L}}\right)$

Peak reverse voltage: $\qquad U_{p\,r} = \sqrt{2}\,U_{0\,\text{r.m.s.}}$

Mean diode current: $\qquad \bar{I}_D = \frac{1}{2}I_o$

Repetitive peak current: $\qquad I_{Dp} = \dfrac{U_{o\,0}}{\sqrt{2R_i R_L}}$

Rated power of transformer: $\qquad P_n = 1.2\,I_o(U_{o\,\infty} + 2U_D)$

Ripple voltage (peak-to-peak): $\qquad U_{r\,pp} = \dfrac{I_o}{2Cf_s}\left(1 - \sqrt[4]{\dfrac{R_i}{2R_L}}\right)$

Lowest value of output voltage: $\qquad U_{o\,\text{min}} \approx U_{o\,\infty} - \frac{2}{3}U_{r\,pp}$

load output voltage:

$$U_{o\,0}=\sqrt{2}\,U_{0\,\text{r.m.s.}}-2U_{\text{D}}=\sqrt{2}\,U_n f_1-2U_{\text{D}},\qquad(6.4)$$

and is only half that of the half-wave rectifier.

For the calculation of the voltage reduction at load, we initially assume an infinitely large smoothing capacitor C. The output voltage is then a pure d.c. voltage which we define as $U_{o\,\infty}$. The more the output voltage decreases due to the load, the longer the recharge time. Steady state is reached when the incoming charge of the capacitor equals the outgoing charge, i.e. that supplied to the load. Hence,

$$U_{o\,\infty}\approx U_{o\,0}\left(1-\sqrt{\frac{R_i}{2R_L}}\right),\qquad(6.5)$$

where $R_L=U_{o\,\infty}/I_o$ is the load resistance. The deduction of this equation is based on calculations involving the approximation of sine waves by parabolas and is omitted here because of its complexity.

To dimension the rectifier correctly, the currents must be known. As no d.c. current flows through the capacitor, the mean forward current of each bridge arm is half the output current. As the forward voltage is only slightly dependent on the current, the power dissipation of a single diode is given by

$$P_{\text{D}}=\tfrac{1}{2}U_{\text{D}}I_o.$$

During every recharge period a peak current I_{Dp} flows, the value of which may be many times that of the output current:

$$I_{\text{Dp}}=\frac{\hat{U}_0-2U_{\text{D}}-U_{o\,\infty}}{R_i}=\frac{U_{o\,0}-U_{o\,\infty}}{R_i}.$$

With Eq.(6.5), it follows that

$$I_{\text{Dp}}=\frac{U_{o\,0}}{\sqrt{2R_iR_L}}.$$

It can be seen that the internal resistance R_i of the a.c. voltage source significantly influences the peak current. If R_i is very small, it may be necessary to series-connect a resistance or inductance so as not to exceed the maximum peak current of the rectifier.

The r.m.s. value of the pulsating charging current is larger than its mean value. Therefore the d.c. power must always be kept smaller than the rated power of the transformer for resistive load, so as not to exceed the thermal transformer rating. The d.c. power is determined by the power supplied to the load ($I_oU_{o\,\infty}$) and the losses in the rectifier (approx. $2U_{\text{D}}I_o$). The rated power of the transformer must therefore be chosen as

$$P_n = \alpha I_o (U_{o\,\infty} + 2 U_D), \qquad (6.6)$$

where α is the form factor allowing for the increased r.m.s. value of the current. For full-wave bridge rectification, $\alpha \approx 1.2$.

For a finite smoothing capacitor, a superposed ripple voltage appears at the output. It can be calculated from the charge supplied by the capacitor during the discharge time t_d,

$$U_{r\,pp} = \frac{I_o t_d}{C}.$$

With Eq. (6.5), approximately

$$t_d \approx \frac{1}{2}\left(1 - \sqrt[4]{\frac{R_i}{2 R_L}}\right) T_s,$$

where $T_s = 1/f_s$ is the reciprocal of the a.c. supply frequency. Hence

$$U_{r\,pp} = \frac{I_o}{2 C f_s}\left(1 - \sqrt[4]{\frac{R_i}{2 R_L}}\right). \qquad (6.7)$$

In the presence of ripple, the lowest instantaneous value of the output voltage is of special interest. It is approximately

$$U_{o\,min} \approx U_{o\,\infty} - \tfrac{2}{3} U_{r\,pp}. \qquad (6.8)$$

The dimensioning of a transformer rectifier is best illustrated by an example. A d.c. voltage supply is required having a minimum output voltage of $U_{o\,min} = 9\,V$ for an output current of $I_o = 1\,A$, and a ripple of $U_{r\,pp} = 3\,V$.

To begin with, we obtain from Eq. (6.8)

$$U_{o\,\infty} = U_{o\,min} + \tfrac{2}{3} U_{r\,pp} = 11\,V,$$

and from Eq. (6.6) the rated power of the transformer

$$P_n = \alpha I_o (U_{o\,\infty} + 2 U_D) = 1.2 \cdot 1\,A\,(11\,V + 2\,V) = 15.6\,W.$$

It can be seen in the table in Fig. 6.1 that an M 55-type core having a loss factor of $f_1 = 1.2$ must be used. To continue the calculation the internal resistance of the transformer must be known; however, it is dependent on the rated voltage, the value of which is not yet known. For its determination, the system of non-linear equations (6.3) to (6.5) must be solved. This is best done by iteration: We set the initial value of U_n to $U_n \approx U_{o\,min} = 9\,V$. With Eq. (6.3), it follows that

$$R_i = R_n(f_1 - 1) = \frac{U_n^2}{P_n}(f_1 - 1) = \frac{(9\,V)^2}{15.6\,W}\cdot(1.2 - 1) = 1.04\,\Omega.$$

Hence, with Eqs. (6.4) and (6.5)

$$U_{o\infty} = (\sqrt{2}\, U_n f_1 - 2 U_D)\left(1 - \sqrt{\frac{R_i}{2R_L}}\right)$$

$$= (\sqrt{2}\cdot 9\,\text{V}\cdot 1.2 - 2\,\text{V})\left(1 - \sqrt{\frac{1.04\,\Omega}{2\cdot\dfrac{11\,\text{V}}{1\,\text{A}}}}\right) = 10.39\,\text{V}.$$

The voltage is about 0.6 V smaller than that initially required. For the first iteration, we increase the rated transformer voltage by this amount and obtain correspondingly

$$R_i = 1.18\,\Omega \quad \text{and} \quad U_{o\infty} = 10.98\,\text{V},$$

which is already the desired value for the output voltage. The data of the transformer are therefore

$$U_n = 9.6\,\text{V}; \qquad I_n = \frac{P_n}{U_n} = 1.6\,\text{A}.$$

Figure 6.1 gives the winding data for a primary voltage of 220 V.

$$w_1 = 2671, \qquad\qquad\qquad d_1 = 0.18\,\text{mm}$$

$$w_2 = 14.62\,\frac{1}{\text{V}}\cdot 9.6\,\text{V} = 140, \quad d_2 = 0.62\,\frac{\text{mm}}{\sqrt{\text{A}}}\sqrt{1.6\,\text{A}} = 0.78\,\text{mm}.$$

The capacitance of the smoothing capacitor is given by Eq. (6.7) as

$$C = \frac{I_o}{2\,U_{r\,pp} f_s}\left(1 - \sqrt[4]{\frac{R_i}{2R_L}}\right) = \frac{1\,\text{A}}{2\cdot 3\,\text{V}\cdot 50\,\text{Hz}}\left(1 - \sqrt[4]{\frac{1.18\,\Omega}{22\,\Omega}}\right) \approx 1700\,\mu\text{F}.$$

The no-load output voltage is 14.3 V. The capacitor must be rated for at least this voltage.

The calculation for transformers having several secondary windings is the same as that above. For P_n, the rated power of the corresponding secondary winding must be inserted. The total power is the sum of the individual powers of the secondary windings. This determines the choice of the core and therefore the loss factor f_1.

6.2.3   Bridge rectifier for two output voltages
symmetrical about ground

For the operation of electronic circuits it is often necessary to have a positive and an equally large negative supply potential. These can be generated using two identical bridge rectifiers where the negative terminal of one rectifier and the positive terminal of the other is

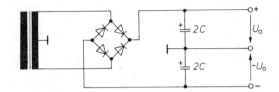

Fig. 6.5 Bridge rectifier for output voltages symmetrical about ground

connected to ground. Should the output currents of the two rectifiers be of the same order of magnitude, it is more convenient to generate both voltages by a single bridge rectifier. Figure 6.5 shows the corresponding arrangement. The centre tap of the transformer winding is connected to ground. In this way, both a positive and a negative transformer voltage are available at any time. During the recharge period, the rectifier connects whichever transformer terminal is positive to the output "plus" and the negative terminal to the output "minus". This effects full-wave rectification for both output voltages.

For the determination of the circuit parameters, the equations of the previous sections can be used in which the output voltage is replaced by the total output voltage, i.e. $2U_o$, and the voltage ripple is $2U_{r\,pp}$. As a result the data of the total secondary winding are obtained. The winding is tapped at half the number of turns. The smoothing capacitor determined by the calculation is realized by two series-connected capacitors of double the capacitance.

6.3 Series regulation

As too large a smoothing capacitor is not desirable, the output voltage of transformer rectifiers usually shows a ripple of a few volts. Furthermore, it is influenced by supply and load variations. These changes can be reduced by inserting a controlled series resistance, this method being known as series loss regulation.

6.3.1 Basic circuit

The simplest voltage regulator is an emitter follower, the base of which is connected to a reference voltage source. This reference voltage can be obtained from the unstabilized input voltage U_i, for example by means of a Zener diode as in Fig. 6.6. Further solutions are shown in Section 6.4. Because of the series feedback, the output voltage assumes the value

$$U_o = U_{ref} - U_{BE}.$$

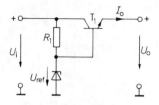

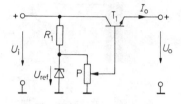

Fig. 6.6 Series regulation
using an emitter follower.

Fig. 6.7 Supplementary circuit
for adjustment of the output voltage

Output voltage: $U_o = U_{ref} - U_{BE}$

$0 \leqq U_o \leqq U_{ref} - U_{BE}$

The output resistance determines the degree to which this voltage changes with load (load regulation). It is given by

$$r_o = -\frac{\partial U_o}{\partial I_o} = \frac{1}{g_m} = \frac{U_T}{I_o}.$$

As $U_T \approx 26\,\text{mV}$, a value of about $0.3\,\Omega$ is obtained for $I_o = 100\,\text{mA}$.

Variations of the input voltage are levelled off by the low incremental resistance r_z of the Zener diode. The change in the output voltage is

$$\Delta U_o = \Delta U_{ref} = \frac{r_z}{R_1 + r_z} \Delta U_i \approx \frac{r_z}{R_1} \Delta U_i.$$

The ratio

$$\frac{\Delta U_i}{\Delta U_o} = \frac{R_1}{r_z}$$

is known as the ripple rejection which for this circuit has a value in the order of $10 \dots 100$.

If an adjustable output voltage is needed, a potentiometer is used to tap the reference voltage as is shown in Fig. 6.7. The resistance of the potentiometer must be chosen small in comparison with r_{BE} so that the increase in the output resistance of the circuit is kept minimal.

6.3.2   Circuit with error amplifier

In the circuits described above, the output resistance is determined by the emitter follower. It can be reduced further by using an error amplifier and feedback across the entire arrangement. We have already discussed such circuits in Chapter 2.1 under the heading "controlled voltage sources". They have the additional advantage that the output voltage can be set with high precision simply by choosing an appropriate resistance ratio, and that the output voltage is independent of U_{BE}.

In principle, the regulator circuits using error amplifiers do not differ from the voltage sources of Chapter 2.1. However, the current which an operational amplifier can normally provide, is not sufficient for the output of a power supply. It is therefore necessary to add a power amplifier which should also be included in the feedback loop. Theoretically, the power amplifiers of Chapter 5 could be used. Since normally only a positive *or* a negative output voltage is required, the amplifiers for power supplies can be simplified and are usually reduced to just one power transistor or a Darlington pair.

Figure 6.8 shows such a circuit for positive output voltages. It consists of a non-inverting amplifier the output current of which is boosted by the emitter follower T_1. The operational amplifier is not connected to the usual two symmetrical-about-ground supply potentials but is supplied only from a positive voltage. In this way, the permissible range of the input and output potentials is limited to positive values. For power supplies, this is not usually a serious restriction, and the need for a negative supply potential for the operational amplifier is therefore avoided. It is also an advantage that the positive supply potential can be raised to twice its normal value without exceeding the absolute maximum ratings. Thus, standard operational amplifiers can be employed to control output voltages of up to about 30 V.

The generation of a separate positive supply voltage for the operational amplifier is superfluous if the input voltage U_i is used directly to supply the amplifier, as in Fig. 6.8. The variations of this voltage have virtually no influence on the output voltage as the supply ripple rejection of an operational amplifier is usually high.

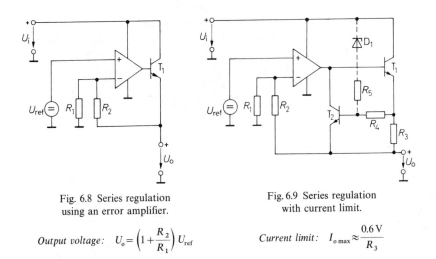

Fig. 6.8 Series regulation
using an error amplifier.

Fig. 6.9 Series regulation
with current limit.

Output voltage: $\quad U_o = \left(1 + \dfrac{R_2}{R_1}\right) U_{ref}$

Current limit: $\quad I_{o\,max} \approx \dfrac{0.6\,V}{R_3}$

Current limit

Integrated operational amplifiers have a built-in current limit which prevents the base current of transistor T_1 in Fig. 6.8 exceeding the value $I_{B\,max} = 10 \ldots 20\,\text{mA}$. It also limits the output current I_o to the value $I_{o\,max} = B \cdot I_{B\,max}$, where B is the current transfer ratio of transistor T_1. As B varies from transistor to transistor, and also increases for rising temperature, this indirect method of limiting the output current is very inaccurate.

Better accuracy is achieved if the current actually flowing at the output is sensed. Resistor R_3 and transistor T_2 in Fig. 6.9 are incorporated for this purpose. If the voltage across R_3 exceeds approx. 0.6 V, transistor T_2 becomes conducting and thereby prevents a further rise in the base current of T_1. As already shown for the power amplifier in Fig. 5.13, this method limits the output current to the value

$$I_{o\,max} \approx \frac{0.6\,\text{V}}{R_3}.$$

When the maximum output current is reached, the power dissipation of the output transistor is

$$P_3 = I_{o\,max}(U_i - U_o). \tag{6.9}$$

At short circuit it is considerably larger than at normal operation, as the output voltage then falls below the rated value and becomes zero. To avoid this increase in power dissipation, the current limit can be reduced for decreasing output voltages. This results in a retrograde current-voltage characteristic, as is shown in Fig. 6.10 (foldback current limit).

A considerable rise in the power dissipation can also occur if the input voltage U_i is raised, as this again increases the voltage difference $(U_i - U_o)$ in Eq. (6.9). Optimum protection of the output transistor T_1 is therefore obtained if the current limit $I_{o\,max}$ is adjusted in accordance to the voltage difference $(U_i - U_o)$. The resistor R_5 and Zener diode D_1 serve this purpose and are shown in Fig. 6.9 by broken lines.

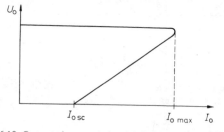

Fig. 6.10 Output characteristic with foldback current limit

When the potential difference $(U_i - U_o)$ is smaller than the Zener voltage U_z of diode D_1, no current flows through resistor R_5. In this case, the current limit remains unchanged at $0.6\,\text{V}/R_3$. When the potential difference exceeds the value U_z, a positive base-emitter bias voltage appears across transistor T_2 because of the voltage divider R_4, R_5. Transistor T_2 therefore becomes conducting for a correspondingly smaller voltage across R_3.

6.3.3   Integrated voltage regulators

The arrangement of Fig. 6.9 is available as a monolithic integrated circuit, sold as a fixed-voltage regulator (e.g. series 7800) for all the usual supply voltages between 5 V and 24 V. This type has only three terminals: input, output and ground. The requirements imposed on the error amplifier are not particularly stringent as the emitter follower is itself quite a useful voltage regulator, as is described in Section 6.3.1. Hence, a simple differential amplifier is sufficient, such as that in Fig. 6.11. There are several ways of providing the reference voltage and these are discussed in detail in Section 6.4. In Fig. 6.11, we have represented them symbolically by the Zener diode D_2. Owing to negative feedback via the voltage divider R_1, R_2, the output voltage assumes the value $U_o = U_{ref}(1 + R_2/R_1)$.

Capacitor C_p effects the necessary frequency compensation. As an additional measure of stabilization, two capacitors each of about $0.1\,\mu\text{F}$ are usually connected one to the output and one to the input.

In addition to the fixed voltage regulators, adjustable voltage regulators are also available (series 78 G). In these circuits, the voltage

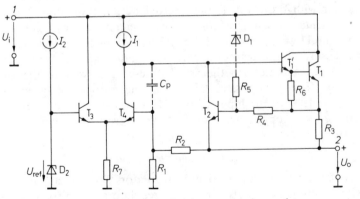

Fig. 6.11 Typical circuit of an integrated series regulator.

$$\textit{Output voltage:}\quad U_o = \left(1 + \frac{R_2}{R_1}\right)U_{ref}$$

divider R_1, R_2 is omitted and the base terminal of T_4 is available at an additional pin, the circuit thus having 4 terminals. With the externally connected voltage divider R_1, R_2, any desired output voltage between $U_{ref} \approx 5\,\text{V} \leqq U_o < U_i$ can be defined. The input voltage must be larger than the output voltage by at least 3 V in order to prevent saturation of the error amplifier. The maximum permissible input voltage is about 40 V.

Even with a fixed-voltage regulator, the output voltage can be varied within certain limits if a Zener diode is inserted in the ground lead, as is shown in Fig. 6.12. The output voltage is then raised by U_z. Resistor R serves to increase the Zener current by the constant amount $\Delta I = (U_o - U_z)/R$. This effects reduction of Zener voltage fluctuations arising from the current variations in the regulator ground lead.

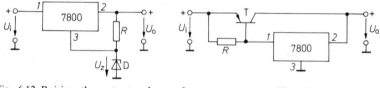

Fig. 6.12 Raising the output voltage of fixed-voltage series regulators

Fig. 6.13
Boosting the output current

Boosting the output current

The maximum output current of integrated voltage regulators of the 7800 series is approximately 1 A. It can be increased with the aid of an additional power transistor, and such a circuit is shown in Fig. 6.13. Together with transistor T, the internal output transistor T_1 forms a complementary Darlington connection, as can be seen by a comparison with Fig. 6.11 which represents the internal circuit of the 7800. The resistor R ensures that transistor T becomes conducting only for larger output current and not for the small quiescent current of the voltage regulator. This prevents, at light load, the output voltage from rising to the level of the input voltage. If transistor T is to become conducting at an output current of 0.3 A, the resistor must have the value $R = 0.6\,\text{V}/0.3\,\text{A} = 2\,\Omega$.

Negative output voltages

So far we have described only regulators for the stabilization of positive output potentials. The same regulators can also be used for negative output potentials if a floating input voltage source is available. Such a circuit is shown in Fig. 6.14. It is obvious that as soon as one of the terminals of the unregulated voltage source is connected to ground,

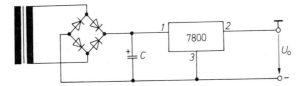

Fig. 6.14 Regulation of a negative voltage

the circuit will no longer function. Either the voltage regulator or the output voltage would then be short-circuited.

Figure 6.5 showed a simplified way of simultaneously providing a positive and a negative supply voltage. The terminal common to both voltages is grounded and for this reason, the circuit of Fig. 6.14 cannot be used for regulating the negative supply potential. In this case, a voltage regulator for negative output voltages is required. Such circuits are also available as integrated circuits (e.g. series 7900 or 79 G) and operate complementarily to the circuit in Fig. 6.11. Their application is shown in Fig. 6.15.

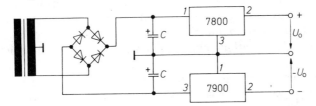

Fig. 6.15 Regulation of two voltages symmetrical about ground, using ICs

6.3.4 Voltage regulator with reduced dropout voltage

The minimum voltage difference $(U_i - U_o)$ required for the correct operation of integrated voltage regulators, is about 3 V. This voltage difference, or dropout voltage, is too high for some applications. However, the design concept used above is such that the dropout voltage cannot be much reduced. As can be seen in Fig. 6.11, the current source I_1 must supply the collector current of the driver transistor T_4 and the base current for the output Darlington pair T_1', T_1. Even with optimum design, it requires a voltage drop of about 1.5 V. A further voltage drop is due to the base-emitter junctions of the output Darlington pair and this is also about 1.5 V.

A considerable reduction in the dropout voltage can be achieved using pnp-transistors in the output stage. The collector current of the driver transistor can then also be used as base current for the output

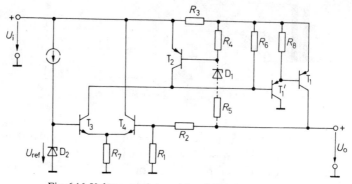

Fig. 6.16 Voltage regulator with reduced dropout voltage.

$$Output\ voltage: \quad U_o = \left(1 + \frac{R_2}{R_1}\right) U_{ref}$$

stage and the current source I_1 may be omitted. The resulting circuit is shown in Fig. 6.16, where it can be seen that the output stage operates in common-emitter connection. As a result, phase inversion occurs, and therefore transistor T_3 is used instead of transistor T_4 to control the output stage. The minimum voltage drop is determined by the collector-emitter saturation voltage of transistor T_1 and is therefore below 1 V.

The internal resistance of the output stage is higher than that of an emitter follower, but this is compensated for by the fact that the output stage contributes to the voltage gain, and therefore increases the loop gain.

The method of limiting the output current is identical to that of the circuit in Fig. 6.11. Resistor R_3 serves to measure the current in the emitter lead of the output transistor T_1. The voltage divider R_4, R_5 is again responsible for the retrograde output characteristic.

For the regulation of negative output potentials the circuit can be built in a complementary manner. The output stage then becomes an npn Darlington pair. In this form, the circuit is particularly suited to monolithic integration and therefore often used for negative-voltage regulation.

6.3.5   Regulation of voltages symmetrical about ground

For the regulation of voltages symmetrical about ground, two independent voltage regulators can be used, as in Fig. 6.15. However, there are cases where it is important to have two voltages of opposite polarity but of exactly the same magnitude, independent of their actual

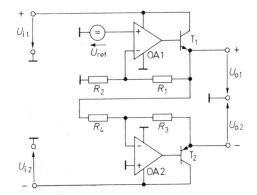

Fig. 6.17 Tracking regulation of voltages symmetrical about ground. IC type e.g. SG 3502 from Silicon General

values. This condition can be fulfilled by the circuit in Fig. 6.17. The positive output voltage U_{o1} is regulated in the usual manner by the operational amplifier OA 1. This voltage serves in turn as a reference for the regulation of the negative voltage U_{o2}. The amplifier OA 2 is operated as an inverting amplifier having the input voltage U_{o1}. When $R_3 = R_4$, the output voltage becomes $U_{o2} = -U_{o1}$.

Since the output potential of OA 2 is always negative and its input potential is zero, the common ground can be used here as the positive supply potential for OA 2. However, the condition for this is that the permissible limit for the positive common-mode signal of the amplifier is equal to the positive supply voltage, as for example is the case for the LM 301 type.

Symmetrical division of a floating voltage source

The problem often arises, especially in battery operated equipment, that two regulated symmetrical-about-ground voltages must be obtained from a floating unstabilized voltage source. For the solution of this problem, the sum of the two voltages can be stabilized to the desired value using one of the circuits previously described. A second circuit is then required to ensure that the voltage is split in the correct ratio. In principle, a voltage divider could be used for this purpose, the node of which is connected to ground. The division of the voltage is kept constant if the internal resistance of the voltage divider is low, but the loss in the voltage divider is then considerably increased. It is therefore better to replace the divider by two transistors. Only the transistor which is connected to the d.c. bus carrying the smaller load current, is turned on. Figure 6.18 shows the circuit.

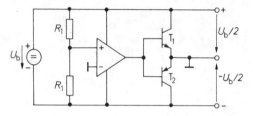

Fig. 6.18 Symmetrical division of a floating voltage source

The voltage divider formed by the two resistors R_1 halves the voltage U_b. It may have a high internal resistance, as its only load is the input bias current of the operational amplifier. If the node of the voltage divider is at zero potential, the voltage U_b is split into a positive and a negative output voltage, their ratio being 1:1. The operational amplifier compares the divider node potential with the ground potential and adjusts its output voltage so that their difference becomes zero. The feedback is realized as follows: If, for instance, the positive output is loaded more than the negative, the positive output potential falls. This also reduces the potential at the P-input of the operational amplifier. Because of the high gain, the amplifier output potential reduces even further, so that T_1 is turned off and T_2 becomes conducting. This counteracts the assumed reduction in the positive output voltage. At steady state, the current through T_2 is just large enough to ensure that the two output voltages share the load equally. The transistors T_1 and T_2 therefore operate as shunt regulators only one of which is conducting at any time.

If the load is only slightly unsymmetrical, the output stage of the operational amplifier can be used directly, instead of the transistors T_1 and T_2. The output of the amplifier is then simply connected to ground.

6.3.6   Voltage regulator with sense inputs

The resistance R_w of the connecting leads between the voltage regulator and the load, including possible contact resistances, makes the low output resistance of the regulator pointless. This effect can be eliminated by including the spurious resistances in the feedback loop, i.e. by measuring the output voltage as near to the load as possible. This is the purpose of the sense inputs, S^+ and S^-, in Fig. 6.19. The currents in the sensing leads must be kept small so that no errors are induced by the lead resistance, i.e. the voltage divider must have a sufficiently large resistance. The sensing lead S^- also carries the negative supply current of the regulating amplifier.

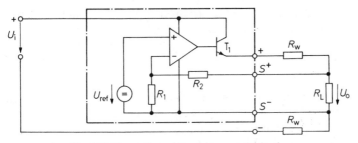

Fig. 6.19 Regulation of the voltage at the load

The described method of four-wire regulation can also be implemented by an integrated adjustable voltage regulator. The voltage divider R_1, R_2 is connected directly to the load, as in Fig. 6.19. The ground terminal of the regulator is also connected to the load, but via a separate lead so that it serves as the negative sense input.

6.3.7 Laboratory power supplies

The output voltage of the voltage regulators described can be adjusted only within a certain range $U_o \geqq U_{ref}$. The current limit serves only to protect the voltage regulator and is therefore fixed to the value I_{max}.

A laboratory power supply must have an output voltage and a current limit which are linearly adjustable between zero and the maximum value. A suitable circuit is shown in Fig. 6.20. Operational

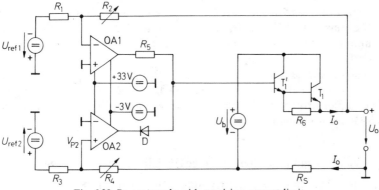

Fig. 6.20 Power supply with precision current limit.

$$\text{Output voltage:} \quad U_o = -\frac{R_2}{R_1} U_{ref\,1}$$

$$\text{Current limit:} \quad I_{o\,limit} = \frac{R_4}{R_S R_3} U_{ref\,2}$$

amplifier OA 1 is operated as an inverting amplifier and is responsible for voltage regulation. Hence, the output voltage

$$U_o = -\frac{R_2}{R_1} U_{\text{ref 1}},$$

which is proportional to the variable resistance R_2. Voltage control of the output is possible by varying the voltage $U_{\text{ref 1}}$. The output current flows from the floating unregulated power voltage-source, U_b, via the Darlington pair T_1, T_1', through the load and the current measuring resistance, R_S, back to the source. The voltage across R_S is therefore proportional to the output current I_o. It is compared with a second reference voltage, $U_{\text{rlf 2}}$, by the operational amplifier OA 2 which is connected as inverting amplifier. As long as

$$\frac{I_o R_S}{R_4} < \frac{U_{\text{ref 2}}}{R_3},$$

$V_{P\,2}$ remains positive. The output voltage of OA 2 is therefore at its positive maximum, and diode D is reverse biased. Under these conditions, the voltage regulation is not affected. If the output current reaches the limit

$$I_{\text{o limit}} = \frac{R_4}{R_S R_3} U_{\text{ref 2}},$$

then $V_{P\,2} = 0$. The output voltage of OA 2 falls, and the diode becomes forward biased. This causes the base potential of the Darlington pair to fall, i.e. the current regulation comes into effect. Amplifier OA 1 tries to prevent a decrease in the output voltage by raising its output potential to the maximum. However, this does not affect the current control, as amplifier OA 2 maintains priority because of the forward biased diode D.

The difference between the currents through the voltage dividers R_1, R_2 and R_3, R_4 causes an additional voltage on R_S. They must therefore have sufficiently high resistance in order not to adversely affect the current measurement.

6.3.8   High-power output stage for laboratory power supplies

Power supplies, the output voltages of which can be adjusted down to zero, may have a particularly large power dissipation. In order to attain a maximum output voltage $U_{\text{o max}}$, the unstabilized voltage U_b must be larger than $U_{\text{o max}}$. Maximum power dissipation in T_1 occurs when maximum output current $I_{\text{o max}}$ is taken at zero output voltage, and is therefore $U_b \cdot I_{\text{o max}}$. The maximum power dissipation is thus

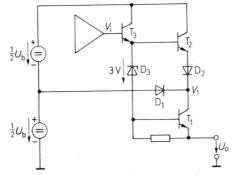

Fig. 6.21 Electronic selection of the output voltage range

nearly as large as the maximum available output power. For values of power dissipation of over 100 W a cooling fan is normally required. To avoid this additional component, the power dissipation must be kept as small as possible. This can be achieved by splitting the total required output voltage range into several successive ranges. A different supply voltage U_b is selected for each range, and in this way the voltage across T_1 remains small.

Figure 6.21 shows one way of electronically selecting the correct U_b for each of the two output voltage ranges [6.3]. At small input potentials V_i, transistor T_2 is turned off and diode D_1 forward biased. The collector potential of T_1 is therefore about $\frac{1}{2}U_b$. The total power dissipation for this mode of operation is then

$$P_{tot} = P_{T1} = I_o(\tfrac{1}{2}U_b - U_o).$$

At zero output voltage, this is only half the power dissipation of a supply having no such voltage selection.

When the input potential V_i rises to values larger than $\frac{1}{2}U_b + 2U_{BE}$, T_2 becomes conducting, and the potential V_1 rises with V_i:

$$V_1 = V_i - 2U_{BE} - U_{D2} \approx V_i - 2\,\text{V}.$$

Hence, diode D_1 is reverse biased, and the output current is supplied by the series connection of the two voltage sources $\frac{1}{2}U_b$. The collector-emitter voltage of transistor T_1 does not become zero but is regulated to the value

$$U_{CE1} = V_1 - U_o = (V_i - 2\,\text{V}) - (V_i - 3\,\text{V} - 1.4\,\text{V}) = 2.4\,\text{V}.$$

The total power dissipation for this mode of operation is now

$$P_{tot} = P_{T1} + P_{T2} = 2.4\,\text{V} \cdot I_o + (U_b - U_o - 2.4\,\text{V})I_o = (U_b - U_o)I_o.$$

The curve of the power dissipation is shown in Fig. 6.22 as a function of the output voltage.

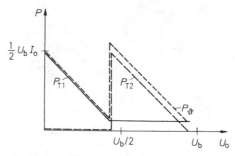

Fig. 6.22 Power dissipation of the transistors T_1 and T_2 against output voltage

Diode D_2 protects the emitter-base junction of transistor T_2 at low output voltages.

6.4   Generation of the reference voltage

Every voltage regulator requires a reference voltage with which the output voltage is compared. The output voltage regulation cannot be better than the stability of the reference. In this section, various aspects of reference voltage generation are discussed in detail.

6.4.1   Reference-voltage sources with Zener diodes

The simplest method of providing a reference voltage is to apply the unregulated input voltage to a Zener diode and a series resistance, as in Fig. 6.23. The quality of the regulation is described by the ripple rejection factor

$$G = \frac{\Delta U_i}{\Delta U_{ref}},$$

which is usually given in dB. For the circuit in Fig. 6.23

$$G = 1 + \frac{R}{r_z} \approx \frac{R}{r_z} \quad \text{(approx. } 10 \ldots 100\text{)},$$

where r_z is the incremental resistance of the Zener diode for the chosen point of operation. In a first order approximation, r_z is inversely proportional to the current flowing in the diode. Increase of the series resistance R for a given input voltage will therefore not result in any improvement of the ripple rejection. An important aspect to consider when defining the diode current, is noise in the Zener voltage which increases quickly for small currents. The resistance R is given such a

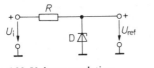

Fig. 6.23 Voltage regulation
by a Zener diode

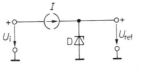

Fig. 6.24 Improvement of the ripple
rejection using a current source

value that the diode carries adequate current for minimum input voltage and maximum output current.

A considerable improvement in ripple rejection can be achieved if the resistor R is replaced by a current source, as in Fig. 6.24. It is easiest to use a FET current source as this has only two terminals. Ripple rejection factors of up to 10000 can then be attained.

For fixed-voltage regulators having an output voltage higher than the reference voltage, high ripple rejection can be achieved even with a series resistance R, if it is connected not to the unregulated input voltage but rather to the output, as in Fig. 6.25. The ripple rejection is then determined mainly by the power supply ripple rejection $D = \Delta V_b / \Delta U_{OFFSET}$ of the operational amplifier. From Fig. 6.25, we take the relationships

$$\Delta V_P = \frac{r_z}{r_z + R} \Delta U_o,$$

$$\Delta V_N = \frac{R_1}{R_1 + R_2} \Delta U_o.$$

Hence, with $\Delta V_b = \Delta U_i$

$$G = \frac{\Delta U_i}{\Delta U_o} = D \left(\frac{r_z}{r_z + R} - \frac{R_1}{R_1 + R_2} \right) \approx |D| \frac{R_1}{R_1 + R_2}.$$

Values of around 10000 can be achieved. If the variations of the input voltage remain below 10 V, the output voltage thus varies by less than 1 mV.

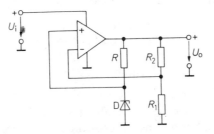

Fig. 6.25 Reference voltage taken from the regulated output voltage

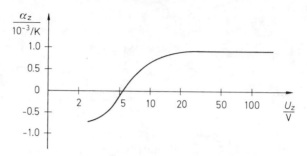

Fig. 6.26 Dependence of the temperature coefficient $\alpha_z = \Delta U_z/(\Delta\vartheta \cdot U_z)$ on the Zener voltage

Considerably larger fluctuations may arise due to shift in temperature. The temperature coefficient of the Zener voltage is in the order of $(+1 \ldots -1) \cdot 10^{-3}/K$. For small Zener voltages, it is negative and for larger ones, positive. Its typical characteristic is shown in Fig. 6.26. It can be seen that the temperature coefficient is smallest for Zener voltages around 6 V. For larger Zener voltages, it can be reduced by series-connecting forward biased diodes. Such elements are available as single components (reference diodes) having temperature coefficients down to $10^{-5}/K$. To guarantee these values, the diode current must be kept constant within approx. 10%. This condition can be easily fulfilled with the circuit in Fig. 6.25.

6.4.2   Generation of small reference voltages

Zener diodes can be obtained for Zener voltages of more than 2.5 V. Smaller voltages can be realized by a series connection of several forward biased silicon diodes. With three diodes, a voltage of about 2 V is obtained, having a temperature coefficient of $-6\,mV/K$, corresponding to $-3 \cdot 10^{-3}/K$. Better values can be achieved using light-emitting diodes, the forward voltages of which have the following typical values. Depending on the wavelength emitted:

infra-red:	1.4 V
red:	1.6 V
light-red:	2.2 V
yellow:	2.2 V
green:	2.4 V

Their temperature coefficient is about $-2\,mV/K$, corresponding to $-1 \cdot 10^{-3}/K$. In addition, their incremental resistance is lower, as can be seen from the comparison of the diode characteristics in Fig. 6.27.

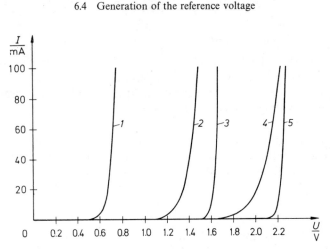

Fig. 6.27 Diode characteristics.
Curve *1*: Single silicon diode. Curve *2*: Two series-connected silicon diodes. Curve *3*: Red
LED. Curve *4*: Three series-connected silicon diodes. Curve *5*: Yellow LED

FET as reference voltage source

The temperature coefficient of the gate-source voltage of field-effect
transistors is positive for large drain currents, and negative for small
drain currents. For a medium drain current I_{DZ}, the temperature
coefficient is zero. At this current, a FET is well suited as a reference
element. The current can be adjusted to the desired value by

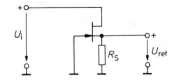

Fig. 6.28 FET as a reference voltage source

series feedback, as in Fig. 6.28 and the gate-source voltage is then the
output voltage. The value of R_S is determined from the transfer
characteristic of the FET

$$I_D = I_{DS}\left(1 - \frac{U_{GS}}{U_p}\right)^2$$

and from the relationship $U_{GS} = -R_S I_D$ as

$$R_S = \frac{U_p}{I_{DZ}}\left(\sqrt{\frac{I_{DZ}}{I_{DS}}} - 1\right).$$

Bipolar transistor as a reference voltage source

In principle, the base-emitter voltage of a transistor can also be used as a reference. However, the temperature coefficient of $-2\,\mathrm{mV/K}$, corresponding to $-3\cdot10^{-3}/\mathrm{K}$, is rather high. It can be compensated for by adding a voltage which has a positive temperature coefficient. Such a voltage can be established by making use of the difference between the base-emitter voltages of two transistors operated at different currents. The transistors T_1 and T_2, in Fig. 6.29, serve this purpose.

Transistor T_1 is operated as a diode. Its collector current is

$$I_{C1} = \frac{U_{ref} - 0.6\,\mathrm{V}}{R/n_1}.$$

Shunt feedback is applied to transistor T_3 by resistor R. The collector of T_2 therefore has a potential of 0.6 V, and its collector current is

$$I_{C2} = \frac{U_{ref} - 0.6\,\mathrm{V}}{R}.$$

The ratio of the collector currents thus becomes independent of U_{ref} and is given as

$$\frac{I_{C1}}{I_{C2}} = n_1.$$

We can now determine the voltage U_1, defined as the difference between the two base-emitter voltages,

$$U_1 = U_{BE1} - U_{BE2} = U_T \ln\frac{I_{C1}}{I_{C2}} = \frac{kT}{e_0}\ln n_1. \qquad (6.10)$$

To ensure that this voltage is positive, we must choose $n_1 > 1$ and therefore $I_{C1} > I_{C2}$, e.g. $n_1 = 10$. We then obtain

$$U_1 = 26\,\mathrm{mV}\ln 10 \approx 60\,\mathrm{mV}.$$

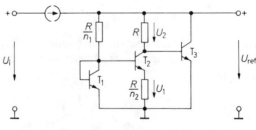

Fig. 6.29 Bandgap reference.

Output voltage: $U_{ref} = 1.205\,\mathrm{V}$
Design parameters: $n_2 \lg n_1 \approx 10$

The temperature coefficient of this voltage is positive and, according to Eq. (6.10), has the value

$$\frac{\partial U_1}{\partial \vartheta} = \frac{k}{e_0} \ln n_1 = \frac{U_T}{T} \ln n_1 = \frac{1}{T} U_1. \tag{6.11}$$

Room temperature, $T \approx 300 \, \text{K}$, gives a temperature coefficient of about $+0.2 \, \text{mV/K}$ for our example. In order to obtain a compensating voltage having the required temperature coefficient of $+2 \, \text{mV/K}$, U_1 must be amplified by a factor of 10. This can be achieved by transistor T_2 if $n_2 = 10$. The voltage U_2 is then $600 \, \text{mV}$ and has the required positive temperature coefficient. The shunt feedback of T_3 adjusts the output voltage to the temperature-compensated value

$$U_{\text{ref}} = U_{\text{BE 3}} + U_2 = U_{\text{BE 3}} + U_T n_2 \ln n_1 \approx 1.2 \, \text{V}.$$

It will be shown that the temperature coefficient becomes exactly zero, if n_1 and n_2 are chosen such that the output voltage has the value

$$U_{\text{ref}} = \frac{E_g}{e_0} = 1.205 \, \text{V},$$

where E_g is the bandgap of silicon. The described circuit is therefore often referred to as "bandgap reference".

For the derivation of the above equation, an expression for the temperature coefficient of the base-emitter voltage is required. From $U_{\text{BE 3}} = U_T \ln(I_{C3}/I_{C0})$, we obtain

$$\left.\frac{\partial U_{\text{BE 3}}}{\partial \vartheta}\right|_{I_{C3} = \text{const}} = \frac{k}{e_0} \ln \frac{I_{C3}}{I_{C0}} - \frac{kT}{e_0} \frac{\partial \ln I_{C0}}{\partial \vartheta}, \tag{6.12}$$

where I_{C0} is the theoretical reverse saturation current. According to [6.3], it follows from the theory for the diode that

$$\frac{\partial I_{C0}/\partial \vartheta}{I_{C0}} = \frac{\partial (\ln I_{C0})}{\partial \vartheta} = \frac{E_g}{kT^2}, \tag{6.13}$$

which, when inserted in Eq. (6.12), gives the temperature coefficient

$$\left.\frac{\partial U_{\text{BE 3}}}{\partial \vartheta}\right|_{I_{C3} = \text{const}} = \frac{1}{T} U_{\text{BE 3}} - \frac{E_g}{e_0 T} \approx -2 \, \text{mV/K}. \tag{6.14}$$

For a complete temperature compensation, the temperature coefficient of U_2 must have the same value as that of $U_{\text{BE 3}}$, but opposite sign. With Eq. (6.11) we obtain

$$\frac{\partial U_2}{\partial \vartheta} = n_2 \frac{\partial U_1}{\partial \vartheta} = \frac{n_2 U_1}{T} = \frac{U_2}{T}. \tag{6.15}$$

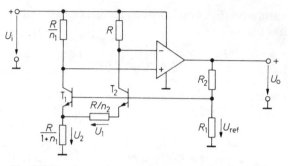

Fig. 6.30 Bandgap reference using an operational amplifier.

Output voltage: $U_o = 1.205\,\mathrm{V}\left(1 + \dfrac{R_2}{R_1}\right)$

Design parameters: $n_2 \lg n_1 \approx 10$

Equating this with Eq. (6.14) gives the result

$$U_2 = \frac{E_g}{e_0} - U_{BE\,3}$$

and therefore

$$U_{ref} = U_{BE\,3} + U_2 = \frac{E_g}{e_0} = 1.205\,\mathrm{V}.$$

A variation of the circuit, operating on the same principle, is shown in Fig. 6.30. The output voltage of the operational amplifier assumes a value so that $I_{C1} = n_1 I_{C2}$, as for the previous circuit. The difference U_1 of the two base-emitter voltages appears, amplified by the factor n_2, across the resistor $R/(1 + n_1)$. The reference voltage is therefore given by

$$U_{ref} = U_{BE\,1} + U_2 = U_{BE\,1} + U_T n_2 \ln n_1.$$

The temperature coefficient is again zero if the factor $n_2 \ln n_1$ is chosen so that $U_{ref} = 1.205\,\mathrm{V}$. The output voltage of the circuit can be adjusted, for constant U_{ref}, by the voltage divider R_1, R_2.

Both circuits are employed in integrated voltage regulators (e.g. in the series 78L00 or in type AD 580). In these implementations, however, the collector currents are chosen as having the same value, the required ratio n_1 being defined instead by the areas of the transistors.

6.5   Switching power supplies

The principle of series regulation requires that a d.c. voltage is provided, the minimum level of which is higher then the required output voltage. The difference between the two voltages appears across the controlling power transistor connected in series to the load. As the power dissipation of the series transistor may be considerable, an efficiency in many cases of only 50% is obtained, especially if small output voltages are to be stabilized.

Far better efficiencies can be attained by replacing the continually controlled series transistor by a switch. The mean value of the output voltage can then be influenced by periodically opening and closing the switch and varying the duty cycle. After the switch a filtering element is needed to eliminate the ripple. An LC filter is most suitable as no loss of power is incurred. Since the switch is situated at the secondary side of the supply transformer, such regulators may be referred to as secondary switching regulators.

The losses of the line transformer in a power supply should not be overlooked. Transformer loss can be considerably reduced if the frequency of the transformer voltage is in the kHz range, as then only a few turns are sufficient to provide the voltage. Therefore, the mains supply voltage can be directly rectified and transistor switches can be used to generate an a.c. voltage of higher frequency which is then applied to an appropriately designed transformer. A bridge rectifier is connected to the secondary winding. To regulate the resulting d.c. voltage, the ON-times of the transistor switches at the transformer primary are varied. Such regulators may therefore be named "primary switching regulators". Their efficiency is the highest and may be over 80%. A further advantage is the small size and weight of the high-frequency transformer.

6.5.1 Secondary switching regulators

Figure 6.31 shows the basic circuit of a secondary switching regulator (buck regulator). Transistor T_1 is periodically turned off (cut-off) and on (saturation), with a frequency of about 20 kHz. The "free-wheeling" diode D prevents high voltages being induced in the reactor when turning off the transistor since it maintains the current flow in the reactor. Thus, during the OFF-time, not only the capacitor but also the reactor contributes to the output current, and in this way, a well smoothed output voltage is obtained without loss of power.

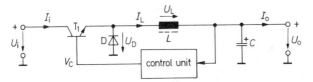

Fig. 6.31 Basic circuit of a secondary switching regulator

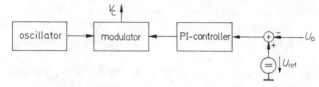

Fig. 6.32 Block diagram of the control unit

The block diagram of the control unit is shown in Fig. 6.32. The controller compares the output voltage with the reference voltage. If the output voltage is too small, the duty cycle t_{ON}/T of the control voltage V_C is increased by the modulator. The frequency, $f = 1/T$, of the control voltage remains constant in this process. It is determined by the oscillator.

For the design of the switching regulator we must initially determine the time dependence of the reactor current. To begin with, we assume the capacitor to be infinitely large so that the output voltage ripple is zero.

Faraday's law of induction gives the expression

$$U_L = L \cdot \frac{dI_L}{dt}. \tag{6.16}$$

During the OFF-time, the reactor voltage is given by

$$U_L = -0.7\,\text{V} - U_o \approx -U_o = \text{const}.$$

Therefore, the reactor current decreases linearly with time

$$\frac{dI_L}{dt} = -\frac{U_o}{L}. \tag{6.17}$$

During the ON-time,

$$U_L = U_i - U_o = \text{const},$$

so that the reactor current rises linearly with time according to

$$\frac{dI_L}{dt} = \frac{U_i - U_o}{L}. \tag{6.18}$$

The time dependence of the reactor current is shown in Fig. 6.33.
From Eqs. (6.17) and (6.18), we obtain the relationship

$$\Delta I_L = I_{L\,max} - I_{L\,min} = \frac{U_o t_{OFF}}{L} = \frac{(U_i - U_o)t_{ON}}{L}. \tag{6.19}$$

Hence

$$\boxed{\frac{U_o}{U_i} = \frac{t_{ON}}{t_{ON} + t_{OFF}} = \frac{t_{ON}}{T}}. \tag{6.20}$$

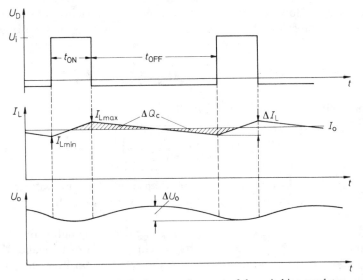

Fig. 6.33 Waveshape of voltages and current of the switching regulator

The output voltage is thus proportional to the duty cycle and independent of the output current as long as $I_o > \frac{1}{2}\Delta I_L$.

During the ON-time, transistor T_1 must supply the output current I_o as well as the charging current for the capacitor. The smaller the inductance L, the larger the ratio

$$\alpha = \frac{I_{L\,max}}{I_o}.$$

It should be restricted to values of about 1.2 so that the rating of transistor T_1 can be kept within reasonable limits. For the determination of L we take from Fig. 6.33 the relationship

$$I_{L\,max} = I_o + \frac{1}{2}\Delta I_L,$$

or with Eqs. (6.19) and (6.20)

$$\boxed{L = \frac{R_{LOAD}(1 - U_o/U_i)}{2f(\alpha - 1)}}, \qquad (6.21)$$

where $R_{LOAD} = U_o/I_o$ is the load resistance.

With a finite value for the capacitance, a ripple voltage occurs at the output. The charging curent is given as

$$I_C = I_L - I_o.$$

The charge supplied to, and taken from the capacitor during one period is represented by the shaded areas in Fig. 6.33, and we obtain the relationship for the ripple

$$\Delta U_o = \frac{\Delta Q_C}{C} = \frac{1}{C} \cdot \frac{1}{2}(\tfrac{1}{2}t_{ON} + \tfrac{1}{2}t_{OFF}) \cdot \tfrac{1}{2}\Delta I_L.$$

Hence, with Eqs. (6.19) and (6.20)

$$\Delta U_o = \frac{U_o}{8LCf^2}\left(1 - \frac{U_o}{U_i}\right). \tag{6.22}$$

The measured ripple voltage is somewhat larger than this value as the unavoidable series resistance of the capacitor can never be completely neglected.

In contrast to the continually operating series regulator, the mean current through the transistor of the switching regulator is smaller than the output current. Neglecting losses, the relationship between power input and output must be

$$U_i \cdot \bar{I}_i \approx U_o \cdot I_o,$$

and therefore

$$\bar{I}_i = \frac{U_o}{U_i}I_o. \tag{6.23}$$

An example is given to clarify the design process for the switching regulator. A regulated power supply of 5 V at 5 A is required. The input voltage is 10 V, and the oscillator frequency 20 kHz. We choose $\alpha = 1.2$. With Eq. (6.21) we then obtain the value of the inductance, $L = 63\,\mu H$. The maximum energy stored in the reactor is $E_{L\,max} = \tfrac{1}{2}LI_{L\,max}^2 = 1.1\,mJ$, which is important for the choice of the reactor core.

The ripple of the output voltage is required not to exceed 30 mV. Equation (6.22) then gives the size of the capacitor as $C > 413\,\mu F$.

Stepping up the voltage

In the described circuit of Fig. 6.31, the output voltage is always lower than the input voltage (buck regulator). By rearranging the circuit elements, higher output voltages can also be obtained (boost

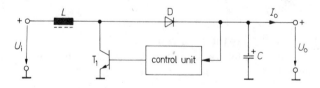

Fig. 6.34 Switching regulator to step up the voltage (boost regulator)

regulator). If transistor T_1 in Fig. 6.34 is turned off, its collector potential rises above the input voltage. The capacitor C is then charged via the diode. With deductions similar to those for Eqs. (6.17) and (6.18), the output voltage can be determined as

$$\frac{U_o}{U_i} = \frac{T}{t_{OFF}}.$$
(6.24)

The remainder of the design equations can be deduced accordingly.

Inverting the voltage

A reactor together with a switching regulator enables the generation of a negative output voltage from a positive input voltage. The corresponding circuit is shown in Fig. 6.35. When transistor T_1 is turned off, its collector potential becomes negative due to the change in current of the reactor. Diode D is then forward biased and the capacitor is charged to a negative voltage. The output voltage is given by the expression

$$\frac{U_o}{U_i} = -\frac{t_{ON}}{t_{OFF}}.$$
(6.25)

The control unit is the same for all three arrangements and is available as a monolithic integrated circuit, e.g. as the type TL 497 (Texas Instr.) or the μA 78S40 (Fairchild).

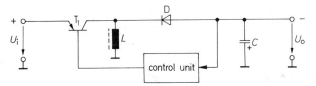

Fig. 6.35 Switching regulator to generate a negative output from a positive input voltage

6.5.2 Primary switching regulators

Figure 6.36 shows the basic circuit of a primary switching regulator [16.5]. The mains supply voltage is rectified directly by means of a diode bridge. The series-connected smoothing capacitors C_1, C_2 then each have a voltage of 150 V. With the transistor switches, T_1 and T_2, the voltages

$$U_1 = \begin{cases} +150\,V, & \text{if } T_1 \text{ is ON} \\ -150\,V, & \text{if } T_2 \text{ is ON} \end{cases}$$

can be applied alternately to the primary winding of the high-frequency transformer. The primary is connected to the mains rectifier in such a

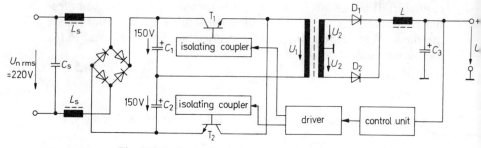

Fig. 6.36 Basic circuit of a primary switching regulator.
L_s, C_s: Filter for noise suppression

way that no d.c. current can flow through it. This prevents the transformer from saturating, should the ON-times of T_1 and T_2 not be the same.

The alternating voltage of the transformer secondary is rectified by a two-pulse midpoint connected circuit. This particular connection is to be preferred for this regulator type as there is only one diode producing loss at any one time. The additional secondary winding normally avoided at 50 Hz operation represents no difficulties at high-frequency operation. These aspects are particularly important for the generation of small output voltages, as the diodes, D_1 and D_2, are the main cause of loss. In order to keep the static and dynamic losses to a minimum, it is recommended to use power Schottky diodes, e.g. the types MBR 3520 ... MBR 7545 from Motorola.

In the same way as for the secondary switching regulator, smoothing of the output voltage is achieved by means of an LC element.

The control unit is in principle identical to that of the secondary switching regulator. However, an additional driver circuit is required to distribute the ON-signal to the appropriate transistor switch. As the transistors are connected to the transformer primary, and the driver circuit and control unit to the secondary, the transistors must be isolated from the driver circuit. Pulse transformers or opto-couplers can be used for transmission of the gating pulses.

In order that the power dissipation of the transistor switches may be kept small, they must be switched on and off as fast as possible and must not even temporarily be ON simultaneously. With optimum design, efficiencies of over 80% can be achieved. The control unit is available as an IC: the SG 3524 from Silicon General or the MC 3420 of Motorola.

The described arrangement can also be directly supplied from a d.c. voltage source instead of the rectified a.c. voltage. It then operates as a highly efficient "d.c. voltage transformer" (d.c./d.c. converter).

7 Analog switches and comparators

An analog switch must be able to switch a continuous input signal ON and OFF. When the switch is ON, the output voltage must closely resemble the input voltage; when it is OFF, the voltage must be zero.

7.1 Principle

There are several switch arrangements which fulfill the above requirements. They are represented in Fig. 7.1 as mechanical switches.

Figure 7.1a shows a single-throw series switch. As long as its contact is closed, $U_o = U_i$. On opening the switch, the output voltage becomes zero, although this is so only for the case of no-load. For capacitive loads, the output voltage will only fall slowly to zero because of the finite output resistance $r_o = R$.

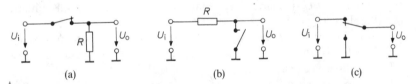

Fig. 7.1 (a) Series switch. (b) Short-circuiting switch. (c) Series/short-circuiting switch

The single-throw short-circuiting switch in Fig. 7.1b overcomes this difficulty. However, in the ON mode, i.e. when the contact is open, the circuit possesses a finite output resistance $r_o = R$.

The double-throw series/short-circuiting switch in Fig. 7.1c combines both advantages, and has a low output resistance for each mode of operation.

7.2 Electronic switches

The switch arrangements of Fig. 7.1 can be realized electronically if the mechanical contact is replaced by a controllable resistor having a small minimum and a large maximum resistance. Field effect transistors, diodes, bipolar junction transistors or other controllable circuit elements can be used for this purpose.

7.2.1   FET switch

The drain-source resistance of most field-effect transistors can be controlled between values of below $100\,\Omega$ and several $G\Omega$ by means of the gate-source voltage U_{GS}. A FET is therefore suitable for switching applications, and Fig. 7.2 shows its use as a series switch. The FET is OFF and the output voltage zero if the control voltage U_C is made more negative than the maximum negative input voltage, and if the difference is at least the pinch-off voltage, U_p, of the FET.

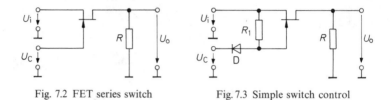

Fig. 7.2 FET series switch   Fig. 7.3 Simple switch control

To make the FET conduct, the voltage U_{GS} must be zero. This condition is not so easy to fulfill as the source potential is not constant. A solution to this problem is shown in Fig. 7.3 where diode D becomes reverse biased if U_C is made larger than the most positive input voltage, and therefore $U_{GS}=0$, as required.

For sufficiently negative control voltages, diode D is forward biased and the FET is turned off. In this mode, a current flows from the input voltage source via resistor R_1 into the control circuit. This can usually be tolerated as the output voltage in this case is zero. However, this effect becomes troublesome if the input voltage is connected to the switch via a coupling capacitor, as the latter becomes charged to a negative voltage during the OFF mode.

These problems do not arise if a MOSFET is used for switching. It can be made to conduct by applying a control voltage which is larger than the most positive input voltage. No current flows from gate to channel, so that diode D and resistor R_1 are no longer necessary. To ensure that the bipolar input voltage range is as large as possible, it is better to use, instead of a single MOSFET, a CMOS switch consisting of two complementary MOSFETs connected in parallel (e.g. type MC 14066 from Motorola).

For the switch to be ON, a positive control voltage is applied to the gate of the enhancement-mode MOSFET T_1 in Fig. 7.4, which must be at least $2U_p$. An equal but opposite voltage is applied to the gate of T_2. At small input voltages U_i, both MOSFETs are conducting. If the input voltage rises to larger positive values, U_{GS1} is reduced and T_1 thereby has a higher effective channel resistance. This does no harm as

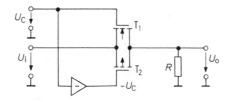

Fig. 7.4   CMOS series switch

at the same time the level of U_{GS2} rises. The channel resistance T_2 therefore falls. At negative input voltages, the effect on T_1 and T_2 is reversed. To turn the switch off, the polarity of the control voltage is reversed.

When reversing the control voltage, a small spike appears in the output voltage, due to the gate-to-channel capacitance, and this may have a detrimental effect, particularly for small input signals. To keep it low, the amplitude of the control voltage is chosen to be no larger than absolutely necessary and additionally, its slew rate is limited. Some improvement is also possible if a low-impedance signal voltage source is used. The maximum switching frequency is in the audio frequency range.

It is most convenient to use CMOS switches for which a level-shift circuit with a TTL-compatible control input is already integrated on the chip. Such ICs usually contain several switches which are often controlled simultaneously, e.g. the types IH 5040...5051 from Intersil or DG 300...307 from Siliconix.

An important special case is an arrangement where all switches have one terminal in common. The switches can be closed individually by applying a binary number to a built-in 1-out-of-n decoder. Such an electronic selector switch is known as an analog multiplexer/demultiplexer (e.g. the DG 506...509 from Siliconix having between 4 and 16 inputs or the MC 14051...14053 from Motorola with 2 to 8 inputs).

7.2.2   Diode switch

Diodes are also suitable for use as switches because of their low forward resistance and high blocking resistance. If a positive control voltage is applied to the circuit in Fig. 7.5, the diodes D_5 and D_6 become reverse biased. The impressed current I then flows through the branches D_1, D_4 and D_2, D_3 from one current source to the other. The potentials V_1 and V_2 thereby take on the values

$$V_1 = U_i + U_D, \qquad V_2 = U_i - U_D.$$

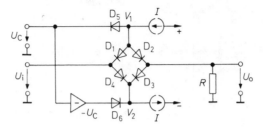

Fig. 7.5 Series switch using diodes

The output voltage then is

$$U_o = V_1 - U_D = V_2 + U_D = U_i,$$

if the forward voltages U_D are the same. Should this not be the case, an offset voltage arises.

If the control voltage is negative, the two diodes D_5 and D_6 are forward biased. Potential V_1 takes on a high negative value and V_2 a high positive value. As can be seen in Fig. 7.5, all diodes of the quartet D_1 to D_4 are then reverse biased. The output is separated from the input, and the output voltage is zero.

To keep switching times short and capacitive transients low, Schottky diodes are usually used. In this way, extremely fast switches can be built, the switching times of which may be below 1 ns.

7.2.3   Bipolar junction transistor switch

To investigate the suitability of a bipolar junction transistor for use as a switch, we take a look at its output characteristic curves around the origin, an expanded view of which is given in Fig. 7.6 for small positive and negative collector-emitter voltages.

The first quadrant contains the familiar output characteristics. If the voltage U_{CE} is made negative without changing the base current, the output characteristics of the third quadrant are obtained. In this reverse mode operation, the current gain of the transistor is considerably reduced and is about $\frac{1}{30}\beta$. The maximum permissible collector-emitter voltage in this mode is the breakdown voltage U_{EB0}, since the base-collector junction is forward biased and the base-emitter junction is reverse biased. This type of operation is known as reverse-region operation, and the accompanying current gain as reverse current gain ratio β_r. The collector current is zero for a collector-emitter voltage of about 10 to 50 mV. If the base current exceeds a few mA, this *offset voltage* increases steeply; for small base currents it remains constant over a large range.

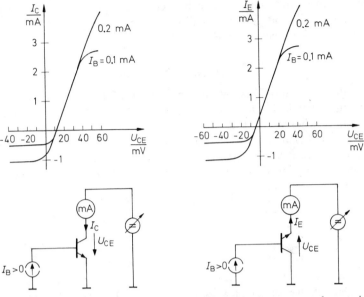

Fig. 7.6 Complete output characteristics of a transistor in common-emitter connection, and test circuit

Fig. 7.7 Complete output characteristics for interchanged emitter and collector, and test circuit

The offset voltage can be considerably reduced by ensuring that the transistor is in reverse-region operation when the output current crosses zero. To achieve this, collector and emitter must be interchanged. The resulting output characteristics are shown in Fig. 7.7. At larger output currents, practically the same curves are obtained as for the normal operation in Fig. 7.6, if U_{CE} is still measured at the correct polarity (collector-to-emitter). The reason for this is that the emitter current, now the output current, is very nearly the same as the collector current.

Near the origin, however, a major difference arises in that the base current can no longer be neglected with respect to the output current. If for normal operation, the output current is made zero, the emitter current is identical to the base current and thus not zero, and an offset voltage of 10 to 50 mV appears at the output. If collector and emitter are interchanged and the output current is made zero, the collector current is the base current. The collector-base junction is then forward biased (reverse operation). The offset voltage for this operation is usually about $\frac{1}{10}$ of that at normal operation, but is also positive since, for the circuit in Fig. 7.7, $U_o = -U_{CE}$. Typical values for the offset voltage are between 1 and 5 mV [7.1], and it is therefore desirable to operate transistor switches with interchanged collector and emitter. If the emitter current is kept small, the transistor operates almost exclusively in the reverse mode.

Short-circuiting switch

Figures 7.8 and 7.9 show the application of a transistor as a short-circuiting switch. In the circuit of Fig. 7.8, the transistor is in normal operation whereas in Fig. 7.9, it is in reverse-region operation. To obtain a sufficiently low transistor resistance, the base current must be in the mA-range. The collector current in Fig. 7.8, and the emitter current in Fig. 7.9, should not be much larger to ensure that the offset voltage remains small.

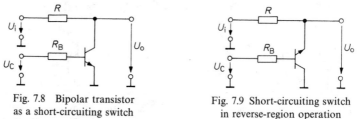

Fig. 7.8 Bipolar transistor
as a short-circuiting switch

Fig. 7.9 Short-circuiting switch
in reverse-region operation

Series switch

Figure 7.10 shows the application of a bipolar transistor as a series switch. A negative control voltage must be applied to cut off the transistor. It must be more negative than the most negative value of the input voltage, but also has a limit, since the control voltage may not be more negative than $-U_{EBO} \approx -6\,\text{V}$. To make the transistor conducting, a positive control voltage is applied which is larger than the input voltage by a value of $\Delta U = I_B R_B$. The collector-base junction is then forward biased and the transistor operates as a switch in reverse-region mode. The disadvantage is that the base current flows into the input voltage source, and unless the internal resistance of the source is kept very small, large errors may occur.

If this condition can be fulfilled, the circuit is particularly suitable for positive input voltages as the emitter current in the ON mode is

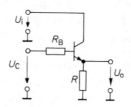

Fig. 7.10 Saturated emitter follower as a series switch

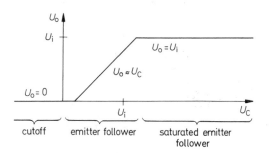

Fig. 7.11 Transfer characteristics for positive input voltages

positive. The offset voltage is therefore reduced and even becomes zero for a particular emitter current, as can be seen in Fig. 7.7. The circuit in this mode of operation is known as a saturated emitter follower, since for control voltages between zero and U_i, it operates as an emitter follower for U_C. This is illustrated in Fig. 7.11 by the transfer characteristics for positive input voltages.

Series/short-circuiting switch

If the saturated emitter follower of Fig. 7.10 is combined with the short-circuiting switch of Fig. 7.9, a series/short-circuiting switch is obtained having a low offset voltage for both modes of operation. It has the disadvantage that complementary control signals are required. The control is particularly simple if a complementary emitter follower is used, as in Fig. 7.12. It is saturated in both the ON and OFF mode if $U_{C\,max} > U_i$ and $U_{C\,min} < 0$. Due to the low output resistance, a fast switchover of the output voltage, between zero and U_i, is possible.

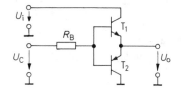

Fig. 7.12 Series/short-circuiting switch

7.3 Analog switch using amplifiers

7.3.1 Improved FET switch

A comparison of the previous sections shows that only the FET switch has no offset voltage. It is therefore well suited as a precision switch, although its relatively large output resistance is a disadvantage.

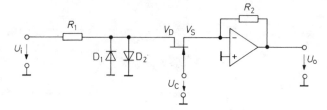

Fig. 7.13 FET switch using an operational amplifier.

$$\textit{Output voltage:} \quad U_{\mathrm{o}} = \begin{cases} 0 & \text{for } U_{\mathrm{C}} < U_{\mathrm{p}} - 0.6\,\text{V} \\ -U_{\mathrm{i}} R_2/(R_1 + r_{\mathrm{DS\,ON}}) & \text{for } U_{\mathrm{C}} = 0. \end{cases}$$

This can be reduced if a voltage follower is connected in series. However, since an amplifier is used, it is better to connect it as in Fig. 7.13, as additional improvements can be achieved in this way. The amplifier is operated in the inverting mode, and the FET is connected as a series switch to the summing point. Hence, the source potential V_{S} is fixed. The series resistance R_1 is chosen such that the voltage across the conducting FET remains small, i.e. $V_{\mathrm{D}} \approx V_{\mathrm{S}} = 0$. Therefore, the FET is switched to ON when $U_{\mathrm{C}} = 0$, independently of the magnitude of the input voltage.

If the FET is switched to OFF, the drain potential V_{D} rises and, depending on the polarity, either diode D_1 or D_2 becomes forward biased. In this way, V_{D} is limited to values between $\pm 0.6\,\text{V}$. The control voltage need therefore not be very much more negative than the pinch-off voltage U_{p} to safely turn off the FET. The voltage transients due to the FET capacitances can be kept small in this way.

Since, in this circuit, the voltages across the FET are small for both modes of operation, input signals of any magnitude can be handled if R_1 is chosen appropriately.

Additional identical FET switches may be connected to the summing point to obtain an analog multiplexer.

7.3.2 FET switch for polarity change

A FET switch can be combined with an amplifier as in Fig. 7.14. The output voltage is then switched not between zero and U_{i}, but between $+U_{\mathrm{i}}$ and $-U_{\mathrm{i}}$. The circuit is based on that for the bipolar coefficient in Fig. 1.5 for $n = 1$. The potentiometer R_2 is replaced by the FET and the fixed resistance R_2.

If $U_{\mathrm{C}} = 0$, the FET is conducting and the circuit operates as an inverting amplifier where

$$U_{\mathrm{o}} = -U_{\mathrm{i}}.$$

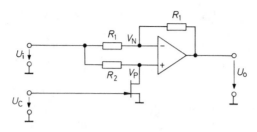

Fig. 7.14 FET switch for polarity change.

Output voltage: $\quad U_o = \begin{cases} U_i & \text{for } U_C < U_{i\,\text{min}} + U_p \\ -U_i & \text{for } U_C = 0 \end{cases}$

If U_C is made more negative than the most negative value of the input voltage, the FET is switched to OFF. No current will then flow through resistor R_2, and $V_P = U_i$. Because of the feedback, $V_N = U_i$, and thus no current flows through the resistors R_1, and

$$U_o = +U_i.$$

These relationships hold only if the following condition is fulfilled

$$r_{DS\,ON} \ll R_2 \ll r_{DS\,OFF}.$$

7.3.3 Differential amplifier as a switch

Figure 1.40 shows the use of a differential amplifier as a multiplier. It can be seen that it could also be used as an analog switch if one considers that switching the input voltage on and off is equivalent to a multiplication by zero or unity. To implement this, the emitter current

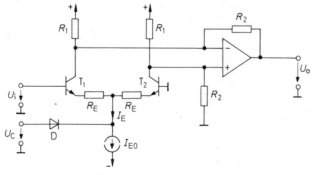

Fig. 7.15 Differential amplifier as a switch.

Output voltage: $\quad U_o = \begin{cases} 0 & \text{for } U_C > U_{i\,\text{max}} \\ g' R_2 U_i & \text{for } U_C < -I_{E0} R_E \end{cases}$

I_E, of the differential amplifier in Fig. 7.15 is made either zero or I_{E0}, by means of diode D.

The operational amplifier evaluates the difference of the collector currents as follows

$$U_o = R_2(I_{C1} - I_{C2}). \tag{7.1}$$

If the control voltage U_C is made positive, diode D is forward biased and the transistors cut off so that $I_{C1} = I_{C2} = 0$ and therefore $U_o = 0$.

If diode D is reverse biased by a negative control voltage, then $I_E = I_{E0}$. The collector currents are then

$$I_{C1} = \tfrac{1}{2}(I_{E0} + g'U_i) \quad \text{and} \quad I_{C2} = \tfrac{1}{2}(I_{E0} - g'U_i), \tag{7.2}$$

where $g' = 1/(R_E + 1/g_m)$ is the reduced transconductance. With Eq. (7.1), we obtain for the output voltage

$$U_o = g'R_2 U_i. \tag{7.3}$$

Connecting a second differential amplifier in parallel, as in Fig. 7.16, gives a very versatile switch. The emitter current I_{E0} is switched from one differential amplifier to the other by means of the transistors T_5 and T_6. This avoids a common-mode step in the collector voltages of T_1 and T_2 during switching, as occurs in the previous circuit. The relationship for the output voltage can be directly derived from Eq. (7.3)

$$U_o = \begin{cases} g'R_2(U_1 - U_2) & \text{for } U_C \approx 1\,\text{V} \\ g'R_2(U_3 - U_4) & \text{for } U_C \approx -1\,\text{V}. \end{cases} \tag{7.4}$$

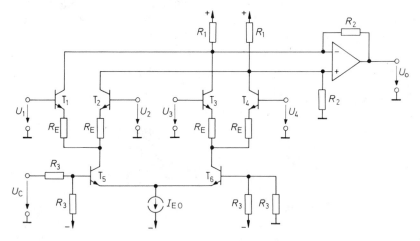

Fig. 7.16 Broadband multiplexer.

Output voltage: $U_o = \begin{cases} g'R_2(U_1 - U_2) & \text{for } U_C \approx 1\,\text{V} \\ g'R_2(U_3 - U_4) & \text{for } U_C \approx -1\,\text{V} \end{cases}$

The circuit can therefore be employed to switch over from one input voltage $U_{i1} = U_1 - U_2$ to a second, $U_{i2} = U_3 - U_4$. If by an appropriate connection $U_3 = U_2$ and $U_4 = U_1$, then $U_{i2} = -U_{i1}$. This is a switch for changing the polarity.

With suitable design parameters, bandwidths of over 100 MHz can be attained and the circuit is thus well suited as a modulator, demodulator or phase detector in communication circuits, or as a chopper in multi-channel broadband oscilloscopes.

The differential amplifier arrangement is available without the output operational amplifier as an integrated circuit (MC 1445 from Motorola).

7.4 Sample-and-hold circuits

As for an analog switch, the output voltage of a sample-and-hold circuit in the ON state should be the same as the input voltage. In the OFF mode, the value of the input voltage at the instant of switching-off must be stored. For this purpose, a capacitor is charged to the value U_i via a series switch, as in Fig. 7.17. When the switch is opened and no discharge current flows, the voltage of the capacitor remains constant. Therefore, the voltage follower OA 2 must be an amplifier having a FET input.

The quality of a sample-and-hold circuit, when in the hold mode, is characterized by the *hold decay rate*

$$\frac{\partial U_o}{\partial t} = \frac{I_L}{C},$$

where I_L is the discharge current. This consists of the leakage current of the capacitor, that of the series switch, and the input current of the amplifier OA 2.

For a given leakage current, the hold decay rate can be reduced by increasing the capacitance C. However, the properties during the sampling mode will then deteriorate. An important factor determining

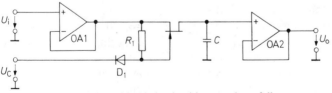

Fig. 7.17 Sample-and-hold circuit with two voltage followers.

$$\text{Output voltage:} \quad U_o = \begin{cases} U_i & \text{for } U_C > U_{i\,max} \\ \text{const} & \text{for } U_C < U_{i\,min} + U_p \end{cases}$$

the quality in this mode is the *acquisition time* t_{Ac}. It defines the time required in the worst case to charge the capacitor to a given percentage of the input voltage.

Amplifier OA 1 serves to reduce the acquisition time if the signal source has a high impedance. Depending on the required acquisition accuracy,

$$t_{Ac} \approx r_{DS\,ON}\, C \cdot \begin{cases} 4.0 & \text{for } 1\,\% \\ 6.9 & \text{for } 0.1\,\%. \end{cases}$$

It is thus proportional to the capacitance C. For large steps in the input voltage, t_{Ac} may be up to three times the given value, since the charging current is limited by either the FET switch or the operational amplifier.

An additional source of error is that, at switching off, some charge is taken from the storage capacitor C by the finite gate-to-channel capacitance of the FET switch. This results in a voltage error (hold step)

$$\Delta U_o = \frac{C_{GD}}{C}\, \Delta U_C.$$

As the gate-to-channel capacitance C_{GD} is in the order of several pF, a storage capacitance of at least 1 nF is required for an accuracy of 0.1 %. The conditions become more favourable if the diode switch of Fig. 7.5 is used instead of the FET switch. The detrimental capacitive effects are thereby compensated to a large extent because of the symmetry of the control. Considerably smaller storage capacitors are then needed, and acquisition times down to about 20 ns can be achieved whereas with FETs it is difficult to get below 500 ns.

A further important characteristic is the *aperture time* t_{Ap}. It defines the delay between switching off the control voltage and the actual cutoff of the series switch. The delay is subject to certain variations known as *aperture jitter* Δt_{Ap}. It defines the degree of uncertainty as to the beginning of the hold mode. For the circuit in Fig. 7.17, the level of the control voltage at which the switch cuts off, is dependent on the instantaneous value of the input voltage. Since the control voltage requires a finite interval for a change from one state to the other, the aperture time varies, giving a systematic aperture jitter which is smaller the steeper the edges of the control signal.

It has already been pointed out that amplifier OA 2 must have a FET input. Its offset voltage can be eliminated by employing feedback across the entire arrangement, including amplifier OA 1, as in Fig. 7.18.

If the switch is ON, the output potential V_1 of amplifier OA 1 assumes a value such that $U_o = U_i$. This eliminates offset errors which

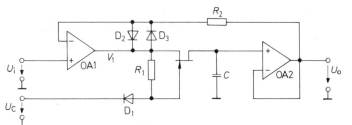

Fig. 7.18 Sample-and-hold circuit with overall feedback.

$$\text{Output voltage:} \quad U_o = \begin{cases} U_i & \text{for } U_C > U_{i\,max} \\ \text{const} & \text{for } U_C < U_{i\,min} + U_p \end{cases}$$

may arise from OA 2 or from the switch. Diodes D_2 and D_3 are cut off in this mode as only the small voltage $V_1 - U_o$ appears across them which is the same as the offset voltage mentioned.

If the switch is OFF, the output voltage remains constant. Resistor R_2 and diodes D_2, D_3 prevent amplifier OA1 becoming saturating in this mode of operation. This is important since the time for recovery after saturation would be large and would increase the acquisition time.

The circuit described is available from National Semiconductors as monolithic integrated circuit LF 398, the data of which are ($C = 10\,\text{nF}$):

hold decay rate: 3 mV/s
acquisition time for 0.1 %:   20 µs
output voltage error ΔU_o: 1 mV

Sample-and-hold circuit with integrator

Instead of a grounded capacitor with voltage follower, an integrator can be used as the analog storing element. Figure 7.19 shows this

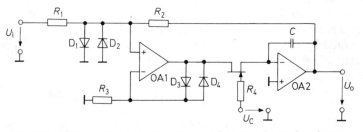

Fig. 7.19 Sample-and-hold circuit with an integrator as the storing element.

$$\text{Output voltage:} \quad U_o = \begin{cases} -(R_2/R_1)\, U_i & \text{for } U_C = 0 \\ \text{const} & \text{for } U_C < U_p - 1.2\,\text{V} \end{cases}$$

version. As in Fig. 7.13, the series switch is connected to the summing point and is therefore easy to control. The voltage across the FET is limited by the diodes to a value of $\pm 1.2\,\text{V}$, and the level at which the switch cuts off is largely independent of the input voltage. This reduces the aperture jitter.

If the switch is ON, the output voltage assumes the value

$$U_o = -\frac{R_2}{R_1} U_i,$$

due to the negative feedback. As for the previous circuit, amplifier OA 1 reduces the acquisition time and also eliminates the offset voltage of the FET input amplifier OA 2.

7.5   Analog comparators

In the circuits described so far, the input voltage is switched on or off, or is stored, depending on a control signal. Another type of analog switch is the comparator. It initiates a switching process if the continuous input voltage exceeds, or falls below a certain predetermined level.

7.5.1   Basic circuit

An operational amplifier without feedback, as in Fig. 7.20, represents the basic circuit of a comparator. Its output voltage is given by

$$U_o = \begin{cases} U_{o\,\text{max}} & \text{for } U_1 > U_2 \\ U_{o\,\text{min}} & \text{for } U_1 < U_2. \end{cases}$$

The corresponding transfer characteristic is shown in Fig. 7.21. Owing to the high gain, the circuit responds to very small voltage differences $U_1 - U_2$. It is thus suitable for the comparison of two voltages, and operates with high accuracy.

At zero crossing of the input voltage difference, the output voltage does not immediately reach the saturation level because the transition is limited by the slew rate. For frequency-compensated standard operational amplifiers it is about $1\,\text{V}/\mu\text{s}$. A rise from $-12\,\text{V}$ to $+12\,\text{V}$

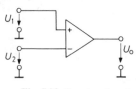

Fig. 7.20 Comparator

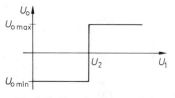

Fig. 7.21 Transfer characteristic

therefore takes 24 µs. An additional delay is incurred due to the
recovery time needed after the amplifier has been saturated.

As the amplifier possesses no feedback, it also does not require any
frequency compensation. Its omission can improve the slew rate and
recovery time by a factor of about 20.

Considerably shorter response times can be attained when using
special comparator amplifiers which are designed for use without
feedback and have especially small recovery times. However, the gain
and hence, the accuracy of the threshold are somewhat lower than for
operational amplifiers. Usually, the amplifier output is directly connected to a level-shift circuit which permits compatible operation with
logic circuits. The integrated circuit LM 339 from National Semicon-
ductor, for example, has an open-collector output and a response time
of 500 ns; the IC NE 521 from Signetics possesses a TTL output and a
response time of 8 ns. A particularly short delay of 5 ns is obtained with
the comparator Am 685 from Advanced Micro Devices, having an ECL
output.

The available range of input voltages for a comparator is limited. If
large voltages are to be compared, the comparators can be used in the
circuit shown in Fig. 7.22. The comparator changes its output when V_P
crosses zero. This is the case when

$$\frac{U_1}{R_1} = -\frac{U_2}{R_2}.$$

The voltages to be compared must therefore have different polarities.
The circuit can be expanded by connecting further resistors to the P-
input. The comparator then determines whether the weighted sum of
the input voltages is larger or smaller than zero. The diodes clamp the
input of the comparator so that it is limited to values between $\pm 0.6\,\text{V}$.

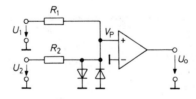

Fig. 7.22 Summing comparator

7.5.2   Comparator having a well defined output voltage

Many applications require that the output voltage of the compara-
tor accurately assumes the two defined levels U_{max} and U_{min}. The fastest

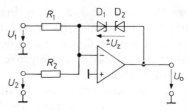

Fig. 7.23 Operational amplifier as a comparator with a well defined output voltage

and most precise method is to use the output voltage of a normal comparator to operate an analog switch.

For low signal frequencies, the problem can be solved as in Fig. 7.23, with a frequency-compensated operational amplifier having suitable feedback elements. The circuit is based on the comparator of Fig. 7.22. When the output voltage reaches the value $\pm(0.6\,\text{V} + U_z)$, the Zener diodes effect feedback for the operational amplifier, and in this way a further rise in voltage is prevented. As the amplifier does not saturate, there is no recovery time.

7.5.3   Window comparator

A window comparator can determine whether or not the value of the input voltage lies between two reference voltage levels. This requires that the output signals of two comparators are connected to a logic circuit, as in Fig. 7.24. The integrated circuit NE 521 mentioned previously is particularly well suited as it contains on a single chip not only two comparators with level-shift circuitry but also two NAND gates. As can be seen in Fig. 7.25, the output is at Boolean "one" only if

$$U_1 < U_i < U_2$$

as then both comparator outputs are also at Boolean "one".

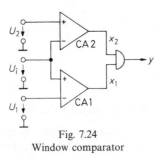

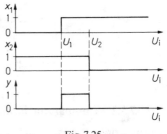

Fig. 7.24
Window comparator

Fig. 7.25
Signals of the window comparator

7.6 Schmitt trigger

The Schmitt trigger is a comparator, for which the positive and negative transitions of the output occur at different levels of the input voltage. Their difference is characterized by the hysteresis ΔU_i. Schmitt triggers can be realized by transistors [1], but in the following section some designs involving comparators are discussed.

7.6.1 Inverting Schmitt trigger

In the Schmitt trigger of Fig. 7.26 the hysteresis is effected by a positive feedback of the comparator, via the voltage divider R_1, R_2. If a large negative voltage U_i is applied, $U_o = U_{o\,max}$. At the P-input, the potential is then given as

$$V_{P\,max} = \frac{R_1}{R_1 + R_2} U_{o\,max}.$$

If the input voltage is changed towards positive values, U_o does not change at first; only when U_i reaches the value $V_{P\,max}$, does the output voltage reduce and therefore also V_P. The difference $U_D = V_P - V_N$ becomes negative. Due to the positive feedback, U_o falls very quickly to the value $U_{o\,min}$. The potential V_P assumes the value

$$V_{P\,min} = \frac{R_1}{R_1 + R_2} U_{o\,min}.$$

U_D is negative and large, this resulting in a stable state. The output voltage changes to $U_{o\,max}$ only when the input voltage has reached $V_{P\,min}$. The corresponding transfer characteristic is represented in Fig. 7.27.

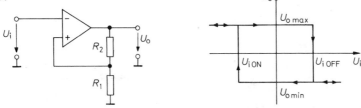

Fig. 7.26 Inverting Schmitt trigger.   Fig. 7.27 Transfer characteristic

Threshold for switch-on: $U_{i\,ON} = \dfrac{R_1}{R_1 + R_2} U_{o\,min}$

Threshold for switch-off: $U_{i\,OFF} = \dfrac{R_1}{R_1 + R_2} U_{o\,max}$

Hysteresis: $\Delta U_i = \dfrac{R_1}{R_1 + R_2}(U_{o\,max} - U_{o\,min})$

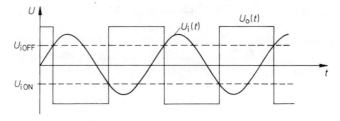

Fig. 7.28 Voltage waveshapes of the inverting Schmitt trigger

The circuit has two stable states only if the loop gain

$$g = \frac{A_D R_1}{R_1 + R_2} > 1.$$

Figure 7.28 shows an important application of the Schmitt trigger. The circuit converts an input voltage of any shape into a square-wave output voltage. This has a defined rise time which is independent of the shape of the input voltage.

7.6.2 Non-inverting Schmitt trigger

If one of the two inputs of the comparator in Fig. 7.22 is connected to the output, the result is the non-inverting Schmitt trigger of Fig. 7.29, the transfer characteristic of which is shown in Fig. 7.30.

If a large positive input voltage U_i is applied, $U_o = U_{o\,max}$. When reducing U_i, U_o does not change until V_P crosses zero. This is the case for the input voltage

$$U_{i\,OFF} = -\frac{R_1}{R_2} U_{o\,max}.$$

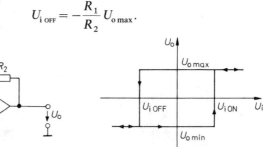

Fig. 7.29 Non-inverting Schmitt trigger.   Fig. 7.30 Transfer characteristic

Threshold for switch-on: $U_{i\,ON} = -\dfrac{R_1}{R_2} U_{o\,min}$

Threshold for switch-off: $U_{i\,OFF} = -\dfrac{R_1}{R_2} U_{o\,max}$

Hysteresis: $\Delta U_i = \dfrac{R_1}{R_2}(U_{o\,max} - U_{o\,min})$

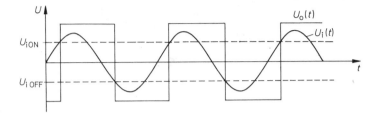

Fig. 7.31 Voltage waveshapes of the non-inverting Schmitt trigger

The output voltage jumps to $U_{o\,min}$ as soon as U_i reaches or falls below this value. The transition is initiated by U_i but is then determined only by the positive feedback via R_2. The new state is stable until U_i returns to the level

$$U_{i\,ON} = -\frac{R_1}{R_2} U_{o\,min}.$$

Figure 7.31 depicts the time function of the output voltage for a sinusoidal input. Since, at the instant of transition, $V_p = 0$, the formulae for the trigger level have the same form as those for the inverting amplifier.

In the same way that an inverting amplifier can be expanded to become a summing amplifier so can the Schmitt trigger be extended to become a summing Schmitt trigger. For this purpose additional resistors are connected to the P-input providing further voltage inputs. This method is shown in Fig. 7.32. The trigger levels can be varied by means of the voltage U_2 but the hysteresis remains unchanged.

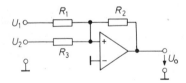

Fig. 7.32 Summing Schmitt trigger.

Threshold for switch-on: $\quad U_{i\,ON} = -\dfrac{R_1}{R_2} U_{o\,min} - \dfrac{R_1}{R_3} U_2$

Threshold for switch-off: $\quad U_{i\,OFF} = -\dfrac{R_1}{R_2} U_{o\,max} - \dfrac{R_1}{R_3} U_2$

7.6.3 Precision Schmitt trigger

The trigger levels of the Schmitt triggers described so far are not as accurate as is normally expected for operational amplifier circuitry. The accuracy can be improved by connecting a comparator to an analog

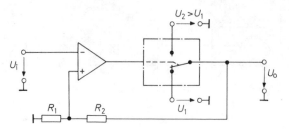

Fig. 7.33 Precision Schmitt trigger with analog switch

Threshold for switch-on: $U_{i\,\mathrm{ON}} = \dfrac{R_1}{R_1 + R_2} U_1$

Threshold for switch-off: $U_{i\,\mathrm{OFF}} = \dfrac{R_1}{R_1 + R_2} U_2$

switch, as in Fig. 7.33. Thus, depending on the state of the comparator, the output voltage assumes two precisely defined values, U_1 or U_2. Due to the positive feedback via the voltage divider R_1, R_2, the trigger levels are given as

$$U_{i\,\mathrm{OFF}} = \frac{R_1}{R_1 + R_2} U_2, \qquad U_{i\,\mathrm{ON}} = \frac{R_1}{R_1 + R_2} U_1.$$

They are therefore no longer dependent on the level at which the operational amplifier saturates, as is the case for the circuit in Fig. 7.26.

The circuit can also be operated as a non-inverting Schmitt trigger if, as in Fig. 7.29, the N-input is connected to ground and the input signal is applied to resistor R_1.

Often a high accuracy is required for the trigger levels but not for the output voltage. This demand can be fulfilled without the use of an analog switch, as Fig. 7.34 shows. The amplifier OA 3 is operated as a summing Schmitt trigger, and the two comparators OA 1 and OA 2 accurately specify the trigger levels.

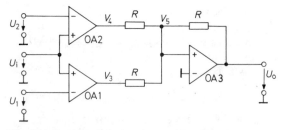

Fig. 7.34 Precision Schmitt trigger with two comparators.

Threshold for switch-on: $U_{i\,\mathrm{ON}} = \mathrm{Max}(U_1, U_2)$

Threshold for switch-off: $U_{i\,\mathrm{OFF}} = \mathrm{Min}(U_1, U_2)$

If the input voltage is larger than the two reference voltages U_1 and U_2, then $U_o = U_{o\,max}$. When the input voltage falls below the larger of the two reference voltages, U_o does not change. This is because one of the two output potentials, V_3 and V_4, is at $U_{o\,max}$ and the other at $U_{o\,min} \approx -U_{o\,max}$. Hence $V_5 \approx \frac{1}{3} U_{o\,max} > 0$. Potential V_5 becomes negative only when the input voltage has fallen below the second reference voltage too. At this instant, the output jumps from $U_{o\,max}$ to $U_{o\,min}$. The trigger level for the $U_{o\,min}$ state is therefore identical to the smaller of the two reference voltages, and the trigger level for the $U_{o\,max}$ state identical to the larger. This is illustrated by the transfer characteristic in Fig. 7.35.

If, instead of the operational amplifiers OA 1 and OA 2, special comparators having level-shift circuits are used, the amplifier OA 3 can be replaced by an RS flip-flop, as shown in Fig. 7.36. The flip-flop is set when the input voltage exceeds the trigger level $U_2 > U_1$, and is reset when it falls below the value U_1.

The IC NE 521 mentioned previously consists of two comparators, each with a NAND gate connected to it. By an appropriate external connection, the desired function can be implemented by a single IC. For low frequencies, a further single-chip solution is available, using the timer 555, and this is discussed in more detail in Section 8.5.1.

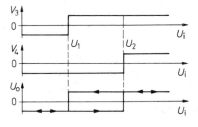

Fig. 7.35 Transfer characteristic
of the circuit in Fig. 7.34

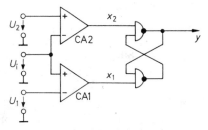

Fig. 7.36 Precision Schmitt trigger
with RS flip-flop.

Threshold for switch-on: $\quad U_{i\,ON} = U_2$
Threshold for switch-off: $\quad U_{i\,OFF} = U_1$

for $U_2 > U_1$

8 Signal generators

A signal generator is a circuit providing an alternating voltage. To begin with, we describe the generation of sine waves, and then that of triangular and square waves.

8.1 *LC* oscillators

The simplest way to generate a sinewave is to use an amplifier to eliminate the damping of an LC resonant circuit. In the following section, we deal with some of the basic aspects of this method.

8.1.1 Condition for oscillation

Figure 8.1 shows the principle of an oscillator circuit. The amplifier multiplies the input voltage by the gain $\underline{A}$, and thereby causes a parasitic phase shift α between $\underline{U}_2$ and $\underline{U}_1$. The load resistance R_L and a frequency-dependent feedback network, for example a resonant circuit, are connected to the amplifier output. The voltage feedback is therefore $\underline{U}_3 = \underline{k}\,\underline{U}_2$, and the phase shift between $\underline{U}_3$ and $\underline{U}_2$ is denoted by β.

To establish whether the circuit can produce oscillations, the feedback loop is opened. An additional resistor R_i is introduced at the output of the feedback network, representing the input resistance of the amplifier. An alternating voltage $\underline{U}_1$ is applied to the amplifier and $\underline{U}_3$ is measured. The circuit is capable of producing oscillations if the output voltage is the same as the input voltage. Hence, the necessary condition for oscillation

$$\underline{U}_1 = \underline{U}_3 = \underline{k}\,\underline{A}\,\underline{U}_1.$$

The loop gain must therefore be

$$\underline{g} = \underline{k}\,\underline{A} = 1, \tag{8.1}$$

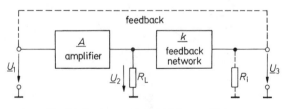

Fig. 8.1 Basic arrangement of an oscillator

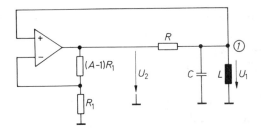

Fig. 8.2 Principle of an *LC* oscillator

from which two conditions can be deduced, i.e.

$$|g| = |\underline{k}| \cdot |\underline{A}| = 1 \tag{8.2}$$

and

$$\alpha + \beta = 0, 2\pi, \dots. \tag{8.3}$$

Equation (8.2) is the *amplitude condition* which states that a circuit can oscillate only if the amplifier eliminates the attenuation due to the feedback network. The *phase condition* of Eq. (8.3) states that an oscillation can arise only if the output voltage is in phase with the input voltage. Details of the oscillation, e.g. frequency and waveform, can be obtained only with additional information on the feedback network. We therefore continue the discussion, taking the *LC* oscillator of Fig. 8.2 as an example.

The non-inverting amplifier multiplies the voltage $U_1(t)$ by the gain A. As the output resistance of the amplifier is low, the resonant circuit is damped by the parallel resistor R. For the calculation of the feedback voltage, we apply KCL to node 1 and obtain

$$\frac{U_2 - U_1}{R} - C\dot{U}_1 - \frac{1}{L} \int U_1 \, dt = 0.$$

As $U_2 = AU_1$, it follows that

$$\ddot{U}_1 + \frac{1 - A}{RC} \dot{U}_1 + \frac{1}{LC} U_1 = 0. \tag{8.4}$$

This is the differential equation of a damped oscillation. To abbreviate,

$$\gamma = \frac{1 - A}{2RC} \quad \text{and} \quad \omega_0^2 = \frac{1}{LC},$$

and therefore

$$\ddot{U}_1 + 2\gamma \dot{U}_1 + \omega_0^2 U_1 = 0,$$

the solution of which is given by

$$U_1(t) = U_0 \cdot e^{-\gamma t} \sin\left(\sqrt{\omega_0^2 - \gamma^2}\, t\right). \tag{8.5}$$

One must differentiate between three cases:

(1) $\gamma > 0$, i.e. $A < 1$.

The amplitude of the output alternating voltage decreases exponentially; the oscillation is damped.

(2) $\gamma = 0$, i.e. $A = 1$.

The result is a sinusoidal oscillation with the frequency $\omega_0 = \dfrac{1}{\sqrt{LC}}$ and with constant amplitude.

(3) $\gamma < 0$, i.e. $A > 1$.

The amplitude of the output alternating voltage rises exponentially.

With Eq. (8.2) we have the necessary condition for an oscillation. This can now be described in more detail: For $A = 1$, a sinusoidal output voltage of constant amplitude and the frequency

$$\omega = \omega_0 = \frac{1}{\sqrt{LC}}$$

is obtained. For tighter feedback, the amplitude rises exponentially and for looser feedback, the amplitude falls exponentially. To ensure that the oscillation builds up after the supply has been switched on, the gain A must initially be larger than unity. The amplitude then rises exponentially until the amplifier begins to saturate. Because of the saturation, A decreases until it reaches the value 1; the output however, is then no longer sinusoidal. If a sinusoidal output is required, an additional gain control circuit should ensure that $A = 1$ before the amplifier saturates. For the high-frequency range, resonant circuits with high Q-factors are usually simple to implement. The voltage of the resonant circuit is still sinusoidal, even if the amplifier saturates. For this frequency range, an extra amplitude control is usually not required, and the voltage across the resonant circuit is then taken as the output voltage.

8.1.2   Meissner oscillator

It is characteristic of the Meissner circuit that the feedback is effected by a transformer. A capacitor C, together with the transformer primary winding, forms the frequency-determining resonant circuit. Figures 8.3 to 8.5 show three Meissner oscillators, each in common-emitter connection. At the resonant frequency

$$\omega_0 = \frac{1}{\sqrt{LC}},$$

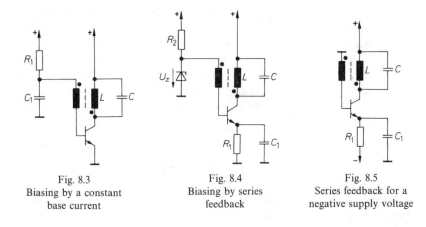

Fig. 8.3
Biasing by a constant
base current

Fig. 8.4
Biasing by series
feedback

Fig. 8.5
Series feedback for a
negative supply voltage

the amplified input voltage appears at the collector with maximum amplitude and a phase shift of 180°. Part of this alternating voltage is fed back via the secondary winding. To fulfill the phase condition, the transformer must effect a further phase displacement of 180°. This is achieved by a.c.-grounding the secondary winding at the end having the same voltage polarity as the collector end of the primary winding. The dots on the two windings indicate which winding ends have the same polarity. The turns ratio is chosen so that the magnitude of the loop gain $\underline{k}\underline{A}$ at resonant frequency is always larger than unity. Oscillation then begins at switch-on of the supply, its amplitude rising exponentially until the transistor saturates. Saturation reduces the mean value of the gain until $|\underline{k}\underline{A}| = 1$ and the amplitude of the oscillation remains constant. Two saturation effects can be distinguished, that at the input and that at the output. The output saturation arises when the collector-base junction is forward biased. This is the case for the circuits in Figs. 8.3 and 8.5 when the collector potential goes negative. The maximum amplitude of the oscillation is therefore $\hat{U}_C = V^+$. The maxima of the collector potential are then $\hat{U}_{CE\,max} = 2V^+$. This influences the choice of the transistor. For the circuit in Fig. 8.4, the maximum amplitude is smaller than V^+, the reduction being due to the Zener voltage.

With tight feedback, input saturation can also occur. Large input amplitudes arise which are rectified at the emitter-base junction. The capacitor C_1 is therefore charged, and the transistor conducts only during the positive peaks of the input alternating voltage.

In the circuit of Fig. 8.3, a few oscillations may be sufficient to charge the capacitor C_1 to such a high negative voltage that the oscillation stops altogether. It restarts only when the base potential has risen to $+0.6\,V$ with the relatively large time constant $R_1 C_1$. In this case, a sawtooth voltage appears across C_1 and the circuit was therefore often used as a sawtooth generator, such a circuit being known as a *blocking oscillator*.

To prevent the circuit from operating in the blocking oscillator mode, the input saturation must be reduced by choosing a correspondingly smaller turns ratio. In addition, the resistance of the base biasing circuit should be kept as low as possible [8.1]. This is difficult to attain for the circuit in Fig. 8.3 as the base current would then be far too large. Bias by series feedback as in Figs. 8.4 and 8.5 is therefore preferable.

8.1.3   Hartley oscillator

The Hartley oscillator resembles a Meissner osillator. The only difference is that the transformer is replaced by a tapped coil (auto-transformer). The inductance of this coil, together with the parallel-connected capacitor, determines the resonant frequency.

Figure 8.6 shows a Hartley oscillator in common-emitter connection. An alternating voltage is supplied to the base via capacitor C_2; it is in phase opposition to the collector voltage so that positive feedback occurs. The amplitude of the feedback voltage can be adjusted to the required value by appropriate positioning of the tap. As for the Meissner oscillation in Fig. 8.5, the mean collector current is determined by the series feedback resistor R_1.

For the Hartley oscillator in Fig. 8.7, the transistor is operated in common-base connection. Therefore, a voltage must be taken from the coil L by means of capacitor C_1, in phase with the collector voltage. If only a small negative supply voltage is available, the emitter resistance R_1 must be fairly low to ensure that the required collector quiescent current

$$I_C = \frac{|V^-| - 0.6\,\text{V}}{R_1}$$

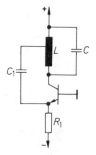

Fig. 8.6 Hartley oscillator
in common-emitter connection

Fig. 8.7 Hartley oscillator
in common-base connection

can flow. The resonant circuit would thereby be damped considerably via capacitor C_1, but this can be reduced by connecting an inductance in series with R_1.

8.1.4   Colpitts oscillator

A characteristic of the Colpitts circuit is the capacitive voltage divider determining the fraction of the output voltage that is fed back. The series connection of the capacitors acts as the oscillator capacitance, i.e.

$$C = \frac{C_a C_b}{C_a + C_b}.$$

The common-emitter circuit of Fig. 8.8 corresponds to the circuit in Fig. 8.6, but requires an additional collector resistor R_3 for the supply of the positive voltage.

The common-base connection is again much simpler, as can be seen in Fig. 8.9. It corresponds to the Hartley oscillator of Fig. 8.7.

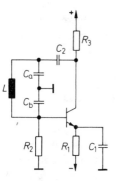

Fig. 8.8 Colpitts oscillator
in common-emitter connection

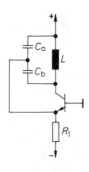

Fig. 8.9 Colpitts oscillator
in common-base connection

8.1.5   Emitter-coupled *LC* oscillator

A simple way of realizing an oscillator is to use a differential amplifier, as in Fig. 8.10. As the base potential of T_1 is in phase with the collector

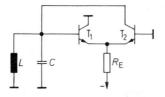

Fig. 8.10 Emitter-coupled oscillator

potential of T_2, positive feedback can be attained by directly connecting the two terminals. The loop gain is proportional to the transconductance of the transistors. It can be adjusted over a wide range by changing the emitter current. As the transistors are operated at $U_{CB}=0$, the amplitude of the output voltage is limited to about 0.5 V.

The amplifier of the emitter-coupled oscillator, together with an output stage and amplitude control, is available as an IC (MC 1648 from Motorola). It is suitable for frequencies of up to about 200 MHz.

8.1.6 Push-pull oscillators

Push-pull circuits are used for power amplifiers to attain higher output power and better efficiency. For the same reason these circuits can also be employed for the design of oscillators. One such design is shown in Fig. 8.11, consisting basically of two Meissner oscillators in which the transistors T_1 and T_2 are alternately conducting.

As the base potential of one transistor is in phase with the collector potential of the other, the secondary winding normally required for phase inversion can be omitted. This version is shown in Fig. 8.12. The positive feedback is effected by the capacitive voltage dividers C_1, C_2. The parallel resistive voltage dividers provide the bias.

Both circuits, in addition to providing larger output power, also generate fewer harmonics than the single-ended oscillators.

Another simple way of designing a push-pull oscillator is to use a bipolar current source [1] to feed the resonant circuit, see Fig. 8.13. The voltage of the resonant circuit is fed back to the current source via the emitter follower T_3. As the damping of the resonant circuit is only very slight, alternating voltages with low distortion can be generated. Resistor R_6 ensures a smooth cut-in of the voltage limit thereby keeping distortion low, even at saturation of the current sources.

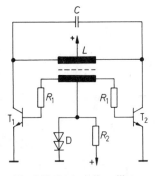

Fig. 8.11 Push-pull oscillator
with inductive feedback

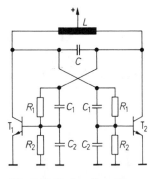

Fig. 8.12 Push-pull oscillator
with capacitive feedback

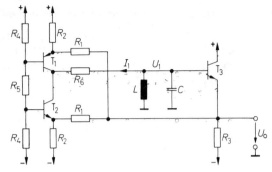

Fig. 8.13 Push-pull oscillator with controlled current sources

The voltage divider R_4, R_5 determines the limit of this saturation and thereby the amplitude of the alternating voltage. The resistances R_2 define the quiescent current of the current source. If a low distortion factor is important, R_2 should be chosen such that the transistors T_1 and T_2 are in class-A operation. The resistances R_1 determine the amount of feedback.

The circuit can be interpreted as a negative resistance which cancels the damping of the resonant circuit. To determine its value, we assume a positive change in voltage, ΔU_1, which effects a drop of $\Delta U_1/R_1$ in the collector current of T_2 and an equally large rise in the collector current of T_1. Current I_1 is therefore reduced by $2\,\Delta U_1/R_1$. This corresponds to a resistor connected in parallel to the resonant circuit, having a value

$$R = \frac{\Delta U_1}{\Delta I_1} = -\tfrac{1}{2}R_1.$$

To ensure that the condition for oscillation is fulfilled, $\tfrac{1}{2}R_1$ must be somewhat smaller than the resistance of the LC circuit at resonance.

8.2   Quartz oscillators

The frequency of the LC oscillators described is not sufficiently constant for many applications as it depends on the temperature coefficients of capacitance and inductance of the resonant circuit. Considerably more stable frequencies can be achieved using quartz crystals. Such a crystal can be excited by electric fields to vibrate mechanically and, when provided with electrodes, behaves electrically as a resonant circuit having a high Q-factor. The temperature coefficient of the resonant frequency is very small. The frequency stability that can be attained by a quartz oscillator is in the order of

$$\frac{\Delta f}{f} = 10^{-6} \ldots 10^{-10}.$$

8.2.1   Electrical characteristics of a quartz crystal

The electrical behaviour of a quartz crystal can be described by the equivalent circuit in Fig. 8.14. The two parameters C and L are well defined by the mechanical properties of the crystal. The resistance R is

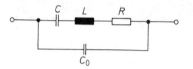

Fig. 8.14 Equivalent circuit of a quartz crystal

small and characterizes the damping. The capacitor C_0 gives the value of the capacitance formed by the electrodes and leads. Typical values for a 4 MHz quartz are

$$L = 100\,\text{mH}, \qquad\qquad R = 100\,\Omega,$$
$$C = 0.015\,\text{pF}, \qquad\qquad Q = 25\,000.$$
$$C_0 = 5\,\text{pF},$$

For the calculation of resonant frequency we initially determine the impedance of the quartz. From Fig. 8.14, by neglecting R

$$\underline{Z}_q = \frac{\text{j}}{\omega} \cdot \frac{\omega^2 L C - 1}{C_0 + C - \omega^2 L C C_0}. \tag{8.6}$$

It can be seen that there is a frequency for which $\underline{Z}_q = 0$ and another for which $\underline{Z}_q = \infty$. The quartz crystal therefore has a series and a parallel resonance. For the calculation of the series resonant frequency f_s, the numerator of Eq. (8.6) is set to zero, and thus

$$f_s = \frac{1}{2\pi\sqrt{LC}}. \tag{8.7}$$

The parallel resonant frequency is calculated by setting the denominator to zero:

$$f_P = \frac{1}{2\pi\sqrt{LC}}\sqrt{1 + \frac{C}{C_0}}. \tag{8.8}$$

As can be seen, the series resonant frequency is dependent only on the well defined product LC, whereas the parallel resonant frequency is influenced by the electrode capacitance C_0 which is far more susceptible to variations.

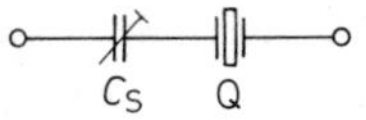

Fig. 8.15 Tuning of the series resonant frequency

The frequency of a quartz oscillator must often be adjustable within a small range. This can be achieved by simply connecting a capacitor C_S in series with the quartz crystal, see Fig. 8.15; C_S must be large in comparison with C.

To calculate the shift in the resonant frequency, we determine the impedance of the series connection. With Eq. (8.6) we obtain

$$\underline{Z}'_q = \frac{1}{j\omega C_S} \cdot \frac{C + C_0 + C_S - \omega^2 LC(C_0 + C_S)}{C_0 + C - \omega^2 LC C_0}. \tag{8.9}$$

Setting the numerator to zero gives the new series resonant frequency

$$f'_s = \frac{1}{2\pi\sqrt{LC}}\sqrt{1 + \frac{C}{C_0 + C_S}} = f_s\sqrt{1 + \frac{C}{C_0 + C_S}}. \tag{8.10}$$

By expanding this into a power series, we arrive at the approximation

$$f'_s = f_s\left[1 + \frac{C}{2(C_0 + C_S)}\right]$$

if $C \ll C_0 + C_S$. The relative shift in frequency is therefore

$$\frac{\Delta f}{f} = \frac{C}{2(C_0 + C_S)}.$$

The parallel resonant frequency is not changed by C_S as the poles of Eq. (8.9) are independent of C_S. A comparison of Eqs. (8.10) and (8.8) shows that for $C_S \to 0$, the series resonant frequency cannot be raised to a value higher than that of the parallel resonant frequency.

8.2.2 Quartz oscillators with *LC* resonant circuits

The resonant frequency of an *LC* oscillator can be stabilized by inserting a quartz crystal in the feedback loop. It is preferable to operate the quartz at series resonance as the frequency stability is then better. However, the external series resistances must be kept as small as possible in relation to the damping resistor R in Fig. 8.14; otherwise the Q-factor of the quartz deteriorates, and the smaller the Q-factor the smaller the slope of the phase-frequency response at resonant frequency. Parasitic phase shifts then have a stronger effect on the resonant frequency.

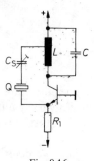

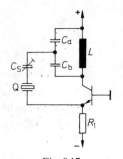

Fig. 8.16   Fig. 8.17
Hartley oscillator using a quartz   Colpitts oscillator using a quartz

The requirement of a small series resistance can be easily fulfilled by operating the oscillator in common-base connection, as in Figs. 8.16 and 8.17. The circuit in Fig. 8.16 is based on the Hartley oscillator of Fig. 8.7, that in Fig. 8.17 on the Colpitts oscillator of Fig. 8.9.

To ensure build-up of the oscillation, the resonant circuit must be tuned to the resonant frequency of the quartz. An integral multiple thereof can also be chosen, as in such a case the appropriate harmonics of the quartz crystal would be excited. This method is preferred for frequencies above 10 MHz.

8.2.3   Quartz oscillators without *LC* resonant circuits

If a quartz crystal is to be operated at its fundamental, an additional resonant circuit is not required. Figure 8.18 shows a suitable circuit for series resonant operation. The circuit around the quartz must have as low a resistance as possible in order not to adversely affect the Q-factor of the quartz. The emitter follower T_1 keeps this resistance low. The current through the quartz is amplified by the current mirror D_1, T_2, and has its

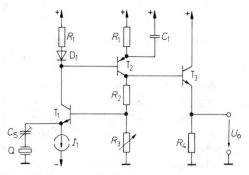

Fig. 8.18 Quartz oscillator without *LC* resonant circuit

maximum at the series resonant frequency. The loop gain is chosen such that the condition for oscillation is fulfilled at this frequency. The resistance of R_3 is made so small that the alternating voltage across the quartz is only a few $10\,\text{mV}$. The power dissipation of the quartz is then sufficiently small so as not to affect the frequency stability. Resistor R_3 is best implemented by an electrically controlled resistor, e.g. a FET, which is adjusted by an amplitude control circuit. This also ensures start-up of the oscillator and that the output voltage is reasonably sinusoidal.

The same circuit can be used to excite the harmonic of the quartz if capacitor C_1 is replaced by an appropriately tuned series resonant circuit.

The entire arrangement, including an amplitude control loop, is available as an IC from Plessey (SL 680 C) with which resonant frequencies of up to $150\,\text{MHz}$ can be realized. The frequency deviation due to the amplifier is $10^{-9}/\text{K}$ or $10^{-7}/\text{V}$.

8.3   *RC* sinewave oscillators

For audio frequencies *LC* oscillators are less suitable since the inductances and capacitances become unwieldy. In this range it is preferable to use oscillators the frequencies of which are determined by *RC* networks.

8.3.1   Wien-Robinson oscillator

In principle an *RC* oscillator can be realized by replacing the resonant circuit in Fig. 8.2 by a passive *RC* bandpass. However, the maximum attainable Q-factor would then be limited to $\frac{1}{2}$, as is shown in Chapter 3.7. Consequently, the frequency stability of the resulting sinewave would be

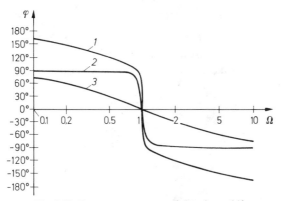

Fig. 8.19 Frequency response of the phase shift.
Curve *1*: Wien-Robinson bridge for $\varepsilon = 0.01$. Curve *2*: Resonant circuit for $Q = 10$.
Curve *3*: Passive bandpass filter for $Q = \frac{1}{3}$

poor, as can be seen from the frequency response of the phase shift shown
in Fig. 8.19. The phase shift of a passive lowpass filter with $Q = \frac{1}{3}$ is 27° at
half the resonant frequency. In the case of the amplifier causing a phase
shift of $-27°$, the circuit would oscillate at half the resonant frequency as
the total phase shift is then zero. For good frequency stability, a feedback
network is thus required for which the gradient of the phase-frequency
response at zero phase is as steep as possible. High-Q resonant circuits and
the Wien-Robinson bridge have this characteristic. However, the output
voltage of the Wien-Robinson bridge is zero at resonant frequency and the
bridge is therefore suitable as a feedback network only under certain
conditions. For the use in oscillators it must be slightly detuned, as in
Fig. 8.20, where ε is positive and small as against unity.

The frequency dependence of the phase shift of a detuned Wien-
Robinson bridge can be determined qualitatively: At high and low
frequencies, $\underline{U}_1 = 0$, and thus $\underline{U}_D \approx -\frac{1}{3}\underline{U}_i$ and the phase shift is $\pm 180°$. At
resonant frequency, $\underline{U}_1 = +\frac{1}{3}\underline{U}_i$ and

$$\underline{U}_D = \left(\frac{1}{3} - \frac{1}{3+\varepsilon}\right)\underline{U}_i \approx \frac{\varepsilon}{9}\underline{U}_i,$$

i.e. $\underline{U}_D$ is in phase with $\underline{U}_i$. To determine the shape of curve 1 in Fig. 8.19
quantitatively, we first calculate the transfer function

$$\frac{\underline{U}_D}{\underline{U}_i} = -\frac{1}{3+\varepsilon} \cdot \frac{(1+P^2)-\varepsilon P}{1+\dfrac{9+\varepsilon}{3+\varepsilon}P+P^2}.$$

Hence, when neglecting higher powers of ε, the frequency response of the
phase is

$$\varphi = \tan^{-1}\frac{3\Omega(\Omega^2-1)(3+2\varepsilon)}{(\Omega^2-1)^2(3+\varepsilon)-9\varepsilon\Omega^2},$$

which is shown in Fig. 8.19 for $\varepsilon = 0.01$. It can be seen that the phase shift of
the detuned Wien-Robinson bridge increases to $\pm 90°$ within a very
narrow frequency range. It is narrower the smaller the value of ε. As far as

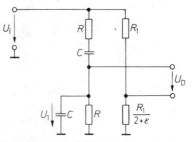

Fig. 8.20 Detuned Wien-Robinson bridge

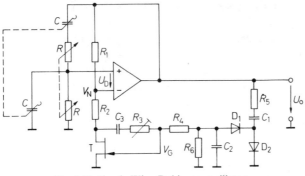

Fig. 8.21 Simple Wien-Robinson oscillator.

Resonant frequency: $f_0 = 1/2\pi RC$

this aspect is concerned, the Wien-Robinson bridge is comparable with high quality resonant circuits, and an additional advantage is that the phase shift is not limited to $\pm 90°$ but to $\pm 180°$. Harmonics are thereby strongly damped. One disadvantage of the Wien-Robinson bridge is that the attenuation at resonant frequency becomes higher the smaller the value of ε. In general, the attenuation at resonant frequency is

$$\frac{\hat{U}_D}{\hat{U}_i} = k \approx \frac{\varepsilon}{9},$$

and, in our example, $k \approx \frac{1}{900}$. To fulfill the amplitude condition of an oscillator, the amplifier must compensate for the attenuation. Such an oscillator circuit is represented in Fig. 8.21.

If the amplifier has the open-loop gain A_D, the detuning factor ε must have the value

$$\varepsilon = 9k = \frac{9}{A_D}$$

to fulfill the amplitude condition $kA_D = 1$. If ε is slightly larger, the oscillation amplitude increases until the amplifier begins to saturate. If ε is too small or negative, there will be no oscillation. However, it is impossible to adjust the resistors R_1 and $R_1/(2+\varepsilon)$ with the required precision. Therefore one of the two resistances must be controlled automatically depending on the output amplitude. This is the purpose of the field-effect transistor T in Fig. 8.21. The channel resistance R_{DS} is dependent only on the voltage U_{GS} if U_{DS} remains sufficiently small. To ensure that U_{DS} does not become too large, only part of V_N is applied to the FET and the rest appears at R_2. The sum of R_2 and R_{DS} must have the value $R_1/(2+\varepsilon)$. The smallest possible value of R_{DS} is R_{DSON}, and hence R_2 must be smaller than

$$\tfrac{1}{2}R_1 - R_{DSON}.$$

When the supply voltage is switched on, V_G is initially zero and therefore $R_{DS} = R_{DS_{ON}}$. If the above design condition is fulfilled, the sum of R_2 and R_{DS} is then smaller than $\frac{1}{2}R_1$. At resonant frequency of the Wien-Robinson bridge, there is therefore a relatively large voltage difference U_D. As a consequence, oscillation begins and the amplitude increases. The output voltage is rectified by the voltage doubler D_1, D_2. The gate potential thereby becomes negative and R_{DS} increases. The amplitude of the output voltage rises until

$$R_{DS} + R_2 = \frac{R_1}{2+\varepsilon} = \frac{R_1}{2+\dfrac{9}{A_D}}.$$

The distortion factor of the output voltage is dependent mainly on the linearity of the FET output characteristic which can be greatly improved if part of the drain-source voltage is superimposed on the gate potential [1]. This is the purpose of the two resistors R_3 and R_4. The capacitor C_3 ensures that no direct current will flow into the N-input of the operational amplifier as this would cause an output offset. In practice, $R_3 \approx R_4$ but, by fine adjustment of R_3, the distortion factor can be reduced to a minimum and values below 0.1 % can be attained.

If R is made adjustable, the frequency can be continually controlled. Good matching of the two resistors is important since the greater their mismatch, the more efficient the amplitude control must be. The maximum value of R should be low enough to ensure that there is no noticeable voltage arising from the input bias current of the operational amplifier. On the other hand, R must not be too low, or the output will be overloaded. For the adjustment of the frequency within a range of $1:10$, fixed resistors with the value $R/10$ are connected in series with the potentiometers R. If, in addition, switches are used to select different values for the capacitances C, a range of output frequencies from 10 Hz to 1 MHz can be covered by this circuit. In order that the amplitude control does not cause distortion even at the lowest frequency, the charge and discharge time constants, $R_5 C_2$ and $R_6 C_2$ respectively, must be larger than the longest oscillation period by a factor of at least 10.

The data of the field-effect transistor T determine the output amplitude. The regulation of the output voltage amplitude is not particularly good since some deviation in amplitude is required to effect a noticeable change in the resistance of the FET. It can be improved by amplifying the gate voltage and such a circuit is shown in Fig. 8.22.

The rectified output alternating voltage is applied to OA 2, connected as a modified PI-controller (Fig. 16.7). This adjusts the gate potential of the FET in such a way that the mean value of $|U_o|$ equals U_{ref}. The controller time constant must be large compared to the oscillation period,

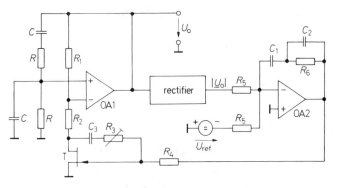

Fig. 8.22 Wien-Robinson oscillator with precise amplitude control.

$$Amplitude: \quad \hat{U}_o = \frac{\pi}{2} U_{\text{ref}}$$

or the gain will change even within a single cycle. This would lead to considerable distortion. A pure PI-controller can therefore not be used; it is better to have a modified one in which a capacitor is connected in parallel to R_6. The alternating voltage at R_6 is thereby short-circuited even at the lowest oscillator frequency. The proportional controller action comes into effect only below this frequency.

8.3.2   Second-order differential equation for sinewave generation

Low-frequency oscillations can be generated by programming operational amplifiers to solve the differential equation of a sinewave. From Section 8.1.1

$$\ddot{U}_o + 2\gamma \dot{U}_o + \omega_0^2 U_o = 0, \tag{8.11}$$

the solution of which is

$$U_o(t) = \hat{U}_o e^{-\gamma t} \sin(\sqrt{\omega_0^2 - \gamma^2}\, t). \tag{8.12}$$

Since operational amplifiers are better suitable for integration than for differentiation, the differential equation is rearranged by integrating it twice:

$$U_o + 2\gamma \int U_o \, dt + \omega_0^2 \iint U_o \, dt^2 = 0.$$

This equation can be simulated using two integrators and an inverting amplifier, and several methods of programming are available. One of these, particularly suitable for an oscillator, is shown in Fig. 8.23. For this circuit, the damping is given by $\gamma = -\alpha/20RC$ and the resonant frequency $f_0 = 1/2\pi RC$. Hence, from Eq. (8.12), the output voltage

$$U_o(t) = \hat{U}_o \, e^{\frac{\alpha}{20RC} t} \sin\left(\sqrt{1 - \frac{\alpha^2}{400}} \frac{t}{RC}\right). \tag{8.13}$$

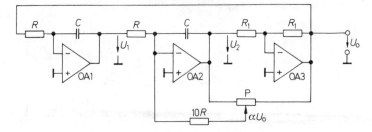

Fig. 8.23 Second-order differential equation for sinewave generation.

Resonant frequency: $f_0 = 1/2\pi RC$

It is evident that the damping of the oscillation is adjusted by α. At the right-hand stop of potentiometer P, $\alpha = 1$. At the left-hand stop

$$\alpha U_o = U_2 = -U_o,$$

i.e.

$$\alpha = -1;$$

in the middle, $\alpha = 0$. The damping can thus be varied between positive and negative values. For $\alpha = 1$, the oscillation amplitude is multiplied, within 20 cycles, by the factor e, for $\alpha = -1$ by the factor $1/e$. For $\alpha = 0$, the oscillation is undamped although this is so only for ideal conditions. In practice, for $\alpha = 0$, a slightly damped oscillation will occur, and to achieve constant amplitude, α must be given a small positive value. The adjustment is so critical that the amplitude can never be kept constant over a longer period and one must therefore introduce automatic amplitude control. As for the Wien-Robinson oscillator in Fig. 8.22, the amplitude at the output can be measured by a rectifier and α controlled, depending on the difference between this amplitude and a reference voltage. As shown previously, the controller time constant must be large compared to the period of oscillation to ensure that the amplitude control causes no distortion. This requirement is increasingly difficult to fulfill for frequencies below 10 Hz.

The difficulties arise from the fact that a whole cycle of the oscillation must be allowed to pass before the correct amplitude is known. Such problems can be eliminated if the amplitude is measured at every instant of the oscillation period. This is possible for the circuit in Fig. 8.23 where, in the case of an undamped oscillation,

$$U_o = \hat{U}_o \sin \omega_0 t$$

and

$$U_1 = -\frac{1}{\tau} \int U_o \, dt = \hat{U}_o \cos \omega_0 t.$$

The amplitude can now be determined at any instant by solving the following expression:

$$U_o^2 + U_1^2 = \hat{U}_o^2(\sin^2 \omega_0 t + \cos^2 \omega_0 t) = \hat{U}_o^2. \tag{8.14}$$

It is obvious that the expression $U_o^2 + U_1^2$ is dependent only on the amplitude of the output signal and not on its phase $\omega_0 t$. One obtains therefore, a pure d.c. voltage which needs no filtering and which can be directly compared with the reference voltage.

A supplementary control based on this principle is shown in Fig. 8.24. The voltage U_o and U_1 are connected to the input of a vector voltmeter as in Fig. 1.50, which produces the voltage $\sqrt{U_1^2 + U_o^2}$ at its output. The PI-controller OA 4 compares this voltage with the reference U_{ref} and adjusts its output voltage U_3 so that

$$\sqrt{U_1^2 + U_o^2} = U_{ref}.$$

With Eq. (8.14),

$$\hat{U}_o = U_{ref}.$$

The voltage $U_o \cdot U_3/E$ appears at the output of the multiplier M. It allows the potentiometer P in Fig. 8.23 to be omitted and is directly applied to the resistor $10R$. Then $\alpha = U_3/E$. If the amplitude increases,

$$\sqrt{U_1^2 + U_o^2} > U_{ref}$$

and both U_3 and α become negative, i.e. the oscillation is damped. If the amplitude decreases, U_3 becomes positive and the oscillation is un-damped.

Apart from a good method of controlling the amplitude, sinewave generation by solving a second-order differential equation offers a further

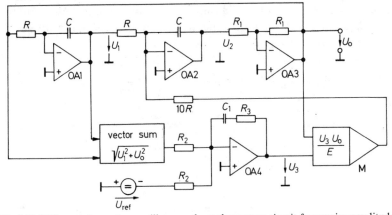

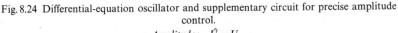

Fig. 8.24 Differential-equation oscillator and supplementary circuit for precise amplitude control.

Amplitude: $\hat{U}_o = U_{ref}$

advantage in that it enables an almost ideal frequency modulation. If this is to be accomplished for LC oscillators, the value of L or C must be varied. However, this will alter the energy of the oscillator and hence the oscillation amplitude, and parametric effects arise. For the differential equation method, however, the resonant frequency can be changed by varying the two resistors R without influencing the oscillator energy.

As each of the two resistors are connected to virtual ground, multipliers connected in front of them can be used to modulate the frequency. The multipliers then produce the output voltages

$$U_o' = \frac{U_c}{E} U_o \quad \text{and} \quad U_1' = \frac{U_c}{E} U_1.$$

This is equivalent to an increase of the resistance R by the factor E/U_c, so that the resonant frequency is given as

$$f_0 = \frac{1}{2\pi RC} \cdot \frac{U_c}{E},$$

i.e. is proportional to the control voltage U_c.

8.4   Function generators

The amplitude control involved in the generation of low-frequency sinewaves is rather cumbersome. It is much easier to use a Schmitt trigger and an integrator to generate a triangular alternating voltage. A sinewave can then be produced if the sine function network of Chapter 1.7.4 is employed. Since with this method, a triangular wave, a square wave and a sinusoidal wave are obtained simultaneously, circuits based on this principle are called function generators. The block diagram of such a circuit is shown in Fig. 8.25.

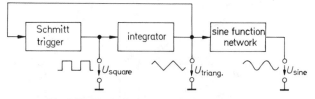

Fig. 8.25 Block diagram of a function generator

8.4.1   Simple triangular/square wave generator

The basis of a triangular/square wave generator is the series-connection of a Schmitt trigger and an integrator, as in Fig. 8.26.

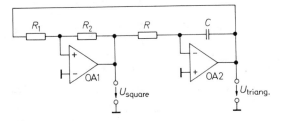

Fig. 8.26 Simple triangular/square wave generator.

$$\text{Frequency:} \quad f = \frac{1}{4RC} \cdot \frac{R_2}{R_1}$$

The Schmitt trigger supplies a constant output voltage which is subsequently integrated. If the output voltage of the integrator reaches the trigger level of the Schmitt trigger, the voltage U_{square} to be integrated instantly changes its polarity. The slope of the output voltage of the integrator therefore also changes polarity, and the voltage rises or falls linearly until the other trigger level is reached. By variation of the integration time constant, the frequency can be adjusted within a wide range. The amplitude of the triangular wave $U_{\text{triang.}}$ is determined solely by the trigger level of the Schmitt trigger. According to Section 7.6.2,

$$\hat{U}_{\text{triang.}} = \frac{R_1}{R_2} U_{\text{max}},$$

where U_{max} is the output voltage limit of the operational amplifier OA 1. The period is twice the time needed by the integrator to integrate from $-\hat{U}_{\text{triang.}}$ to $+\hat{U}_{\text{triang.}}$. Therefore

$$T = 4RC \frac{R_1}{R_2}.$$

The frequency is thus independent of the output saturation level U_{max} of the operational amplifier.

If high stability of the amplitude is required, the simple non-inverting Schmitt trigger can be replaced by one of the precision circuits of Section 7.6.3.

8.4.2 Function generator with controllable frequency

It is relatively easy to modulate the frequency of function generators by controlling the input amplitude of the integrator. Figure 8.27 shows this method, using the circuit of Fig. 7.14. Depending on the polarity of the Schmitt trigger output voltage, the input of the integrator

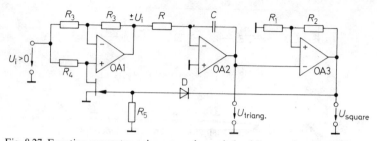

Fig. 8.27 Function generator using an analog switch with operational amplifier

has either the voltage U_i or $-U_i$. The slope of $U_{triang.}$ is therefore given by

$$\frac{\Delta U_{triang.}}{\Delta t} = \mp \frac{U_i}{RC}.$$

According to Section 7.6.1, the inverting Schmitt trigger changes state when the triangular voltage reaches the value $\pm U_{max} R_1/(R_1 + R_2)$, giving the oscillation frequency

$$f = \frac{R_1 + R_2}{4R_1 RC} \cdot \frac{U_i}{U_{max}}.$$

This is proportional to the input voltage, U_i, so that the circuit is suitable for voltage-to-frequency conversion. If

$$U_i = U_{i0} + \Delta U_i,$$

a linear frequency modulation is obtained.

If high stability of amplitude and frequency is required, one of the precision circuits in Section 7.6.3 can be used instead of the Schmitt trigger OA 3.

The analog switch of Fig. 8.27 can be used only for relatively low frequencies. For frequencies above 10 kHz, it is better to use a transistor switch as in Fig. 8.28. Depending on the state of the Schmitt trigger, either

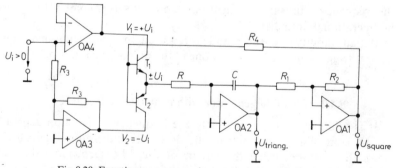

Fig. 8.28 Function generator using a transistor analog switch

the voltage $+U_i$ or $-U_i$ is applied via the transistors T_1 or T_2 to the input of the integrator. As is shown in Section 7.2.3, it is necessary that $|U_i| < U_{max}$ for the two transistors to operate as saturated complementary emitter followers. They then cause a voltage error of only a few mV.

Variable duty cycle

To generate a square-wave voltage of variable duty cycle, the triangular wave can be compared with a d.c. voltage using a comparator circuit, as is shown in Section 1.8.1. The situation becomes slightly more complicated if not only the square wave, but also the triangular wave is to be unsymmetrical, as in Fig. 8.29.

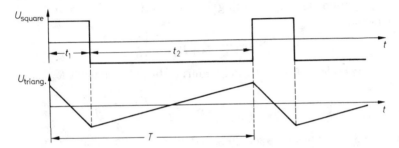

Fig. 8.29 Voltage waveshapes for a duty cycle of $t_1/T = 0.2$

The circuit in Fig. 8.28 offers a solution to the problem if the magnitude of the potential V_1 is made different from that of V_2. The rise and fall times of the triangular wave between $\pm U_{max}$ are then

$$t_1 = \frac{2RCU_{max}}{V_1}, \qquad t_2 = \frac{2RCU_{max}}{|V_2|}.$$

If the symmetry is to be changed without affecting the frequency, the magnitude of one potential must be increased and that of the other decreased so that T remains constant where

$$T = t_1 + t_2 = 2RCU_{max}\left(\frac{1}{V_1} + \frac{1}{|V_2|}\right). \tag{8.15}$$

This condition can be easily fulfilled if the circuit in Fig. 8.30 is used [8.2]. For its output potentials, we obtain the relationship

$$\frac{1}{V_1} + \frac{1}{|V_2|} = \frac{1}{U_iR_3}[R_3 + (1-\alpha)R_4 + R_3 + \alpha R_4] = \frac{1}{U_iR_3}[2R_3 + R_4].$$

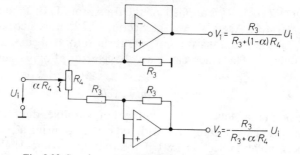

Fig. 8.30 Supplementary circuit for a variable duty cycle

As required, this expression is independent of the symmetry factor α set by the potentiometer. By inserting it into Eq. (8.15), we obtain for the frequency

$$f = \frac{R_3}{2RC[2R_3+R_4]} \cdot \frac{U_i}{U_{max}}.$$

The duty cycle t_1/T or t_2/T can be adjusted by potentiometer R_4 between the values

$$\frac{R_3}{2R_3+R_4} \quad \text{and} \quad \frac{R_3+R_4}{2R_3+R_4}.$$

If $R_4 = 3R_3$, values between 20% and 80% are attainable.

8.4.3   Function generator for higher frequencies

It is difficult to build an integrator which operates satisfactorily at frequencies above $100\,\text{kHz}$. In this frequency range, it is better to generate the triangular voltage by charging a capacitor with a constant current and, by discharging it with a current of opposite polarity. For this purpose, the current can, as in Fig. 8.31, be reversed by a Schmitt trigger-controlled diode bridge.

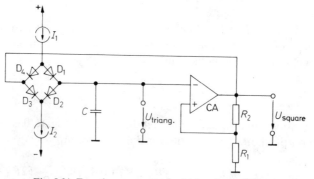

Fig. 8.31 Function generator for higher frequencies

When the triangular voltage reaches the upper trigger level the output voltage of the comparator CA drops to the value $U_{o\,min}$. Diode D_4 becomes forward biased and D_1 reverse biased. Therefore the charging current I_1 no longer flows into the capacitor C, but into the output of the comparator. Simultaneously, diode D_3 becomes reverse biased so that current I_2 flows through diode D_2 and discharges the capacitor C.

When the triangular wave reaches the lower trigger level, the output voltage of the comparator jumps to $U_{o\,max}$ whereby the diodes D_2 and D_4 are reverse biased. The current I_2 flows through D_3 into the comparator output, and I_1 charges the capacitor via diode D_1. Depending on whether the square wave voltage U_{square} is high or low, either the positive or the negative current flows into the capacitor. The current not required flows into the output of the comparator CA.

The output levels $U_{o\,max}$ and $U_{o\,min}$ need not necessarily be symmetrical about ground; TTL output levels, for instance, could also be used. However, it is important that the diode bridge is correctly biased, as is the case if the trigger levels of the Schmitt trigger are between

$$U_{o\,min}+0.6\,V \quad \text{and} \quad U_{o\,max}-0.6\,V.$$

To be able to fulfill this condition for any output level, resistor R_1 in Fig. 8.31 must be connected to the potential $\frac{1}{2}(U_{o\,max}+U_{o\,min})$ instead of to ground. However, this d.c. voltage is then superimposed on the triangular voltage.

8.5 Multivibrators

If the generation of the square-wave voltage is of predominant importance and the linearity of the triangular voltage is not so critical, the described function generators can be simplified. They are then called multivibrators. Simple circuits involving two transistors are described in the literature [1]. In this section, we deal with some arrangements which can be easily realized by integrated circuits.

8.5.1 Multivibrators for low frequencies

The multivibrator of Fig. 8.32 consists of an inverting Schmitt trigger, the feedback path of which includes a lowpass filter.

When the potential at the N-input exceeds the trigger level, the circuit changes state, and the output voltage jumps to the opposite output limit. The potential at the N-input attempts to follow the transition in the output voltage but the capacitor prevents it from changing rapidly. The potential therefore rises or falls gradually until the other trigger level is reached. The circuit then flips back to its initial state. The voltage wave

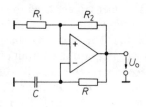

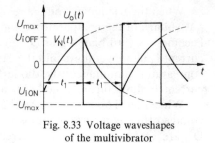

Fig. 8.32 Multivibrator using an
operational amplifier.

Fig. 8.33 Voltage waveshapes
of the multivibrator

Frequency: $f = 1/2RC \ln(1 + 2R_1/R_2)$

shapes are shown in Fig. 8.33. From Section 7.6.1, the trigger levels for
$U_{o\,max} = -U_{o\,min} = U_{max}$ are given by

$$U_{i\,ON} = -\alpha U_{max}$$

and

$$U_{i\,OFF} = \alpha U_{max},$$

where $\alpha = R_1/(R_1 + R_2)$.

The differential equation for V_N can be taken directly from the circuit
diagram as

$$\frac{dV_N}{dt} = \frac{\pm U_{max} - V_N}{RC}.$$

With the initial condition $V_N(t=0) = U_{i\,ON} = -\alpha U_{max}$, we obtain the
solution

$$V_N(t) = U_{max}[1 - (1+\alpha)\,e^{-\frac{t}{RC}}].$$

The trigger level $U_{i\,OFF} = \alpha U_{max}$ is reached after a time

$$t_1 = RC \ln \frac{1+\alpha}{1-\alpha} = RC \ln \left(1 + \frac{2R_1}{R_2}\right).$$

The period is therefore

$$T = 2t_1 = 2RC \ln \left(1 + \frac{2R_1}{R_2}\right). \tag{8.16}$$

For $R_1 = R_2$,

$$T \approx 2.2RC.$$

Multivibrator using a precision Schmitt trigger

A versatile multivibrator giving a very stable frequency can be built
using the precision Schmitt trigger of Fig. 7.36. For frequencies up to
200 kHz, the timer 555 can be used for this purpose. For higher
frequencies, the dual comparator NE 521 is more suitable and, by adding a

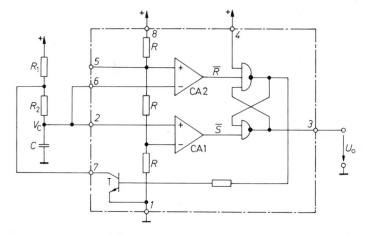

Fig. 8.34 Multivibrator using a timer circuit.

Frequency: $f = 1.44/(R_1 + 2R_2)C$

few external components can be operated in the same way, as has already been shown in Section 7.6.3.

The operation as a precision multivibrator is illustrated in Fig. 8.34 where the pin numbers refer to the package of the timer 555. The internal voltage divider R defines the trigger levels as $\frac{1}{3}V^+$ and $\frac{2}{3}V^+$, but these can be adjusted within certain limits using terminal 5. When the capacitor potential exceeds the upper threshold, $\overline{R} = L$ (low). The output voltage of the flip-flop assumes the L-state and transistor T is turned on. Capacitor C is then discharged by the resistor R_2 until the lower threshold $\frac{1}{3}V^+$ is reached, this process requiring the time

$$t_2 = R_2 C \ln 2 \approx 0.693 R_2 C.$$

On reaching the threshold, $\overline{S} = L$ and the flip-flop resumes its former state. The output voltage assumes the H (high)-state and transistor T is turned off. The charging of the capacitor takes place via the series connection of R_1 and R_2. The time interval needed to reach the upper trigger level is

$$t_1 = (R_1 + R_2) C \ln 2.$$

Hence the frequency

$$f = \frac{1}{t_1 + t_2} \approx \frac{1.44}{(R_1 + 2R_2) C}.$$

The waveshapes of the voltage U_o and V_c are shown in Fig. 8.35. The reset input 4 allows interruption of the oscillation.

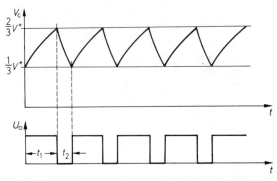

Fig. 8.35 Voltage waveshapes of the circuit in Fig. 8.34

When supplying a voltage to pin 5, the trigger thresholds can be shifted. In this way, the charging time t_1 and therefore the frequency of the multivibrator can be varied. A change in the potential $V_5 = \frac{2}{3}V^+$ by the amount ΔV_5 results in a relative frequency shift of

$$\frac{\Delta f}{f} \approx -3.3 \cdot \frac{R_1 + R_2}{R_1 + 2R_2} \cdot \frac{\Delta V_5}{V^+}.$$

As long as the voltage deviation is not too large, the frequency modulation is reasonably linear.

The timer 555 is also useful for generating single pulses (one-shot), and pulse times of a few μs to several minutes can be achieved. The corresponding external wiring is shown in Fig. 8.36.

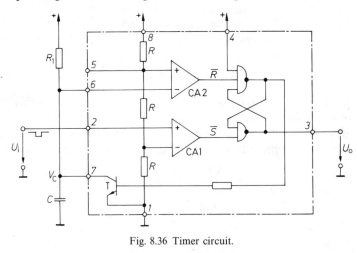

Fig. 8.36 Timer circuit.

ON-*time:* $t_1 = 1.1RC$

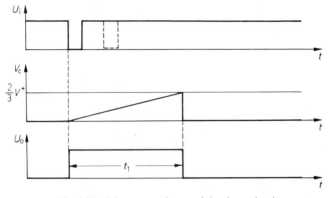

Fig. 8.37 Voltage waveshapes of the timer circuit

When the capacitor potential exceeds the upper trigger threshold, the flip-flop is reset, i.e. the output voltage resumes the L-state. Transistor T becomes conducting and discharges the capacitor. As the lower comparator is no longer connected to the capacitor, this state remains unchanged until the flip-flop is set by an L-pulse at trigger input 2. The ON-time t_1 is equal to the time required by the capacitor potential to rise from zero to the upper threshold $\frac{2}{3}V^+$. It is given by

$$t_1 = RC \ln 3 \approx 1.1\,RC.$$

If a new trigger pulse occurs during this time interval, the flip-flop remains set and the pulse is ignored. Figure 8.37 shows the voltage diagrams.

The discharging of capacitor C at the end of t_1 is not as fast as could be wished, as the collector current of the transistor is limited. The discharge time is known as *recovery time*. If a trigger pulse occurs during this interval, the ON-time is curtailed and is therefore no longer precisely defined.

Retriggerable timer

There are cases where the ON-time is not to be counted from the first pulse of a pulse train, as in the previous circuit, but from the last pulse of the train. Circuits having this characteristic are named retriggerable timers. The appropriate connection of the timer 555 is shown in Fig. 8.38, where use is made only of its function as a precision Schmitt trigger.

When the capacitor potential exceeds the upper trigger threshold, the output voltage of the flip-flop assumes the L-state. The capacitor will not discharge as the transistor T is not connected. The capacitor potential therefore rises to V^+, this being the stable state. The capacitor must be

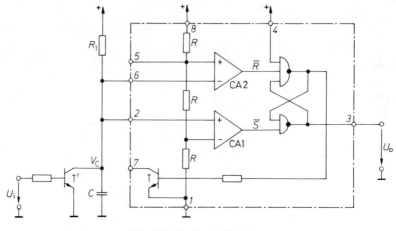

Fig. 8.38 Retriggerable timer.

ON-*time:* $t_1 = 1.1RC$

discharged by a sufficiently long positive trigger pulse applied to the base of the external transistor T'. The flip-flop is set by the lower comparator and the output voltage assumes the H-state. If a new trigger pulse occurs before the end of the ON-time, the capacitor is discharged again and the output voltage remains in the H-state. It flips back only if no new trigger pulse occurs for at least a time interval of

$$t_1 = RC \ln 3.$$

The circuit is therefore also called the "missing pulse detector". The waveshapes of voltages within the circuit are shown in Fig. 8.39 for several consecutive trigger pulses.

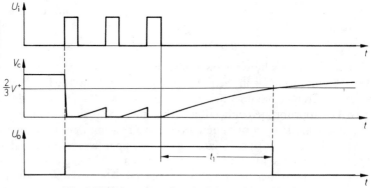

Fig. 8.39 Voltage waveshapes of the retriggerable timer

8.5.2 Multivibrators for higher frequencies

The principle illustrated by Fig. 8.32 can also be used to generate square-wave voltages of more than 200 kHz if, for the comparator, the operational amplifier is replaced by a differential amplifier, as in Fig. 8.40.

The positive feedback necessary for the operation as a Schmitt trigger is effected by connecting the output directly to the input instead of to the usual voltage divider, i.e. $R_2 = 0$. According to Eq. (8.16) this would result in an infinite oscillation period. However, when deducing this equation, we assumed that the comparator had an infinitely large gain, i.e. that it would change state whenever the differential input voltage is zero. The trigger threshold would therefore be equal to the output voltage. This value is indeed reached only after a very long time, because of the exponential charging of the capacitor C.

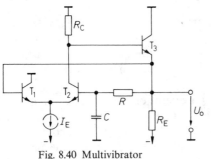

Fig. 8.40 Multivibrator
using a differential amplifier

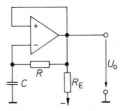

Fig. 8.41 Multivibrator
using an ECL line receiver.

Frequency: $f \approx 0.32/RC$

However, the differential amplifier in Fig. 8.40 has a relatively low gain. The circuit therefore changes state before the difference between the input voltages has reached zero. If the circuit is realized as in Fig. 8.41 by an ECL line-receiver (e.g. MC 10116), this difference is about 150 mV. For a typical change in output voltage of 850 mV, a period of

$$T \approx 3.1 RC$$

is obtained. Frequencies of over 50 MHz can be achieved.

A similar oscillator can be constructed very simply using TTL circuitry, for which Schmitt trigger gates (e.g. 7414 or 74132) are particularly suitable as they already have internal positive feedback. One such circuit is shown in Fig. 8.42. The input current of the gate must flow through resistor R, and R must therefore not be larger than $470\,\Omega$ to ensure that the lower trigger level can be reached. The minimum value of R is determined by the fan-out of the gate and is about $100\,\Omega$. The trigger

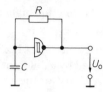

Fig. 8.42 Multivibrator with TTL Schmitt trigger.

Frequency: $f \approx 0.7/RC$

levels of the gate are 0.8 V and 1.6 V. For a typical change in output voltage
of 3 V, an oscillation frequency of

$$f \approx \frac{0.7}{RC}$$

is obtained and the maximum attainable value is about 10 MHz.

The highest frequencies can be reached using special emitter-coupled
multivibrators (e.g. 74S124 or MC 1658), the basic circuit of which is
represented in Fig. 8.43. The integrated circuits have additional TTL or
ECL output stages.

For an explanation of the circuit operation, we initially assume that
the amplitude of each alternating voltage within the circuit does not
exceed the value $U_{pp} \approx 0.5$ V. When T_1 is turned off, its collector potential
is practically the same as the supply voltage. Hence, the emitter potential
of T_2 is $V^+ - 1.2$ V, and the emitter current is $I_1 + I_2$. To attain the
desired oscillation amplitude across R_1, its resistance must be chosen
as $R_1 = 0.5$ V$/(I_1 + I_2)$. In this state, the emitter potential of T_4 is therefore $V^+ - 1.1$ V. As long as T_1 is OFF, the current of the left-hand current
source flows through capacitor C and this effects a decrease in the

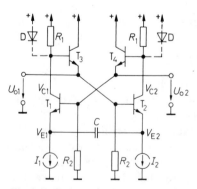

Fig. 8.43 Emitter-coupled multivibrator

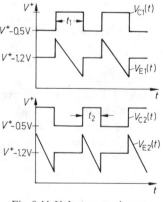

Fig. 8.44 Voltage waveshapes

emitter potential of T_1 at the rate of

$$\frac{\Delta V_{E1}}{\Delta t} = -\frac{I_1}{C}.$$

T_1 becomes conducting when the emitter potential falls below $V^+ - 1.7\,\text{V}$. The base potential of T_2 then falls by $0.5\,\text{V}$ and T_2 is turned off. Its collector potential rises to V^+. The base potential of T_1 also increases due to the emitter follower action of T_4. This makes the emitter potential of T_1 jump to the value $V^+ - 1.2\,\text{V}$ and the jump is transferred via capacitor C to the emitter of T_1, so that the potential there rises from $V^+ - 1.2\,\text{V}$ to $V^+ - 0.7\,\text{V}$.

As long as T_2 is OFF, the current I_2 flows through the capacitor and makes the emitter potential T_2 fall at the rate of

$$\frac{\Delta V_{E2}}{\Delta t} = -\frac{I_2}{C}.$$

Transistor T_2 remains turned off until its emitter potential has fallen from $V^+ - 0.7\,\text{V}$ to $V^+ - 1.7\,\text{V}$. Hence, the switching interval

$$t_2 = \frac{1\,\text{V}\cdot C}{I_2} \quad \text{or, more generally,} \quad t_2 = 2\left(1 + \frac{I_1}{I_2}\right)R_1\,C.$$

Correspondingly,

$$t_1 = \frac{1\,\text{V}\cdot C}{I_1} \quad \text{or, more generally,} \quad t_1 = 2\left(1 + \frac{I_2}{I_1}\right)R_1\,C.$$

The waveshapes of the voltages within the circuit are illustrated in Fig. 8.44. It can be seen that, with the initial assumption of $U_{pp} = 0.5\,\text{V}$, none of the transistors ever becomes saturated. Therefore, higher frequencies can be obtained with this circuit than with the multivibrators described previously, and 100 MHz can be achieved without great effort.

The circuit is particularly suited to frequency modulation. For this purpose, the current sources are designed so that $I_1 = I_2 = I$ and are controlled by a common modulating voltage. To ensure in this case that the amplitude across R_1 remains constant, a diode may be connected in parallel to each of the resistors R_1, as is shown by the broken lines in Fig. 8.43. The oscillation frequency is then

$$f = \frac{I}{4U_D C},$$

where U_D is the forward voltage of the diodes.

9 Combinatorial logic circuitry

The term "combinatorial logic network" describes an arrangement of digital circuits which contains no storage elements for the logic variables. The output variables y_j are defined by the input variables x_i alone, as illustrated by Fig. 9.1. In *sequential logic circuits* on the other hand, the output variables are also dependent on the state of the system at any time and hence on its previous history.

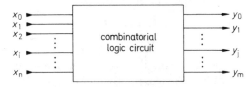

Fig. 9.1 Block diagram of a combinatorial logic network

The correlation between the output and input variables is described by truth tables or Boolean functions. For their implementation, read-only memories (ROMs) can be employed in which the truth table is stored. The input variables are then used as ROM address variables. A second possibility is the use of gates to synthesize the Boolean function.

If the truth table contains only a few cases for which the output variables assume logic "1", the realization by gates is more favourable, as only a small number of logic elements is required. Even for a large number of input variables, a single integrated circuit may be sufficient if a programmable logic array (PLA) is employed. If there are only a few cases of logic "0", the same method is used to implement the inverted function. Figure 9.2 illustrates the different methods of implementing Boolean functions.

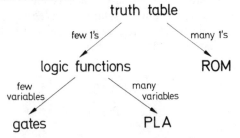

Fig. 9.2 Different methods of implementing Boolean functions

The following sections describe the operation and construction of the most common combinatorial circuits, with particular emphasis on the processing of numbers (arithmetic operations). If the numbers are to be represented by logic variables, they must be converted into *binary* information, i.e. a series of binary digits. One such binary digit is known as a *bit*. A special binary notation is the straight binary (natural binary) notation in which the bits are assigned weightings in ascending powers of 2. The symbol 1 identifies the logic "1", the symbol 0 logic "0". Logic variables characterize the individual bits, and we shall identify them by lower-case letters; the entire number will be symbolized by a capital letter. Therefore, an *N*-bit number in straight binary code is represented as follows:

$$X_N = x_{N-1} \cdot 2^{N-1} + x_{N-2} \cdot 2^{N-2} + \cdots + x_1 \cdot 2^1 + x_0 \cdot 2^0.$$

Obviously, one must always differentiate clearly between an arithmetic operation involving numbers, and a Boolean operation of logic variables. An example should clarify the difference. The expression $1+1$ is to be calculated. If we interpret the operational sign $(+)$ as an instruction to add in the decimal system, we arrive at

$$1+1=2.$$

The addition in the straight binary system gives

$$1+1=10_2 \quad \text{(read: One-Zero)}.$$

On the other hand, if we interpret the sign $(+)$ as the Boolean sum of logic variables we obtain

$$1+1=1.$$

9.1   Coding circuits

Coding circuits convert a number from one notation to another. The most important binary notation is the straight binary code but in some cases, other codes are easier to process. We therefore deal in this section with combinatorial networks for the conversion of the straight binary code into other codes and vice versa.

9.1.1   1-out-of-*n* code

The 1-out-of-*n* code assigns to each number *J* between 0 and $(n-1)$ a logic variable y_J which assumes the value "1" only when the number *J* occurs, and otherwise is "0". Figure 9.3 shows the truth table for the 1-out-of-10 code. The variables x_0 to x_3 represent the straight binary

J	x_3	x_2	x_1	x_0	y_9	y_8	y_7	y_6	y_5	y_4	y_3	y_2	y_1	y_0
0	0	0	0	0	0	0	0	0	0	0	0	0	0	1
1	0	0	0	1	0	0	0	0	0	0	0	0	1	0
2	0	0	1	0	0	0	0	0	0	0	0	1	0	0
3	0	0	1	1	0	0	0	0	0	0	1	0	0	0
4	0	1	0	0	0	0	0	0	0	1	0	0	0	0
5	0	1	0	1	0	0	0	0	1	0	0	0	0	0
6	0	1	1	0	0	0	0	1	0	0	0	0	0	0
7	0	1	1	1	0	0	1	0	0	0	0	0	0	0
8	1	0	0	0	0	1	0	0	0	0	0	0	0	0
9	1	0	0	1	1	0	0	0	0	0	0	0	0	0

Fig. 9.3 Truth table of a 1-out-of-10 encoder

$$y_0 = \bar{x}_0\,\bar{x}_1\,\bar{x}_2\,\bar{x}_3 \qquad y_5 = x_0\,\bar{x}_1\,x_2\,\bar{x}_3$$
$$y_1 = x_0\,\bar{x}_1\,\bar{x}_2\,\bar{x}_3 \qquad y_6 = \bar{x}_0\,x_1\,x_2\,\bar{x}_3$$
$$y_2 = \bar{x}_0\,x_1\,\bar{x}_2\,\bar{x}_3 \qquad y_7 = x_0\,x_1\,x_2\,\bar{x}_3$$
$$y_3 = x_0\,x_1\,\bar{x}_2\,\bar{x}_3 \qquad y_8 = \bar{x}_0\,\bar{x}_1\,\bar{x}_2\,x_3$$
$$y_4 = \bar{x}_0\,\bar{x}_1\,x_2\,\bar{x}_3 \qquad y_9 = x_0\,\bar{x}_1\,\bar{x}_2\,x_3$$

Fig. 9.4 Boolean functions of a 1-out-of-10 encoder

coded number J. The canonical products (disjunctive normal form, standard product terms) of the coding functions can be taken directly from the truth table and are shown in Fig. 9.4. When using monolithic integrated circuits for the implementation of these equations, NAND functions are often chosen rather than AND functions so that the output variables are specified by "negative true logic" levels.

Types of IC:

1-out-of-10 code: SN 7442 (TTL, open coll.),   SN 7445 (TTL),
                   MC 14028 (CMOS);

1-out-of-16 code: SN 74159 (TTL, open coll.),   SN 74154 (TTL),
                   MC 14514 (CMOS).

Application in sequential controller circuits

The 1-out-of-n encoder is suitable for the construction of sequential controllers. For this purpose it is connected to the output of a straight binary counter, thereby ensuring that all possible input combinations come up in succession. At any instant, only one output variable is "1", and an event which is to happen at the instant t_J can be triggered off by the output variable y_J. If the same event is to occur at different instants, the appropriate output variables are combined using an OR operation.

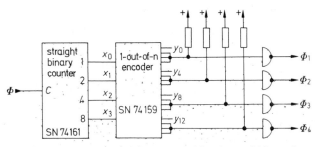

Fig. 9.5 Generation of a 4-phase clock with a 1-out-of-16 encoder

This is particularly easy for open-collector types with negative true logic outputs, as a wired-OR connection can then be used.

As an example, Fig. 9.5 shows the generation of a non-overlapping four-phase clock with defined delays between each phase. For three successive intervals of Φ, one of the four clock outputs Φ_1 to Φ_4 is "1". The terminals y_0, y_4, y_8 and y_{12} are not connected, and therefore a pause of one period of Φ is incurred between the individual clock pulses. This can be seen in the diagram of Fig. 9.6.

The length of the time intervals is determined by the input clock pulse Φ and can be set to any desired value. This is a considerable advantage over other designs involving one-shots.

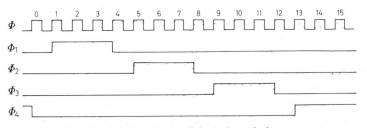

Fig. 9.6 Waveshapes of the 4-phase clock

Conversion of 1-out-of-n to straight binary code

For the conversion of the 1-out-of-n code to the straight binary code, a *priority encoder* can be used. At its outputs, a straight binary number occurs which is equivalent to the highest index number of all those inputs at which a "1" is found. The logic states of the inputs below this highest number are irrelevant, hence the name *priority encoder*. Because of this property, the circuit can be used not only for the conversion of the 1-out-of-n code but also for a code where all less significant bits are also "1". The truth table of the priority encoder is represented in Fig. 9.7.

J	x_9	x_8	x_7	x_6	x_5	x_4	x_3	x_2	x_1	y_3	y_2	y_1	y_0
0	0	0	0	0	0	0	0	0	0	0	0	0	0
1	0	0	0	0	0	0	0	0	1	0	0	0	1
2	0	0	0	0	0	0	0	1	×	0	0	1	0
3	0	0	0	0	0	0	1	×	×	0	0	1	1
4	0	0	0	0	0	1	×	×	×	0	1	0	0
5	0	0	0	0	1	×	×	×	×	0	1	0	1
6	0	0	0	1	×	×	×	×	×	0	1	1	0
7	0	0	1	×	×	×	×	×	×	0	1	1	1
8	0	1	×	×	×	×	×	×	×	1	0	0	0
9	1	×	×	×	×	×	×	×	×	1	0	0	1

Fig. 9.7 Truth table of a priority encoder. × may be either 0 or 1 (don't-care)

Types of IC:

1-out-of-10 code: 　　　　　　 SN 74147 (TTL)

1-out-of-8 code, extendable: SN 74148 (TTL),　　 MC 10165 (ECL),

　　　　　　　　　　　　　　 MC 14532 (CMOS).

9.1.2　BCD code

Since one is used to thinking in the decimal system, straight binary numbers are unsuitable for input and output operations of digital equipment. The binary coded decimal numbers (BCD numbers) have therefore been introduced. In this notation, each indivual decimal digit is represented by a binary number, for instance by the corresponding straight binary number as in the following example:

$$218_{\text{dec.}} \triangleq 0010 \ 0001 \ 1000.$$

A decimal number coded in this way is described more precisely as a 8421-coded BCD number or as a straight BCD number. The individual decimal digit can also be represented by other 4-or-more-bit combinations. As the straight BCD code is the most widely employed, it is often referred to simply as the BCD code. We shall also use this term and indicate whenever a code other than the straight BCD code is involved.

With a 4-bit binary number (tetrad), any decimal number between 0 and $15_{\text{dec.}}$ can be represented. In the BCD code, only ten of these combinations are employed, and for this reason, the BCD representation requires more bits than the straight binary equivalent.

For arithmetic operations in BCD code, results may arise containing the decimal "digits" $10_{\text{dec.}}$ to $15_{\text{dec.}}$. Such unintended digits are called *pseudo tetrads* (pseudo combinations). To correct for them, they must be reduced by $10_{dec.} = 1010_2$ and the next higher tetrad increased by 1. This correction can also be achieved by adding $6 = 0110_2$ to the pseudo tetrad, as shown by the following example:

	tens	units		tens	units
pseudo 13:	0000	1101		0000	1101
$-10_{dec.}$:	0000	1010	+6:	0000	0110
$+10_{dec.}$:	0001	0000			
correct 13:	0001	0011		0001	0011

Conversion of straight binary to BCD code

The above example illustrates the rule for converting a 4-bit straight binary number to a BCD number.

Numbers up to and including 9 remain unchanged.
Numbers over 9 must undergo the pseudo-tetrad correction.

Binary numbers of more than 4 bits are treated accordingly: beginning with the most significant bit (MSB), the straight binary number is shifted from right to left into a BCD "frame", as shown in Fig. 9.8. If

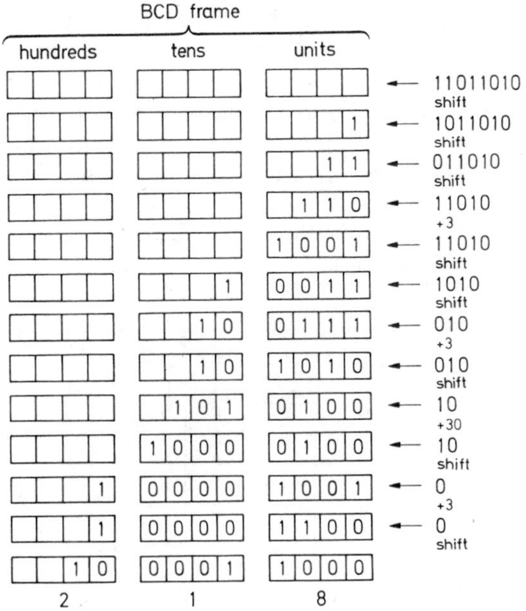

Fig. 9.8 Straight binary-to-BCD conversion; $218_{dec.}$ as an example

a 1 crosses the boundary between the units- and the tens-column, an error is incurred. This is because, for the straight binary number, the weighting changes due to the shift from 8 to 16, whereas for the BCD number it changes only from 8 to 10. After such a shift, the BCD number has become too small by 6. To allow for this, a 6 must be added whenever a 1 crosses the boundary. Similarly, a 6 must be added in the tens-column when a 1 is transferred to the hundreds-column. The resulting BCD number then has the correct value but may still contain pseudo tetrads. To avoid this, any pseudo tetrads that do occur are immediately corrected after every shifting step by adding 6 to the decade in question and carrying 1 to the next decade. Both corrections require the same arithmetic operation, that is the addition of 6.

Instead of adding 6 after the shifting, one can equally well add 3 beforehand, since it can be ascertained before the shift whether or not a correction will be necessary: if the value of a tetrad is smaller than or equal to $4 = 0100_2$, the subsequent shift will result neither in a 1 crossing the column boundary nor in a pseudo tetrad. The tetrad can therefore be shifted to the left unchanged. If the value of the tetrad before the shift is 5, 6 or 7, there again is no boundary crossing as the most significant bit is zero. However, the pseudo tetrads ten, twelve, fourteen, or eleven, thirteen, fifteen may arise depending on whether the next bit entering the frame is 0 or 1. In these cases a pseudo-tetrad correction is necessary, i.e. 3 must be added before the shift. If the value of the tetrad before the shift is 8 or 9, the boundary crossing of the 1 must be corrected, giving the correct tetrads six or seven, or eight or nine, after the shift. Because of the immediate pseudo-tetrad correction, values higher than nine cannot arise. All possibilities are now taken care of, and we obtain the correction table in Fig. 9.9.

decimal	input				output				function
I	x_3	x_2	x_1	x_0	y_3	y_2	y_1	y_0	Y
0	0	0	0	0	0	0	0	0	X
1	0	0	0	1	0	0	0	1	X
2	0	0	1	0	0	0	1	0	X
3	0	0	1	1	0	0	1	1	X
4	0	1	0	0	0	1	0	0	X
5	0	1	0	1	1	0	0	0	$X+3$
6	0	1	1	0	1	0	0	1	$X+3$
7	0	1	1	1	1	0	1	0	$X+3$
8	1	0	0	0	1	0	1	1	$X+3$
9	1	0	0	1	1	1	0	0	$X+3$

x_3 x_2 x_1 x_0

straight binary $\longrightarrow$ BCD

y_3 y_2 y_1 y_0

Fig. 9.9 Correction network for the straight binary-to-BCD conversion

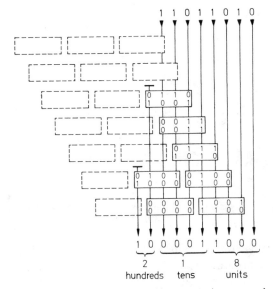

Fig. 9.10 Straight binary-to-BCD conversion using correction networks. The values entered refer to the example of $218_{dec.}$

The conversion of a straight binary to the corresponding BCD number can be implemented by left shifting the straight binary into a shift register divided into 4-bit blocks (decades). A correction circuit is connected to each decade, which alters the contents of the register according to the truth table in Fig. 9.9 before the next shifting step is carried out. Instead of a solution involving sequential logic circuits, combinatorial logic circuitry can be employed if the shift operation is replaced by a suitable connection of the individual circuit elements. This possibility is shown in Fig. 9.10. Rather than shifting the number from right to left, the BCD frame may be shifted from left to right and each tetrad corrected according to the table in Fig. 9.9. In order that the frame shift can be effected by the staggered arrangement alone, an individual correction network is necessary for each decade and each shifting "step". The total circuit may be simplified by omitting a correction network whenever fewer than 3 bits are applied at the inputs, as in this case a correction is definitely not required. Figure 9.10 shows the combinatorial logic circuit for the conversion of an 8-bit binary number, where the omitted elements are drawn with broken lines. The circuit can be extended to deal with larger numbers by expanding its characteristic pattern. The numbers in the diagram apply to the example of Fig. 9.8 and illustrate the conversion procedure.

The correction logic network is available as an integrated circuit. On a single chip three such elements are combined, as in Fig. 9.11, to

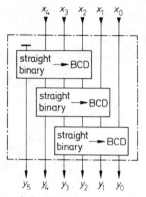

Fig. 9.11 Functional diagram of the integrated straight binary-to-BCD converter network SN 74185

make a circuit with 5 inputs and 6 outputs (SN 74185). It is in fact a
manufacturer-programmed 32-byte ROM. An 8-bit converter can be
constructed with such a circuit using only 3 ICs.

Conversion of BCD to straight binary code

In many cases the BCD code can be generated directly in a simple
way, e.g. with the help of BCD counters. As will be discussed later,
many arithmetic operations can also be performed in BCD code, but
sometimes the conversion to straight binary numbers is necessary. This
can be achieved by a repeated division by 2. To begin with, the binary
coded decimal number is divided by 2, and if it is an odd number, a
remainder of 1 is obtained, i.e. the 2^0-bit has the value 1. The quotient
is again divided by 2. If the remainder is 0, the 2^1-bit is 0, and if the
remainder is 1, the 2^1-bit has the value 1. The more significant bits of
the straight binary number are obtained accordingly.

The division of a BCD number by 2 can be carried out simply by
shifting it one bit to the right, since the individual decades are already
straight-binary coded. In each case, the remainder is the bit which has
been "pushed out" of the BCD frame. If, at shifting, a 1 crosses the
boundary between two columns, an error is incurred: when crossing
from the tens to the units, the bit weighting of the shifted 1 must be
halved from 10 to 5. However, for a straight binary number, the
assigned weighting would be 8 so that, for a correction, 3 must be
subtracted. Therefore the following correction rule applies: If the MSB
of a column (decade) after shifting is 1, a subtraction of 3 is required in
this particular decade. The truth table in Fig. 9.12 of the correction network can thus be directly determined. The conversion is finished when
the BCD number has been completely "pushed out" of the frame.

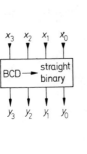

decimal	input				output				function
I	x_3	x_2	x_1	x_0	y_3	y_2	y_1	y_0	Y
0	0	0	0	0	0	0	0	0	X
1	0	0	0	1	0	0	0	1	X
2	0	0	1	0	0	0	1	0	X
3	0	0	1	1	0	0	1	1	X
4	0	1	0	0	0	1	0	0	X
5	0	1	0	1	0	1	0	1	X
6	0	1	1	0	0	1	1	0	X
7	0	1	1	1	0	1	1	1	X
8	1	0	0	0	0	1	0	1	$X-3$
9	1	0	0	1	0	1	1	0	$X-3$
10	1	0	1	0	0	1	1	1	$X-3$
11	1	0	1	1	1	0	0	0	$X-3$
12	1	1	0	0	1	0	0	1	$X-3$
13	1	1	0	1	1	0	1	0	$X-3$
14	1	1	1	0	1	0	1	1	$X-3$
15	1	1	1	1	1	1	0	0	$X-3$

Fig. 9.12 Correction network for the BCD-to-straight binary conversion

Figure 9.13 shows a combinatorial logic network for the conversion of a $2\frac{1}{2}$-digit BCD number. In analogy to Fig. 9.10, shifting of the BCD frame is achieved by a staggered arrangement of identical correction

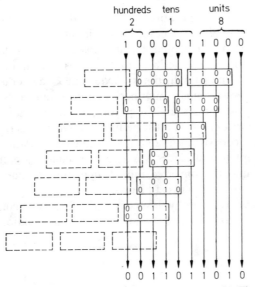

Fig. 9.13 BCD-to-straight binary conversion by correction networks. The values entered refer to the example of $218_{\text{dec.}}$.

networks. To illustrate the basic structure, all three correction networks required for each shift are shown. If the MSB is not used, no correction is necessary as can be seen in Fig. 9.12, and the corresponding correction networks can be omitted. In Fig. 9.13, they are drawn with broken lines.

Blocks such as those in Fig. 9.14, containing two correction networks, are available as manufacturer-programmed 32-byte ROMs, with 5 inputs and 5 outputs (SN 74184).

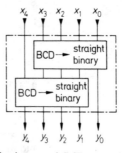

Fig. 9.14 Block diagram of the integrated BCD-to-straight binary converter network SN 74184

9.1.3   Gray code

There are many applications where a code is required in which two consecutive numbers differ in only a single bit (unit-distance code). The Gray code, or reflected binary code, possesses this characteristic in contrast to all those described so far. It is shown in Fig. 9.15 and compared with the straight binary code. The Gray code is derived by reflecting all lower binary numbers about certain axes of symmetry when adding on one bit, and by putting a 1 in front of these reflected numbers. Unwritten zeros must be filled in when reflecting.

The Gray code is unsuitable for arithmetic operations. It is therefore only employed in places where its particular property is essential, and is subsequently converted into straight binary code. The required code converters are discussed below.

As can be verified by inspecting Fig. 9.15, the conversion from straight binary to Gray code is described by the equations

$$g_1 = d_1 \oplus d_2,$$
$$g_2 = d_2 \oplus d_3$$
$$\vdots$$

or, more generally,

$$g_i = d_i \oplus d_{i+1}.$$

decimal code	straight binary code	Gray code
0	0	0
1	1	$\dfrac{1}{11}$ axis of symmetry
2	10	
3	11	$\dfrac{10}{110}$ axis of symmetry
4	100	
5	101	**111**
6	110	**101**
7	111	$\dfrac{\textbf{100}}{\textbf{1100}}$ axis of symmetry
8	1000	
9	1001	**1101**
10	1010	**1111**
11	1011	**1110**
12	1100	**1010**
13	1101	**1011**
14	1110	**1001**
15	1111	$\underline{\textbf{1000}}$ axis of symmetry
⋮	⋮	⋮
	$\ldots d_4 \quad d_3 \quad d_2 \quad d_1$	$\ldots g_4 \quad g_3 \quad g_2 \quad g_1$

Fig. 9.15 Comparison of Gray code and straight binary code

For N-bit numbers, a simplification is obtained for the highest bit g_N, as $d_{N+1}=0$. Hence

$$g_N = d_N \oplus 0 = d_N.$$

This relationship can be verified from Fig. 9.15. To convert N-bit straight binary numbers, $(N-1)$ exclusive-OR gates are required which must be connected as in Fig. 9.16.

The conversion from Gray code to straight binary code can also be carried out using exclusive-OR gates, but is rather more difficult. For N-bit numbers

$$d_N = g_N$$

and

$$d_i = g_N \oplus g_{N-1} \oplus g_{N-2} \cdots \oplus g_i \quad \text{for } i < N.$$

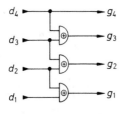

Fig. 9.16 Straight binary-to-Gray encoder

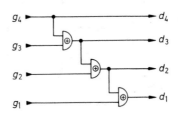

Fig. 9.17 Gray-to-straight binary decoder

The simplest way to implement these expressions is a serial exclusive-OR operation, as in the circuit of Fig. 9.17. As above, $(N-1)$ gates are required. However, the propagation delay is increased and, for the least significant bit (LSB), is $(N-1)\,t_{pd}$ where t_{pd} is the propagation delay of a single gate. For a large number of digits, it is therefore preferable to perform the operation in parallel as far as possible. This is achieved by combining pairs of inputs by means of exclusive-OR gates, the outputs of which are again collected in groups of two.

9.2   Multiplexer and demultiplexer

It is often necessary to sample the logic states of different variables one after the other and relay them to a common output. In such cases, a multiplexer as in Fig. 9.18 is employed. Depending on the state of the address inputs (data select inputs), the output y is connected to one of the data inputs x_0 to x_3. The circuit is constructed in such a way that the output is always linked to the input, the index number of which equals the binary number defined by the two data select variables a_0 and a_1. As can be seen from Fig. 9.18,

$$y = \bar{a}_1 \bar{a}_0 x_0 + \bar{a}_1 a_0 x_1 + a_1 \bar{a}_0 x_2 + a_1 a_0 x_3.$$

It follows that the Boolean product of the data select variables is 1 only for the input variable for which the index number corresponds to the chosen value. If, for instance, $a_1 = 1$ and $a_0 = 0$,

$$y = 0 \cdot 1 \cdot x_0 + 0 \cdot 0 \cdot x_1 + 1 \cdot 1 \cdot x_2 + 1 \cdot 0 \cdot x_3 = x_2.$$

Using this principle, the circuit may be enlarged to any size. With N address inputs, 2^N data inputs can be selected.

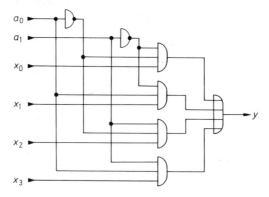

Fig. 9.18 Multiplexer having 4 inputs

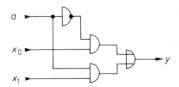

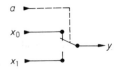

Fig. 9.19 Multiplexer
as a controlled double-throw switch

Fig. 9.20 Mechanical
double-throw switch for comparison

As only one input is assigned to each address, the multiplexer can also be used to implement any desired Boolean function of the address variables. This is done by keeping the data inputs permanently at either 0 or 1, as defined by the truth table. The circuit then operates in the same way as a ROM.

One special and important kind of multiplexer is that having two data inputs, as in Fig. 9.19. Such multiplexers require only one control (address) input, with which the output may be connected to one or the other of the two inputs. This performance as a logic double-throw switch is illustrated by the equivalent circuit in Fig. 9.20.

Single-chip multiplexers are available in the following configurations:

	TTL:	ECL:	CMOS:
16 inputs:	SN 74150		
8 inputs:	SN 74151	MC 10164	MC 14512
2×4 inputs:	SN 74153	MC 10174	MC 14539
4×2 inputs:	SN 74157		MC 14519

The logic double-throw switch of Fig. 9.19 may be simplified by replacing the OR gate by a wired-OR connection. As shown in Fig. 9.21, NAND gates with open-collector outputs are then employed instead of the AND gates, and their outputs are connected in parallel. The disadvantage of this method is that relatively large rise times are incurred since charging of the load capacitance can take place only through the common collector resistor.

Normal push-pull ("totem-pole") outputs do not have this disadvantage as their output resistances are low in both the L- and the H-

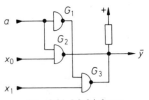

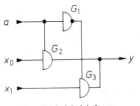

Fig. 9.21 Multiplexer
with open-collector gates

Fig. 9.22 Multiplexer
with three-state gates

state. However, paralleling of the outputs is then impossible. If multiplex operation is to be achieved by a parallel-connection, circuits with "three-state" outputs must be employed. An additional input (output-enable) then allows the totem-pole output to have a third, indifferent i.e. high-impedance state. A logic double-throw switch employing such gates is represented in Fig. 9.22.

Demultiplexer

Occasionally it is necessary to distribute input information to different addresses. Such a circuit is known as a demultiplexer and is represented in Fig. 9.23. The data input x is always connected to the particular output, the address of which is defined by the inputs a_0, a_1. The logic circuit for addressing is identical to that of the multiplexer in Fig. 9.18.

If $x = \text{const} = 1$, the demultiplexer operates as an 1-out-of-n encoder.

IC types:	TTL:	ECL:	CMOS:
16 outputs:	SN 74154		MC 14514
10 outputs:	SN 7442		
8 outputs:	SN 74S138	MC 10162	
2 × 4 outputs:	SN 74155	MC 10172	MC 14555

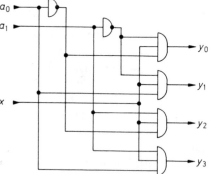

Fig. 9.23 Demultiplexer with 4 outputs

9.3 Unclocked shift register

For many arithmetic operations, a bit sequence must be shifted by one or more binary digits. This operation is usually carried out by means of series-connected D flip-flops which are controlled by a common clock pulse. As will be shown in Section 10.4, a single clock pulse results in a shift by one bit. There is a disadvantage, however, in that a sequential controller is necessary to organize loading of the bit sequence into the shift register and the subsequent shifting by a given number of binary digits.

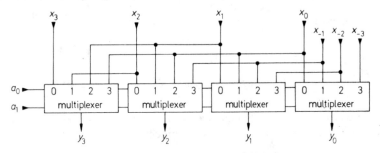

Fig. 9.24 Unclocked shift register made up of multiplexers

The same operation may be carried out without the need for clocked sequential control by employing instead a combinatorial network involving multiplexers, as is illustrated in Fig. 9.24. If the address $A = 0$ is applied, then $y_3 = x_3$, $y_2 = x_2$ etc., but if $A = 1$, $y_3 = x_2$, $y_2 = x_1$, $y_1 = x_0$ and $y_0 = x_{-1}$, due to the special arrangement of the multiplexers. The bit sequence X therefore appears at the output left-shifted by one digit. As with a normal shift register, the MSB is lost. If multiplexers with N inputs are used, a shift of $0, 1, 2 \ldots (N-1)$ bits can be executed. For the example in Fig. 9.24, $N = 4$; the corresponding function table is shown in Fig. 9.25.

a_1	a_0	y_3	y_2	y_1	y_0
0	0	x_3	x_2	x_1	x_0
0	1	x_2	x_1	x_0	x_{-1}
1	0	x_1	x_0	x_{-1}	x_{-2}
1	1	x_0	x_{-1}	x_{-2}	x_{-3}

Fig 9.25 Function table of the unclocked shift register

If the loss of MSBs is to be avoided, the register may be extended by adding identical elements, as is illustrated in Fig. 9.26. For the chosen example, where $N = 4$, a 5-bit number X can be shifted in this way by a maximum of 3 bits without loss of information. The shifted number then appears at the outputs y_3 to y_7.

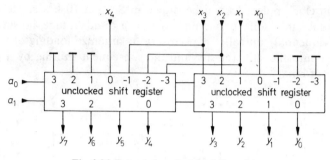

Fig. 9.26 Extended unclocked shift register

The circuit of Fig. 9.24 can also be operated as a ring shift register if the extension inputs x_{-1} to x_{-3} are connected to the inputs x_1 to x_3, as in Fig. 9.27.

IC types:
Am 25 S 10 (TTL, three-state) from Advanced Micro Devices; 8243 (TTL) from Signetics.

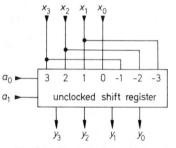

Fig. 9.27 Unclocked ring shift register

9.4   Digital comparators

Comparators check two numbers, A and B, against one another, the relations of interest being $A = B$, $A > B$ and $A < B$. Initially we deal with comparators which determine whether two numbers are equal (identity comparators). The criterion for this is that all corresponding bits of the two numbers are identical. The comparator is to give at its output a logic 1 if the numbers are equal, otherwise a logic 0. In the simplest case, the two numbers consist of only one bit each; to compare them, the EQUALITY operation (COINCIDENCE, exclusive-NOR) may be used. Two N-bit numbers are compared bit by bit with an EQUALITY circuit for each binary digit and the outputs are combined by an AND gate, as is illustrated in Fig. 9.28.

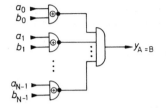

Fig. 9.28 Identity comparator for two N-bit numbers

IC type with 2×6 inputs: DM 8131 (TTL) from National Sem., with 2×8 inputs: Am 25LS2521 (TTL) from AMD.

Comparators have a wider range of application if, in addition to indicating equality, they can also determine which of two numbers is larger. Such circuits are known as magnitude comparators. To enable a comparison of the magnitude of two numbers, their code must be known, and for the following, we assume that both numbers are straight binary coded, i.e. that

$$A = a_N \cdot 2^N + a_{N-1} \cdot 2^{N-1} + \cdots + a_1 \cdot 2^1 + a_0 \cdot 2^0.$$

The simplest case is again that of comparing two single-bit numbers. The setting-up of the logic functions is based on the truth table of Fig. 9.29. From these, we can directly obtain the circuit in Fig. 9.30.

The following algorithm is used for the comparison of numbers consisting of more than one bit: To begin with, the most significant bit (MSB) of A is compared with the MSB of B. If they are different, these bits are sufficient to determine the result. If they are equal, the next to most significant bit must compared, etc. For each bit therefore, a circuit as in Fig. 9.30 can be used, and the result of the comparison of the highest non-equal pair of bits is transferred to the output via a multiplexer. Such circuits are available as monolithic ICs for the comparison of 4- and 5-bit numbers: SN 7485 (TTL), MC 10166 (ECL), MC 14585 (CMOS). The circuits can be cascaded in series or in parallel, Fig. 9.31 showing the serial method. When the 3 most signifi-

a	b	$y_{a>b}$	$y_{a=b}$	$y_{a<b}$
0	0	0	1	0
0	1	0	0	1
1	0	1	0	0
1	1	0	1	0

Fig. 9.29 Truth table of a 1-bit magnitude comparator

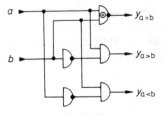

Fig. 9.30
1-bit magnitude comparator

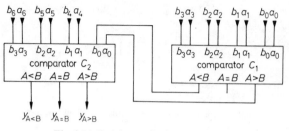

Fig. 9.31 Serial magnitude comparator

cant bits are the same, the outputs of comparator C_1 determine the result, as they are connected to the LSB inputs of comparator C_2.

When comparing many-bit numbers it is better to employ parallel cascading, as in Fig. 9.32, since the propagation delay time is shorter.

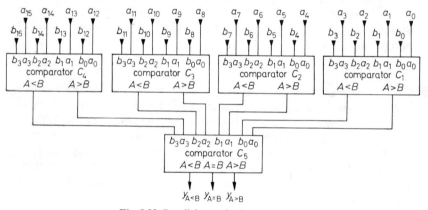

Fig. 9.32 Parallel magnitude comparator

9.5   Adders

9.5.1   Half-adder

Adders are circuits which give the sum of two binary numbers. To begin with, adders for straight binary numbers are described. The simplest case is the addition of two single-bit numbers. To devise the logic circuit, all possible cases must first be investigated so that a logic function table can be compiled. The following cases can occur if two single-bit numbers A and B are to be added:

$$0+0=0,$$
$$0+1=1,$$
$$1+0=1,$$
$$1+1=10.$$

a_0	b_0	s_0	c_0
0	0	0	0
0	1	1	0
1	0	1	0
1	1	0	1

Fig. 9.33 Truth table of the half-adder

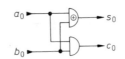

Fig. 9.34 Circuit of the half-adder

If both A and B are 1, a carry to the next higher bit is obtained. The adder must therefore have two outputs, one for the sum and one for the carry to the next higher bit. The truth table in Fig. 9.33 can be deduced by expressing the numbers A and B by the logic variables, a_0 and b_0. The carry is represented by the variable c_0, and the sum by the variable s_0.

By setting up the canonical products, the Boolean functions

$$c_0 = a_0 b_0$$

and

$$s_0 = \bar{a}_0 b_0 + a_0 \bar{b}_0 = a_0 \oplus b_0$$

are obtained. The carry thus represents an AND operation, the sum an exclusive-OR operation. A circuit implementing both operations is known as a half-adder and is shown in Fig. 9.34.

9.5.2 Full-adder

If two straight binary numbers of more than one bit are to be added, the half-adder can only be used for the LSB. For all other binary digits, not two but three bits must be added as the carry from the next lower binary digit must be included. In general, each bit requires a logic circuit with three inputs, a_i, b_i and c_{i-1}, and two

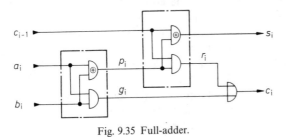

Fig. 9.35 Full-adder.

Sum: $\quad s_i = a_i \oplus b_i \oplus c_{i-1}$

Carry: $\quad c_i = a_i b_i + a_i c_{i-1} + b_i c_{i-1}$

input			internal			output	
a_i	b_i	c_{i-1}	p_i	g_i	r_i	s_i	c_i
0	0	0	0	0	0	0	0
0	1	0	1	0	0	1	0
1	0	0	1	0	0	1	0
1	1	0	0	1	0	0	1
0	0	1	0	0	0	1	0
0	1	1	1	0	1	0	1
1	0	1	1	0	1	0	1
1	1	1	0	1	0	1	1

Fig. 9.36 Truth table of the full-adder

outputs, s_i and c_i. Such circuits are called full-adders and can be realized as in Fig. 9.35, using two half-adders. Their truth table is given in Fig. 9.36.

For each bit, a full-adder is required, but for the LSB a half-adder is sufficient. Figure 9.37 shows a circuit suitable for the addition of two 4-bit numbers, A and B. Such circuits are available as ICs, but usually a full-adder is used for the LSB also to enable extension of the circuit (SN 74LS83).

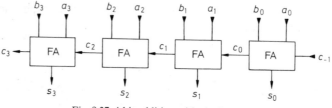

Fig. 9.37 4-bit addition with ripple carry

9.5.3 Look-ahead carry logic

The computing time of the adder in Fig. 9.37 is considerably longer than that of the individual stages because the carry c_3 can assume its correct value only after c_2 has been determined. The same goes for all the previous carries (ripple carry). To shorten the computing time for the addition of many-bit straight binary numbers, a look-ahead carry generator (simultaneous carry logic) can be used. In this method, all carries are determined directly and simultaneously from the input variables. From the truth table in Fig. 9.36, the general relationship for the carry of the stage i can be deduced:

$$c_i = \underbrace{a_i b_i}_{g_i} + \underbrace{(a_i \oplus b_i)}_{p_i} c_{i-1}. \tag{9.1}$$

The quantities g_i and p_i are introduced for brevity and appear as intermediate variables in the full-adder of Fig. 9.35. Their calculation therefore requires no extra effort. These variables can be interpreted as follows: the quantity g_i indicates whether or not the input combination a_i, b_i of the stage i gives rise to a carry, and is therefore called the generate variable. The quantity p_i indicates whether the input combination causes a carry of the previous stage to be absorbed or handed on. It is therefore called the propagate variable.

From Eq. (9.1), we obtain successively the individual carries

$$c_0 = g_0 + p_0 c_{-1},$$
$$c_1 = g_1 + p_1 c_0 = g_1 + p_1 g_0 + p_1 p_0 c_{-1},$$
$$c_2 = g_2 + p_2 c_1 = g_2 + p_2 g_1 + p_2 p_1 g_0 + p_2 p_1 p_0 c_{-1}, \quad (9.2)$$
$$c_3 = g_3 + p_3 c_2 = g_3 + p_3 g_2 + p_3 p_2 g_1 + p_3 p_2 p_1 g_0 + p_3 p_2 p_1 p_0 c_{-1}$$
$$\vdots \qquad \vdots$$

It can be seen that the expressions become ever more complicated but that they can be computed from the auxiliary variables within the propagation delay time of two gates.

Figure 9.38 shows the block diagram of a 4-bit adder with look-ahead carry logic. The Eqs. (9.2) are implemented in the carry generator. The complete circuit is available on a single chip.

IC types:

SN 74181 (TTL), MC 10181 (ECL), MC 14581 (CMOS).

Adder networks for more than 4 bits can be realized by cascading several 4-bit blocks. The carry c_3 would then be applied as c_{-1} to the next block up. This method however, is somewhat inconsistent because the carry is parallel-processed within the blocks but serially between the blocks.

To obtain short computation times, the carries from block to block must therefore also be parallel-processed. The relationship for c_3 in

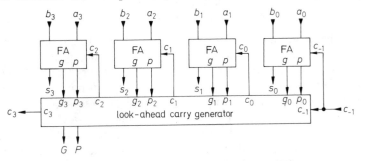

Fig. 9.38 4-bit addition with look-ahead carry logic

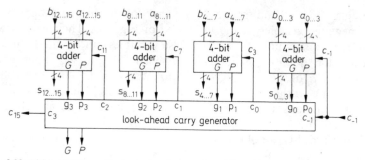

Fig. 9.39 16-bit addition with look-ahead carry from bit to bit and from block to block

Eq. (9.2) is thus reconsidered:

$$c_3 = \underbrace{g_3 + p_3 g_2 + p_3 p_2 g_1 + p_3 p_2 g_0}_{G} + \underbrace{p_3 p_2 p_1 p_0}_{P} c_{-1}. \qquad (9.3)$$

To abbreviate, the block-generate variable G and the block-propagate variable P are introduced, and

$$c_3 = G + P c_{-1}$$

is obtained. The form of this equation is the same as that of Eq. (9.1). Within the individual 4-bit adder blocks, only the additional auxiliary variables G and P need be computed; when these are known, the algorithm given in Eq. (9.2) and used for the bit-to-bit carries, can also be used for the carries from block to block. The result is the block diagram given in Fig. 9.39 for a 16-bit adder with look-ahead carry logic. The carry generator is identical to that of the 4-bit adder. It can be obtained as a separate IC having the type number SN 74182 (TTL), MC 10179 (ECL) or MC 14582 (CMOS). When performing a 16-bit addition with TTL circuits, the computation time is 36 ns; for Schottky-TTL circuits it is reduced to 19 ns.

9.5.4   Addition of BCD numbers

For the addition of two BCD numbers, a 4-bit straight binary adder can be used for each decade. After the addition, a correction is necessary as has already been discussed for the case of straight binary-to-BCD conversion: if a carry arises in a decade, 6 must be added to allow for the difference in the bit weighting. The result obtained is a BCD number already having the correct value, although it may still contain pseudo tetrads. It must therefore be checked whether a number larger than 9 has occurred in any decade. If this is the case, 6 must again be added to eliminate the pseudo tetrad. The carry so obtained is

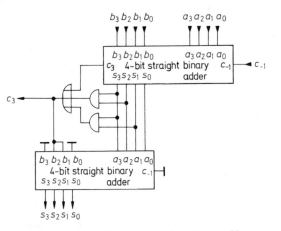

Fig. 9.40 BCD addition with straight-binary adders

transferred to the next higher decade in the same way as the ordinary carry. The described operations are effected most simply by using a second adder for each decade, as shown in Fig. 9.40. The entire arrangement is available as a monolithic integrated circuit.

IC types:
82 S 82 and 82 S 83 (TTL) from Signetics.

9.5.5 Subtraction

The subtraction of two numbers can be reduced to an addition, since

$$D = A - B = A + (C - B) - C. \tag{9.4}$$

In this equation, C must be chosen such that the operation $(C-B)$ (complementation) and the subtraction of C can be performed without the help of a subtractor. For N-bit straight binary numbers, A_N and B_N, this is possible both for $C = 2^N$ and for $C = 2^N - 1$. The expression $C - B_N$ is called the 2's-complement $B_N^{(2)}$ of the subtrahend for $C = 2^N$, or the 1's-complement $B_N^{(1)}$ for $C = 2^N - 1$. Therefore,

$$B_N^{(2)} = 2^N - B_N$$

and

$$B_N^{(1)} = (2^N - 1) - B_N. \tag{9.5}$$

Hence,

$$B_N^{(2)} = B_N^{(1)} + 1. \tag{9.6}$$

A 1's-complement is obtained simply by inverting each bit of B_N. To demonstrate this, we assume that the largest straight binary number which can be represented by N bits, is

$$1111\ldots \cong 2^N - 1.$$

If, for the 1's-complement representation, a straight binary number B_N is subtracted from this number, it can be seen that the result is the same as that obtained by inverting each individual bit. The representation of the 2's-complement is more difficult as 1 must be added to the inverted number, according to Eq. (9.5).

Consider therefore the case of the 1's-complement notation. For $C = 2^N - 1$, it follows from Eq. (9.4):

$$A_N - B_N = A_N + (2^N - 1 - B_N) - (2^N - 1) = A_N + B_N^{(1)} - 2^N + 1. \quad (9.7)$$

The subtraction can thus be carried out by adding the inversion of the number B_N, adding 1 and subtracting 2^N. The subtraction of 2^N is effected simply by inverting the most significant carry bit. For the addition of 1, the free carry input c_{-1} is permanently connected to "1", so that no additional adder is needed. This results in the circuit of Fig. 9.41.

Consider now the case of the 2's-complement. It follows from Eq. (9.4) that

$$A_N - B_N = A_N + (2^N - B_N) - 2^N = A_N + B_N^{(2)} - 2^N. \quad (9.8)$$

If the subtrahend B_N is available in 2's-complement notation, it can be added with a normal adder, the (most significant) carry bit of which is inverted. If B_N is available as a normal positive number, however, the 2's-complement must be computed from the 1's-complement with Eq. (9.6). Hence, from Eq. (9.8)

$$A_N - B_N = A_N + \underbrace{B_N^{(1)} + 1}_{B_N^{(2)}} - 2^N,$$

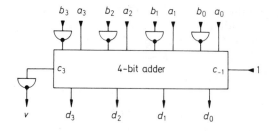

Fig. 9.41　Subtraction of two 4-bit numbers.

$$A - B = \begin{cases} D & \text{for } D \geq 0 \\ D^{(2)} & \text{for } D < 0 \end{cases} \qquad v = \begin{cases} 0 & \text{for } D \geq 0 \\ 1 & \text{for } D < 0 \end{cases}$$

straight binary	decimal		straight binary	decimal
1010	10		1000	8
-1000	-8		-1010	-10
[0] 0010	2		[1] $\underbrace{1110}_{Z_N}$	$-16+14=-2$

Fig. 9.42 Example of the representation of the difference D, for a positive and a negative result

which is the same as Eq. (9.7); the same circuit as in Fig. 9.41 is thus obtained. The only difference between the two methods is the time of the add-1 operation: for the 1's-complement, it is carried out *after* the addition of A_N and B_N, for the 2's-complement, *before*. However, this is of no significance when using a combinatorial adder network.

The gates for the inversion of B_N are already incorporated in the arithmetic units of the 181-type mentioned in the previous section, and can be activated by additional control inputs.

It must not be forgotten that the difference D_N can also be negative. In this case, a 1 appears at the carry bit output v, and this can be interpreted as a negative carry of -2^N. The example of Fig. 9.42 illustrates this, i.e.

$$D_N = -2^N + Z_N,$$

and

$$|D_N| = 2^N - Z_N.$$

The number Z_N appearing at the output thus represents the 2's-complement of the difference. The notation is well suited to further processing, but for a display, the representation of sign and magnitude is preferable. Such a representation is obtained by evaluating the 2's-complement of D_N whenever a 1 occurs at the carry output v. For this purpose, N controllable inverters are required which can be implemented by exclusive-OR gates. However, as Fig. 9.43 shows, a second adder network is needed for the required addition of 1.

This second adder can be omitted if the 1's-complement representation is used. To prove this, we refer back to Eq. (9.4) and determine the magnitude of the negative difference D in its general form:

$$|D| = -D = [C - D] - C \quad \text{for } D < 0.$$

Again with Eq. (9.4),

$$|D| = C - [A + (C - B) - \mathcal{C}] - \mathcal{C},$$
$$|D| = C - [A + (C - B)]. \tag{9.9}$$

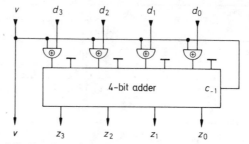

Fig. 9.43 Sign-dependent determination of the 2's-complement.

$$Z = \begin{cases} D & \text{for } v=0 \text{ (positive)}, \\ 2^N - D = D^{(2)} & \text{for } v=1 \text{ (negative)} \end{cases}$$

The exclusive-OR operation and the addition of 1 can both be carried out by the arithmetic unit type SN 74181

Unlike in the previous case of a positive difference, the correction terms $C = 2^N$ or $2^N - 1$ cancel each other. The repeated addition of 1 is thus avoided if the calculation is carried out with the 1's-complement. Hence,

$$|D| = [A + B^{(1)}]^{(1)}.$$

The result can be given automatically in sign-and-magnitude representation, if the numbers A and $B^{(1)}$ are applied to the adder network and the resulting non-inverted carry bit c_3 is examined. If it is 1, the difference is positive and 1 must be added by making $c_{-1} = 1$, as shown above. This can be realized quite simply by applying the most significant carry bit c_3 to the c_{-1} input, this method being known as end-around carry.

If the most significant carry bit is 0, the difference is negative, and the addition of 1 is automatically skipped. The result need only be inverted to obtain the magnitude representation.

The special case of $A = B$ must not be forgotten. All inputs of the adder then have the bit combination 01, i.e. the block-propagate variable P is 1. Thus $c_3 = c_{-1}$, and a positive feedback arises within the combinatorial network. The state of c_3 therefore no longer depends on the input variables, but is left to chance.

This difficulty can be avoided by employing an adder network with look-ahead carry logic. The end-around carry can then be connected to the block-generate output G instead of to the carry output c_3, as is illustrated in Fig. 9.44. As shown in Section 9.5.3 by Eq. (9.3), the state of G is the same as that of c_3 as long as the c_{-1}-term is ignored. This property prevents positive feedback. The output G also suffices for the determination of the sign as this cannot be changed by the addition of 1.

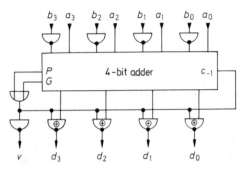

Fig. 9.44 Subtraction of two 4-bit straight binary numbers, using the end-around carry method. The result is given in sign-and-magnitude representation.

$$D = |A - B|; \qquad v = \begin{cases} 0 & \text{for } A \geq B \text{ (positive)}, \\ 1 & \text{for } A < B \text{ (negative)} \end{cases}$$

A flaw in this method is that zero is given as a negative number, which is not wrong but unusual. This effect can be avoided by generating an additional end-around carry if $A = B$, and by preventing the complementation at the output. For this purpose, the block-propagate output P is combined with the end-around carry using an OR operation, as can be seen in Fig. 9.44.

9.5.6 Addition of numbers of either sign

When a negative difference occurs in the subtractor of Fig. 9.41, the magnitude is given in 2's-complement representation. The most significant carry bit may be directly interpreted as the sign: for $v = 0$, the result is positive, for $v = 1$ negative. This representation is of advantage for arithmetic operations involving numbers of either sign (2's-complement number representation). An 8-bit number can be within the following range:

v	2^6	2^5	2^4	2^3	2^2	2^1	2^0	
b_7	b_6	b_5	b_4	b_3	b_2	b_1	b_0	
0	1	1	1	1	1	1	1	$+127$
0	0	0	0	0	0	0	1	$+\ 1$
0	0	0	0	0	0	0	0	0
1	1	1	1	1	1	1	1	$-\ 1$
1	0	0	0	0	0	0	0	-128

B_7

This representation may also be interpreted by reading the sign bit v as a normally weighted straight binary digit having, however, a negative sign. In the case of an 8-bit positive number B_8, $v=b_7=0$, and therefore

$$B_8 = -b_7 2^7 + B_7 = B_7 > 0.$$

In the case of a negative 8-bit number B_8, $v=b_7=1$, and therefore

$$B_8 = -b_7 \cdot 2^7 + B_7 = -2^7 + B_7 < 0.$$

Hence, the magnitude of B_8,

$$|B_8| = -B_8 = 2^7 - B_7 = B_7^{(2)} > 0;$$

i.e. the magnitude of B_8 is the 2's-complement of B_7.

We shall now examine how an adder network must operate so that positive and negative numbers in 2's-complement representation are correctly processed. Following the above example, we consider the processing of two numbers A and B, each having a word length of 7 bits plus a sign bit. The adder network calculates the sum

$$S_7 = A_7 + B_7.$$

In the case of $A>0$ and $B>0$, the numbers are correctly processed. However, a restriction must be imposed which is not required for an adder of positive numbers only: since the eighth bit now represents the sign, it can no longer be used as the carry output. The sum must therefore not exceed the 7-bit range, i.e. must not be larger than $2^7 - 1 = 127$, as it might otherwise be wrongly interpreted as a negative number.

The case where one of the two numbers A_7, B_7 is given in 2's-complement representation, is discussed in the previous section since there, subtraction is treated as an addition in 2's-complement notation. In order that the difference has the correct sign, the carry output must be inverted if either A or B is negative.

In the third case, A and B are both negative. The sum of the 2's-complements is given as

$$S_7 = A_7^{(2)} + B_7^{(2)} = 2^7 - A_7 + 2^7 - B_7 = 2^8 - A_7 - B_7.$$

It can be seen that owing to the addition of the two complement terms 2^7, the carry bit (eighth bit b_7) is not changed; the carry output therefore directly indicates the sign. The function table in Fig. 9.45 summarizes the determination of the sign v_S from the highest carry c_6, for all possible cases. The Boolean function

$$v_S = v_A \oplus v_B \oplus c_6 \tag{9.10}$$

is derived and implemented in the circuit of Fig. 9.46.

v_B	v_A	v_S
0	0	c_6
0	1	$\bar{c}_6$
1	0	$\bar{c}_6$
1	1	c_6

Fig. 9.45 Function table for the determination of the sign bit

As a comparison with Fig. 9.35 shows, the expression of Eq. (9.10) is identical with the sum term of a full-adder. The entire 8-bit number S_8 including the sign can therefore be calculated using an 8-bit adder as in Fig. 9.47, so that special treatment of the sign bits is avoided.

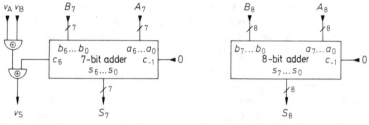

Fig. 9.46 Addition of two straight binary numbers of any sign if negative numbers are represented as 2's-complements

Fig. 9.47 Inclusion of the sign logic in the arithmetic unit

Overflow monitor

When adding two numbers having the same sign, an erroneous result may arise if the available range of the adder is exceeded. This situation is easily recognized: when A and B are positive, the sum must also be positive. If a negative sign v_S occurs, there must be a positive overflow (OV^+).

Hence,
$$OV^+ = \bar{v}_A \, \bar{v}_B \, v_S.$$

If both numbers are negative, the sum must also be negative. The occurrence of a positive sign is, in this case, the criterion for a negative overflow (OV^-).

Hence,
$$OV^- = v_A \, v_B \, \bar{v}_S.$$

For any overflow, we therefore obtain
$$OV = OV^+ + OV^- = \bar{v}_A \, \bar{v}_B \, v_S + v_A \, v_B \, \bar{v}_S. \tag{9.11}$$

This variable can be generated separately to indicate an invalid result. The 4-bit arithmetic unit Am 25 LS 2517 from Advanced Micro Devices has an output for this variable.

9.6 Multiplier

Multiplication of two straight binary numbers is best illustrated by an example. The product $13 \cdot 11 = 143$ is to be calculated:

$$
\begin{array}{r}
1101 \ \cdot\ 1011 \\
\hline
1101 \\
+ \quad 1101 \\
+ \quad 0000 \\
+ \ 1101 \\
\hline
10001111
\end{array}
$$

The calculation is particularly easy because only multiplications by either 1 or 0 occur. The product is obtained by consecutive shifting of the multiplicand to the left by 1 bit at a time, and by adding or not adding, depending on whether the corresponding multiplier bit is 1 or 0. The individual bits are processed in order, and this method is therefore known as serial multiplication.

Multiplication can be implemented by combining a shift register and an adder, although such a circuit would require a sequential controller. As is described for the straight binary-to-BCD conversion, the shifting process can also be carried out by a combinatorial network if N adders are suitably staggered and interconnected. This method requires a large number of adders, but the shift register and the sequential controller are no longer needed. The main advantage, however, is the considerably reduced computation time since, instead of the time-consuming clock control, only gate propagation delays are incurred.

Figure 9.48 shows a suitable circuit for a combinatorial 4×4-bit multiplier. For the additions, we employ the unit SN74181 previously mentioned, the mode of operation of which can be influenced by several controlling variables. The multiplicand X is applied in parallel to the four adder inputs, b_0 to b_3, of all arithmetic units. Each bit of the multiplier Y is entered at an appropriate control input m, producing the following effect

$$
S = \begin{cases} A + 0 & \text{for } m = 0 \\ A + B & \text{for } m = 1. \end{cases}
$$

To begin with, we assume that the number $K = 0$. We then obtain the expression

$$
S_0 = X \cdot y_0
$$

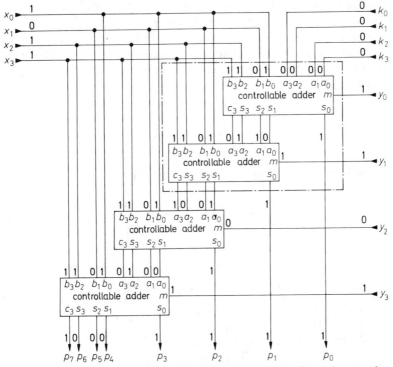

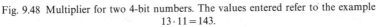

Fig. 9.48 Multiplier for two 4-bit numbers. The values entered refer to the example
$13 \cdot 11 = 143$.

Result: $P = X \cdot Y + K$

at the output of the first element, corresponding to the first partial
product in the above multiplication algorithm. The LSB of S_0 represents the LSB of the product P; it is transferred directly to the output.
The more significant bits of S_0 are added in the second element to the
expression $X \cdot y_1$. The resulting sum is the subtotal of the partial
products in the first and second line of the multiplication algorithm.
The LSB of this sum gives the second lowest bit of P and is therefore
transferred to the output p_1. The subsequent subtotals are treated
accordingly. We have entered in Fig. 9.48 the numbers of the above
example to demonstrate this process.

The additional inputs k_0 to k_3 can be used to add a 4-bit number K
to the total product P, so that the multiplier function becomes

$$P = X \cdot Y + K.$$

The method of expansion for larger numbers can now be understood.
For each additional bit of the multiplier Y, a further arithmetic unit is

added at the bottom left of the circuit in Fig. 9.48. If the multiplicand X is to be increased, the word length is enlarged by stringing an appropriate number of arithmetic units at each stage.

Blocks containing two controllable adder networks, as marked in Fig. 9.48 by a chain-dotted line, are available as 4×2 bit multiplier elements.

IC types:

Am 25 S 05 (TTL) from Advanced Micro Devices,
    93 S 43 (TTL) from Fairchild.

Comparison of computing times when employing these ICs:

configuration	number of ICs	computing time
4×4 bit	2	35 ns
8×8 bit	8	75 ns
16×16 bit	32	155 ns

More complex monolithic integrated multipliers:

8×8 bit:	MM 67558	(TTL) from MMI,	80 ns;
8×8 bit:	MPY- 8	(TTL) from TRW,	70 ns;
12×12 bit:	MPY-12	(TTL) from TRW,	80 ns;
16×16 bit:	MPY-16	(TTL) from TRW,	100 ns.

All the IC types mentioned also permit the multiplication of negative numbers in 2's-complement notation.

In the multiplication method described, the new partial product is always added to the previous subtotal. This technique requires the fewest elements and results in straightforward and easily extensible circuitry. The computing time can be shortened if as many summations as possible are carried out simultaneously and if the individual subtotals are added afterwards by a fast adder circuit. Several procedures are available which differ only in the adding sequence [9.1].

9.7   Digital function networks

The function $Y = f(X)$ can be implemented directly in tabular form using ROMs. For a high resolution, many-bit numbers and therefore large memories are required. The latter can be considerably reduced if only part of the table is stored, and if the remaining values of the function are interpolated from it using simple arithmetic operations. Special characteristics of the particular function can then often be used to advantage.

9.7.1 Sine function

One advantage of the sine function is its periodicity, so that only the function values for $0 \le \theta \le \dfrac{\pi}{2}$ need be stored. The input is a straight binary fraction, $0 \le X \le 1$, according to

$$X = x_1 \cdot 2^{-1} + x_2 \cdot 2^{-2} + \cdots + x_N \cdot 2^{-N}$$

and

$$\theta = \frac{\pi}{2} X.$$

For an input word length of up to 9 bits (0.2%) and an output word length of up to 8 bits, the mask-programmed ROM MM 5232 (AEI mask) from National can be used. An extension to an output word length of 16 bits is possible with a second ROM MM 5232, programmed with mask AEJ.

For larger input word lengths the required memory capacity quickly rises to unrealistic proportions. For an input and output of 16 bits, it will be already 1 Mbit.

The memory required can be reduced if the input variable X is divided into a part M containing the more significant bits and a part L containing the lesser significant bits. If

$$X = M + L,$$

then

$$\sin \theta = \sin \frac{\pi}{2}(M+L) = \sin \frac{\pi}{2} M \cos \frac{\pi}{2} L + \cos \frac{\pi}{2} M \sin \frac{\pi}{2} L. \quad (9.12)$$

The part L is made small so that, for a given output accuracy,

$$\cos \frac{\pi}{2} L = 1. \quad (9.13)$$

Then

$$\sin \frac{\pi}{2} L = \frac{\pi}{2} L, \quad (9.14)$$

and Eq. (9.12) reduces to

$$\sin \theta = \sin \frac{\pi}{2} M + \underbrace{\frac{\pi}{2} L \cos \frac{\pi}{2} M}_{K}. \quad (9.15)$$

The sine and cosine functions for the calculation of this expression require only a short input word length.

We shall now illustrate the method by an example. The sine function is to be calculated with a resolution of 16 bits at the input and the output. To begin with, we determine the word length of L. The error in Eq. (9.13) is largest if L assumes its maximum value. With the condition

$$(\Delta \sin \theta)_{max} < 0.5 \cdot 2^{-16} = 2^{-17},$$

we obtain from Eq. (9.13)

$$1 - \cos \frac{\pi}{2} L_{max} < 2^{-17},$$

i.e.

$$L_{max} < 2^{-9}.$$

In order not to exceed this limit, we can use for L no more than the last 7 bits, i.e. the bit weightings $2^{-10} \ldots 2^{-16}$. Therefore, the 9 MSBs of the straight binary fraction are left for the representation of M, i.e. the bit weightings $2^{-1} \ldots 2^{-9}$. This splitting is illustrated in Fig. 9.49. For the storage of the coarse-increment values, a sine ROM having 2^9 words at 16 bits, is required.

The computation of the interpolation values K in Eq. (9.15) can be carried out using a multiplier. However, at its output, the full word length of 16 bits is not needed. Eight bits are sufficient (binary weighting $2^{-9} \ldots 2^{-16}$), as the largest possible interpolation value is

$$K_{max} = \frac{\pi}{2} L_{max} \cos \left(\frac{\pi}{2} \cdot 0 \right) \approx 3.1 \cdot 10^{-3} < 2^{-8}.$$

For the cosine ROM, a resolution of 9 bits is sufficient to keep the error in the product below 2^{-17}. The total capacity of the memories is 13 kbit i.e. only about 1 % of that of a direct ROM implementation. The

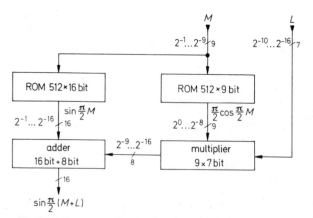

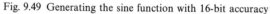

Fig. 9.49 Generating the sine function with 16-bit accuracy

9×7-bit multiplier requires the most components of the circuit, but if resolution is reduced it can be omitted by incorporating the multiplication in the cosine ROM. For a desired input resolution of 12 bits and an output resolution of 14 bits, a mask-programmed correction ROM (MM 5232, mask AEK) is available and can be used in connection with the sine ROMs mentioned, as illustrated in Fig. 9.50. Although the output resolution is still 16 bits, the accuracy is reduced to 0.7×2^{-14}, as only the 6 MSBs of M are connected to the interpolation ROM.

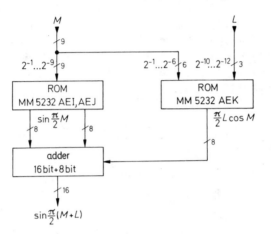

Fig. 9.50 Simplified generation of the sine function, with an input word length of 12 bits and output accuracy of 14 bits

10 Sequential logic circuitry

A sequential network is an arrangement of digital circuits, which can carry out logic operations and, in addition, store the states of individual variables. In contrast to a combinatorial logic network, the output variables y_j are not only dependent on the input variables x_i, but also on the state S_Z of the system at any time. This state is represented by the state vector $Z = (z_1, z_2, ..., z_n)$, the value of which is stored for one clock period in n flip-flops. The basic arrangement is shown in Fig. 10.1.

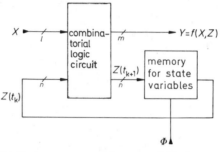

Fig. 10.1 Basic arrangement of a sequential logic circuit.

Input vector:	X	State vector:	Z
Output vector:	Y	Clock pulse:	Φ

The new state $S(t_{k+1})$ is determined by the old state $S(t_k)$ and by the input variables x_i (qualifiers). The sequence of the states can therefore be influenced by the qualifiers. The appropriate correlation is taken care of by a combinatorial circuit: if the old state vector $Z(t_k)$ is applied to its inputs, the output gives the new state vector $Z(t_{k+1})$. This state is to remain until the next clock pulse arrives. The state vector $Z(t_{k+1})$ may therefore only be transferred to the outputs of the flip-flops at the next clock pulse, and for this reason edge-triggered flip-flops must be used.

There are several important special types of sequential logic circuits. One such type is that where the state variables are identical to the output variables. A second type is that where the sequence of the states is always the same, and input variables are then not needed. Both these simplifications are found in *counters* which are therefore the simplest type of sequential circuits and easiest to understand.

The most important standard counter circuits are described in the following sections. In Section 10.7, we deal with a method of systematic design of sequential circuits, which can also be used to advantage for the development of special counters.

10.1   Straight binary counters

So far, we have used logic circuits for computing and for decoding/encoding. Another important application is the counting of pulses. A counter may be any circuit which, within certain limits, has a defined relationship between the number of input pulses and the state of the output variables. As each output variable can have only two values, there are for n outputs, 2^n possible output combinations although often only some of these are used. It is unimportant which number is assigned to which combination, but it is useful to choose a representation which can subsequently be easily processed. The simplest circuits are obtained for the straight binary notation.

Figure 10.2 shows the relationship between the number, Z, of input pulses and the values of output variables z_i, for a 4-bit straight binary counter. If this table is read from top to bottom, two patterns emerge:
(1) an output variable z_i always changes state when the next lower value z_{i-1} changes from 1 to 0.

Z	z_3	z_2	z_1	z_0
	2^3	2^2	2^1	2^0
0	0	0	0	0
1	0	0	0	1
2	0	0	1	0
3	0	0	1	1
4	0	1	0	0
5	0	1	0	1
6	0	1	1	0
7	0	1	1	1
8	1	0	0	0
9	1	0	0	1
10	1	0	1	0
11	1	0	1	1
12	1	1	0	0
13	1	1	0	1
14	1	1	1	0
15	1	1	1	1
16	0	0	0	0

Fig. 10.2 State table of a straight binary counter

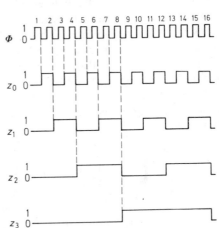

Fig. 10.3 Output states of a straight binary up-counter, as a function of time

(2) an output variable z_i always changes state when all lower variables $z_{i-1}, ..., z_0$ have the value 1 and a new pulse arrives.

These patterns can also be seen in the sequence of Fig. 10.3. Pattern (1) is the basis of an asynchronous counter (ripple counter), whereas pattern (2) yields the synchronous counter.

Occasionally, counters are required, the output state of which is reduced by 1 for each count pulse. The operational principle of such a *down-counter* can also be inferred from the table in Fig. 10.2 by reading it from the bottom up. It follows that

(1a) an output variable z_i of a down-counter changes state when- ever the next lower variable z_{i-1} changes from 0 to 1.

(2a) an output variable z_i of a down-counter always changes state when all lower variables $z_{i-1}, ..., z_0$ have the value 0 and a new clock pulse arrives.

10.1.1 Straight binary asynchronous counter

A straight binary asynchronous (ripple) counter can be realized by joining flip-flops in a chain, as in Fig. 10.4, and by connecting each clock input C to the output Q of the previous flip-flop. If the circuit is to be an up-counter, the flip-flops must change their output states when their clock inputs C change from 1 to 0. Edge-triggered flip-flops are therefore required, e.g. JK master-slave flip-flops where $J = K = 1$. The counter may be extended to any size. Using this principle, one can count up to 1023 with only 10 flip-flops.

Flip-flops triggering at the positive edge of the clock pulse can also be employed. If they are connected in the same way as in Fig. 10.4, down-counter operation is obtained. For up-counter operation, their clock pulse must be inverted. This is achieved by connecting each clock input to the $\bar{Q}$-output of the previous flip-flop.

Every counter is also a frequency divider. The frequency at the output of flip-flop F_0 is half the counter frequency. A quarter of the input frequency appears at the output of F_1, an eighth at the output of F_2 etc. This property of frequency division can be seen clearly in Fig. 10.3.

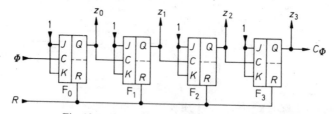

Fig. 10.4 Asynchronous straight binary counter

IC types:

4 bit: SN 7493 (TTL), MC 10178 (ECL);
8 bit: SN 74393 (TTL), MC 14520 (CMOS);
12 bit: MC 14040 (CMOS).
Divider by 2^{18} to 2^{24}: MC 14521 (CMOS).

10.1.2 Straight binary synchronous counters

It is characteristic of an asynchronous counter that the clock pulse is applied only to the input of the first flip-flop, while the remaining flip-flops are indirectly controlled. This means that the input signal of the last flip-flop arrives only after all previous stages have changed state. Each change of the output states z_0 to z_n is therefore delayed by the set-up time of a flip-flop. For long chains and high counter frequencies, this may result in z_n changing with a delay of one or more clock cycles. After the last clock pulse, it is therefore necessary to wait for the delay time of the entire counter chain before the result can be evaluated. If evaluation of the counter state is required during counting, the period of the clock pulse must not be smaller than the delay time of the counter chain.

Synchronous counters do not have these drawbacks as the clock pulses are applied simultaneously to all clock inputs C. In order that the flip-flops do not all change state at every clock pulse, the transitions are additionally controlled by the J- and K-inputs, as shown in Fig. 10.5.

According to the table in Fig. 10.2, the flip-flop F_0 must change its output state with every clock pulse, therefore $J = K = 1$. In an up-counter, flip-flop F_1 may change state only if z_0 was 1 before the arrival of the clock pulse. This can be achieved simply by connecting

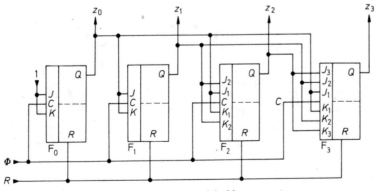

Fig. 10.5 Synchronous straight binary counter

the J- and K-inputs of F_1 to z_0. The flip-flop then retains its old state as long as $z_0 = 0$ and changes state only when $z_0 = 1$ and a new clock pulse occurs. The table in Fig. 10.2 shows further that the flip-flop F_2 may change state only when both z_0 and z_1 were 1 before the arrival of the clock pulse. To fulfill this condition, one JK pair is connected to z_0 and a second pair to z_1. Correspondingly, for flip-flop F_3, three JK pairs are used, each connected to the output of one of the previous flip-flops.

If identical JK flip-flops having three J- and three K-inputs are used, several inputs of F_0, F_1 and F_2 remain free. Since the individual J-inputs and also the K-inputs are combined internally by means of AND gates, the free inputs must be set to logic 1 to obtain the desired operation.

If the flip-flops in question have only one J- and one K-input, the necessary extension can be implemented by joining the two inputs and connecting them to the output of an AND gate.

It is obvious that extension to a very large counter will cause problems as a high number of AND inputs would then be required. For this reason, four flip-flops are usually combined to make the 4-bit counter stage shown in Fig. 10.6. The stages are joined via the carry output C_E and the enable input E by which the entire counter stage can be blocked.

The carry output should be 1 if the counter state is 1111 and all lower stages produce a carry. To achieve this, the operation

$$C_E = z_0 z_1 z_2 z_3 E$$

must be executed in each stage. In this manner, many counter stages may be cascaded without the need for additional logic circuitry, as is shown in Fig. 10.7.

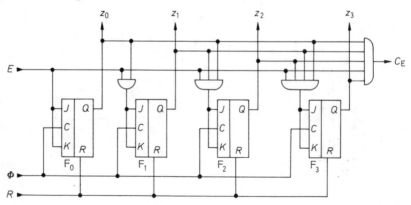

Fig. 10.6 Synchronous counter with carry logic

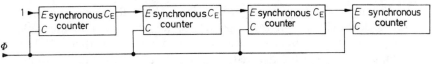

Fig. 10.7 Cascading of synchronous counter stages

IC types:

SN 74161 (TTL), MC 14161 (CMOS), asynchronous reset;

SN 74163 (TTL), MC 14163 (CMOS), synchronous reset.

Counters with up-down control

Straight binary synchronous counters can be easily switched over to down-counter operation by using a double-throw switch which connects the J- and K-inputs to the $\bar{Q}$-outputs instead of to the Q-outputs. As the switching affects the J- and K-inputs only, and not the C-inputs, the state of the counter does not change at switchover. This is a great advantage over asynchronous counters and a reason for implementing up-down counters mainly as synchronous counters. The logic double-throw switch as described in Fig. 9.19 can be employed for the switchover. The circuit in Fig. 10.8 implements these ideas. It involves trailing edge triggered flip-flops.

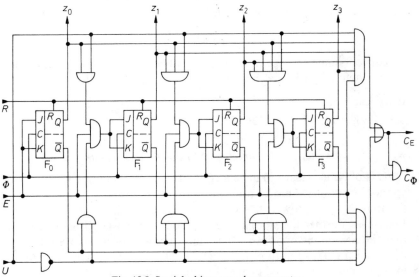

Fig. 10.8 Straight binary up-down counter.

$$\text{Counting direction:} \quad U = \begin{cases} 1 & \text{up} \\ 0 & \text{down} \end{cases}$$

As long as the up-down control variable U is 1, the lower AND gates are blocked. The circuit then operates as an up-counter in the same way as the straight binary synchronous counter in Fig. 10.6. If $U = 0$, the upper AND gates are blocked and the JK pairs are connected to the $\overline{Q}$-outputs. In this case, the circuit operates as a down-counter. As a change in the up-down variable affects the states of J and K, it may be altered only when the clock is 0.

A carry to the next higher counter stage can arise in two cases: when the counter state is 1111 and $U = 1$ (up-counter operation), and when the counter state is 0000 and $U = 0$ (down-counter operation). The following function is therefore obtained for the carry variable C_E

$$C_E = [z_0 z_1 z_2 z_3 U + \overline{z}_0 \overline{z}_1 \overline{z}_2 \overline{z}_3 \overline{U}] E.$$

This variable is connected, as in Fig. 10.7, to the enable input E of the next higher counter stage. Since the counting direction is changed simultaneously for all stages, the carry variable is always interpreted with the correct sign.

If speed is not too important, the individual counter stages can also be cascaded asynchronously by using the carry output as the clock pulse for the next higher stage. However, there is a danger that, owing to differing set-up times, C_E is briefly 1 at the wrong time. This spurious pulse would be counted by the higher stage but can be suppressed by implementing the Boolean product

$$C_\Phi = C_E \Phi.$$

Since we assume the use of trailing edge triggered flip-flops, the stable state is definitely reached by the time the clock pulse is 1.

IC types:

SN 74191 (TTL),   MC 10136 (ECL),   MC 14516 (CMOS).

Counter with up-clock and down-clock inputs

It is often necessary to have a counter with two inputs. Clock pulses applied to input Φ_U are to raise the state of the counter, those applied to input Φ_D are to lower it. A circuit having this behaviour is shown in Fig. 10.9. Consider the clock pulses applied to input Φ_U. It can be seen that the AND gates prevent a clock pulse getting through to a flip-flop unless all previous flip-flops are at $Q = 1$. This is precisely the condition for up-counter operation, deduced from the table in Fig. 10.2.

In the previous circuits, the clock pulse affected all flip-flops. Those flip-flops, the output states of which were not to change, were blocked by making the signal at the J- and K-inputs zero. Here, the clock pulses

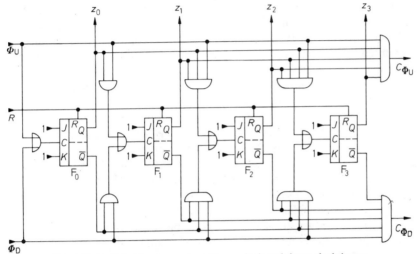

Fig. 10.9 Straight binary counter with up-clock and down-clock inputs

are prevented from reaching a particular flip-flop by blocking the AND gate in front of the clock input. For this purpose, exactly the same logic operations are required, hence the close resemblance of the control logic circuitry in Figs. 10.8 and 10.9.

Next consider the effect of pulses applied to the Φ_D input. The $\bar{Q}$-outputs determine which of the AND gates allow the clock pulse to pass. For this reason only those flip-flops receive a clock pulse for which all preceding outputs are at $Q=0$. This is the condition for a down-counter.

Those flip-flops which are to change state, receive the clock pulse almost simultaneously. The flip-flops for the more significant bits and those for the less significant bits therefore change state at the same time. The circuit thus operates as a synchronous counter. The AND gates at the output determine the carry for up-counter operation and that for down-counter operation. They can be connected to a further identical counter stage which may itself be synchronously controlled but delayed with respect to its preceding stage and therefore operating asynchronously. This operation is known as a semisynchronous operation.

IC type:

SN 74193 (TTL).

Coincidence cancellation

The interval between two count pulses and their duration must not be smaller than the set-up time t_{su} of the counter, or the second pulse would be wrongly processed. Counters with only one clock input can therefore count at a maximum possible frequency of $f_{max} = 1/2t_{su}$. For the counter in Fig. 10.9, the situation is more complicated. Even if the counter frequencies at the up-clock and at the down-clock input are considerably lower than f_{max}, the interval between an up- and a down-clock pulse may, in asynchronous systems, be smaller than t_{su}. Such close or even coinciding pulses result in a spurious counter state. This can be avoided only by preventing these pulses reaching the counter inputs. The state of the counter then remains unchanged as would also be the case after one up- and one down-clock pulse.

Such a coincidence cancellation circuit can, for example, be designed as in Fig. 10.10, where one-shots are used [10.1]. The one-shots (monostable multivibrators) M_1 and M_2 convert the counter pulses Φ_U and Φ_D into the signals x_U and x_D, each having a defined length t_1. Their trailing edges are used to trigger the two one-shots M_4 and M_5 which in turn generate the output pulses. Gate G_1 decides whether the normalized input impulses x_U and x_D overlap. If this is the case, a falling edge is found at its output which triggers one-shot M_3. Both output gates G_2 and G_3 are then blocked for a time t_2, and no pulses can appear at the output. In order that pulses are safely suppressed, the following relationship must hold:

$$t_2 > t_1 + t_3.$$

Time t_3 defines the duration of the output pulses. The interval between them is shortest just before coincidence is detected, i.e. $\Delta t = t_1 - t_3$. For a correct operation of the counter, the additional conditions

$$t_3 > t_{su} \quad \text{and} \quad t_1 - t_3 > t_{su}$$

must therefore be fulfilled. The shortest permissible ON-times of the one-shots are thus $t_3 = t_{su}$, $t_1 = 2t_{su}$ and $t_2 = 3t_{su}$. The maximum counter frequency at the two inputs of the coincidence detector is then

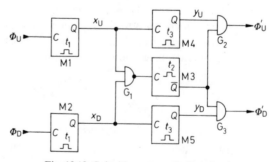

Fig. 10.10 Coincidence cancellation circuit

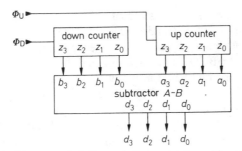

Fig. 10.11 Straight binary up-clock/down-clock counter insensitive to coinciding clock pulses

$f_{max} = 1/t_2 = 1/3t_{su}$. The coincidence cancellation circuit therefore reduces f_{max} by a factor of 1.5.

The described method for the suppression of overlapping pulses can no longer be used at very high frequencies. The up- and down-clock pulses must then be counted separately, and the desired result is obtained by computing the difference between the two counter states using of a combinatorial subtractor, as is illustrated in Fig. 10.11. Coinciding counter pulses are then harmless. A further advantage is that the simpler logic circuit of an up-counter inherently permits a higher clock frequency.

One must do without the representation of the mathematical sign, however, since a positive difference would be wrongly interpreted as being negative should one of the two counters overflow. If positive and negative differences are to be dealt with, the midrange, e.g. 1000, must be chosen as zero point (offset binary representation). To achieve this, the most significant bit of the up-counter is inverted; the initial difference is then 1000, as required.

For the output of the difference in the usual 2's-complement representation, the offset is compensated by also inverting the most significant bit of the difference.

10.2   BCD counters in 8421 code

10.2.1   Asynchronous BCD counter

The table in Fig. 10.2 shows that a 3-bit counter can count up to 7 and a 4-bit counter up to 15. In a counter for straight BCD numbers, a 4-bit straight binary counter is needed for each decimal digit, known as a decade counter. This decade counter differs from the ordinary straight binary counter in that it is reset to zero after every tenth count pulse and produces a carry. This carry bit controls the decade counter for the next higher decimal digit.

A decimal display of the state of BCD counters is achieved much more easily than for the straight binary counter as each decade can be separately decoded and displayed as a decimal digit.

In the straight BCD code, each decimal digit is represented by a 4-bit straight binary number, the bit weightings of which are 2^3, 2^2, 2^1 and 2^0. It is therefore also known as the 8421 code. The state table of a decade counter operating in 8421 code is shown in Fig. 10.12. By definition, it must be identical with that in Fig. 10.2 up to the number 9, but the number $10_{dec.}$ is represented again by 0000. The corresponding output signals are illustrated in Fig. 10.13.

Obviously, additional logic circuitry is required to reset the counter at every tenth count pulse. However, gates may be saved if JK flip-flops are used. It may then often be sufficient to use the J-inputs only and set the unused inputs to 1.

Consider a JK flip-flop of which only two J-inputs are in operation. If $J = J_1 \cdot J_2 = 1$, the flip-flop is a normal toggle flip-flop, since K is also 1 and constant. If $J = 0$, this input state will be transferred to the output at the next clock pulse, as $J = \bar{K}$. Q will then become or remain 0. These two modes are made use of in the 8421 decade counter of Fig. 10.14. In contrast to the normal straight binary counter, the operation must, according to Fig. 10.2, be as follows: flip-flop F_1 may not change state at the tenth count pulse, although z_0 changes from 1 to 0. From Fig. 10.12, we deduce a simple criterion for this case: z_1 must be kept at zero if z_3 is 1 before the clock pulse. To achieve this, the J-input of F_1

Z	z_3	z_2	z_1	z_0
	2^3	2^2	2^1	2^0
0	0	0	0	0
1	0	0	0	1
2	0	0	1	0
3	0	0	1	1
4	0	1	0	0
5	0	1	0	1
6	0	1	1	0
7	0	1	1	1
8	1	0	0	0
9	1	0	0	1
10	0	0	0	0

Fig. 10.12
State table for the 8421 code

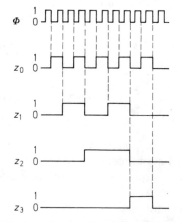

Fig. 10.13 Output states of an 8421-coded counter, as a function of time

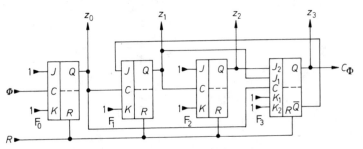

Fig. 10.14 Asynchronous BCD counter

is connected to $\bar{z}_3$. The condition that z_2 must remain zero at the tenth count pulse is therefore automatically fulfilled.

The second deviation from a straight binary counter is that z_3 changes from 1 to 0 at the tenth clock pulse. However, if the clock input of F_3 were connected to z_2 as in a straight binary counter, z_3 would be unable to change after the eighth count pulse, since flip-flop F_1 is blocked by the feedback. The clock input of F_3 must therefore be connected to the output of the flip-flop which is not blocked by feedback, in this case z_0.

On the other hand, the J-inputs must be controlled so that they prevent flip-flop F_3 from changing state too early. Table 10.12 shows that z_3 must not be 1 unless both z_1 and z_2 are 1 before the clock pulse. This may be achieved by connecting the two J-inputs of F_3 to z_1 and z_2. Then, at the eighth count pulse, $z_3 = 1$. Since z_1 and z_2 become zero simultaneously, z_3 resumes the state $z_3 = 0$ as soon as possible, i.e. at the tenth count pulse when z_0 has its next transition from 1 to 0.

IC types:
SN 7490 (TTL),   MC 10138 (ECL).

10.2.2   Synchronous BCD counter

Decade counters can also be operated synchronously. For the development of the circuit, we take the synchronous 4-bit straight binary counter of Fig. 10.5. In order to convert it to a decade counter in the 8421 code, feedback must be introduced to reset the counter at every tenth count pulse. As for the asynchronous counter in Fig. 10.14, the output $\bar{z}_3$ is connected for this purpose to a J-input of F_1, this resulting in the circuit of Fig. 10.15. Up to the ninth count pulse, the counter operates in the same way as the synchronous straight binary counter. At the tenth count pulse, however, F_1 remains at $z_1 = 0$ due to the feedback at J_2, as is required by the state table in Fig. 10.12.

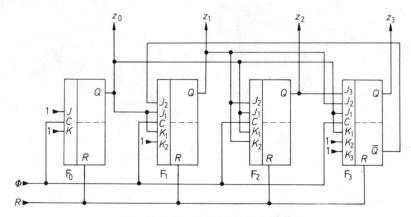

Fig. 10.15 Synchronous BCD counter

If the JK inputs of flip-flop F_3 were connected as for the synchronous straight binary counter, F_3 could never flip back, as z_1 and z_2 remain zero due to the feedback. For this reason, these two outputs are connected only to the J-inputs of F_3 but not to the K-inputs. This ensures that F_3 flips back to $z_3 = 0$ at the tenth count pulse since before that pulse, $K = 1$ and $J = 0$.

For the first seven count pulses, the operation is not influenced by these measures, as only the state $z_3 = 0$ is stored. For a decade counter in 8421 code, flip-flop F_3 must be prevented only once from changing state while $z_3 = 1$. It is therefore sufficient to connect z_0 to a J- as well as a K-input of F_3.

IC types:

SN 74160 (TTL),   MC 14160 (CMOS),   asynchronous reset;
SN 74162 (TTL),   MC 14162 (CMOS),   synchronous reset.

Synchronous BCD up-down counters

BCD counters can be extended to become up-down counters in the same way as straight binary counters. The required logic circuitry is closely related to that described in Section 10.1.2, and a detailed description is therefore omitted. Only the appropriate IC types are given.

BCD counters with up-down control:

SN 74190 (TTL),   MC 10137 (ECL),   MC 14510 (CMOS).

BCD counters with up-clock and down-clock input:

SN 74192 (TTL).

10.3 Presettable counters

Presettable counters are circuits which produce an output signal when a number of input pulses equals a predetermined number M. The output signal can be used to trigger any desired process and is employed to stop the counter or reset it to its initial state. If the counter is allowed to continue counting after reset, it operates as a modulo-m counter, the counting cycle of which is determined by the chosen number M.

Most synchronous counters have additional inputs, as in Fig. 10.16, by which numbers can be read-in in parallel, and synchronously with the clock. Hence, the described presetting operation can be easily realized: the counter is loaded with the number $P = Z_{max} - M$ by setting the load-enable input L to 1 and applying a clock pulse Φ to the counter. For a straight binary counter, the number P is particularly easy to calculate as it is the 1's-complement of M. After M clock pulses, the state Z_{max} of the counter is reached. This is recognized without additional decoders by the change of the carry output C_E from 0 to 1. With the carry, the desired process in another circuit can be triggered. If the other circuit is not in synchronism with the clock Φ, the carry C_E must be combined with Φ to make the signal $C_\Phi = C_E \cdot \Phi$ which is in turn employed as a trigger pulse. In this manner, erroneous triggering by transitionary states is avoided.

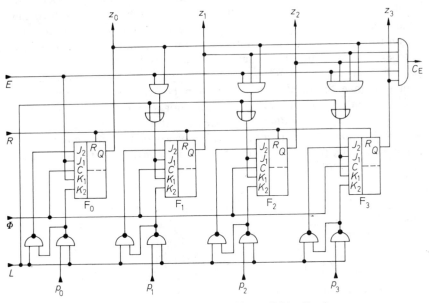

Fig. 10.16 Straight binary counter with parallel loading inputs

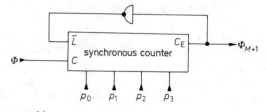

Fig. 10.17 Counter with programmable counting cycle. Suitable IC type e.g. SN 74161

If the counter is to operate in a cyclic mode, it is sufficient to connect C_E to the input L. Clock pulse $M+1$ then resets the counter. This implementation of a programmable modulo-$(M+1)$ counter is illustrated in Fig. 10.17.

10.4 Shift registers

10.4.1 Basic circuit

Flip-flops can store binary information. With master-slave flip-flops, the input information is transferred to an internal store at one edge of the clock pulse, and afterwards transferred to the output. New input information is taken in only at the next edge of the same type.

If, as in Fig. 10.18, several of these flip-flops are cascaded, the information is shifted with each clock pulse from one flip-flop to the next. Such an arrangement is thus called a clocked shift register. Any non-transparent flip-flop is suitable for this circuit, i.e. the master-slave types as well as the single-edge triggered types can be used.

For an explanation of the circuit operation we consider a JK master-slave flip-flop. If $\Phi=1$, flip-flop F_1 transfers the input states $J=D_1$ and $K=\bar{D}_1$ to its internal memory. If Φ becomes 0 again, the output $Q_1, \bar{Q}_1$ assumes this state so that after the first clock pulse, $Q_1=D_1$. The next data bit, D_2, is now applied to the input. After the next clock pulse, F_2 has assumed the former output state of F_1, and

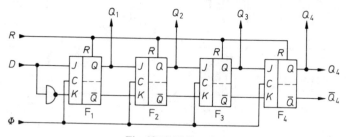

Fig. 10.18 Shift register

Φ	Q_1	Q_2	Q_3	Q_4
1	D_1	–	–	–
2	D_2	D_1	–	–
3	D_3	D_2	D_1	–
4	D_4	D_3	D_2	D_1
5	D_5	D_4	D_3	D_2
6	D_6	D_5	D_4	D_3
7	D_7	D_6	D_5	D_4

Fig. 10.19 Function table of a 4-bit shift register

F_1 the new input state, i.e. $Q_2 = D_1$ and $Q_1 = D_2$. Correspondingly after the third pulse, $Q_3 = D_1$, $Q_2 = D_2$, $Q_1 = D_3$ and finally after the fourth pulse $Q_4 = D_1$, $Q_3 = D_2$, $Q_2 = D_3$ and $Q_1 = D_4$. It can be seen that at each clock pulse, the information in the register is shifted by one flip-flop and the new input information is read in.

Since the shift register in Fig. 10.18 contains four flip-flops, it can store only 4 bits. There are two possible methods of reading the information from the register: After the fourth clock pulse, the states D_1 to D_4 are available simultaneously at the outputs Q_4 to Q_1. It is thus possible to read out, in parallel, data which were read in serially. However, a serial read-out is also possible. At the fourth to the seventh clock pulse, the data D_1 to D_4 may be retrieved one after the other at output Q_4. The new data D_5 to D_7 can be read in simultaneously. Figure 10.19 gives a summary of the states of the individual flip-flops.

IC types:

8 bit: SN 74164 (TTL),   MC 14021 (CMOS).

10.4.2   Ring shift register

It is sometimes necessary that data are read out serially without erasing them. In such cases, the data of the last flip-flop must be read in again at the first flip-flop. Figure 10.20 shows one such circuit. As long as logic 1 is applied to the control input U, $D = D_i$, and the

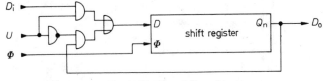

Fig. 10.20 Ring shift register

feedback is ineffective; the shift register operates as previously described. With the first n clock pulses, n bits can thus be stored. In order that the contents of the shift register are not lost at the following clock pulses, the control signal U is made zero. Therefore $D = Q_n$, and each bit of D_o is read in again at the input. After every n clock pulses the shift register is once more at its initial state. The logic state of the control input U thus determines whether new information is to be read in, or old data are to be shifted round the register.

10.4.3   Shift register with parallel input

The read-out of data in the shift register of Fig. 10.18 may be either serial or parallel whereas the read-in of data can only be serial. However, there is often parallel data which must be read into the shift register, e.g. for sequential addition and multiplication.

Loading of parallel data is best effected with D flip-flops, the inputs of which are connected, via a logic double-throw switch, either to the output of the next flip-flop to the left or to the parallel input p_i, see Fig. 10.21. The control variable L defines the mode of operation. At $L = 0$, the next clock pulse effects a shift to the right. At $L = 1$, the next clock pulse effects parallel loading of data.

IC types:

4 bits: SN 74179 (TTL),   MC 14035 (CMOS);
8 bits: SN 74199 (TTL).

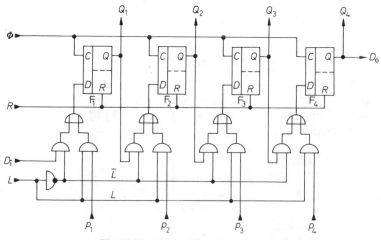

Fig. 10.21 Shift register with parallel input

10.4.4 Bidirectional shift register

The operation of the shift register in Fig. 10.21 can be changed from the right-shifting to the parallel-input mode. Each parallel input could be connected to the output of the next flip-flop to the right. In the parallel-input mode ($L=1$), this would result in a shift of data to the left. The control variable L can therefore be used to change the direction of the shift.

Because of the additional connections, the parallel inputs p_i are no longer available. If they are required in addition to the left-right shift, the logic double-throw switches at the D-inputs of the flip-flops must be replaced by multiplexers having three inputs. Each multiplexer is then connected to the output of the next flip-flop to the left, the output of the next flip-flop to the right and the external parallel input. For control of the shift register operation, two variables are then required.

IC types:

4 bits: SN 74194 (TTL), MC 10141 (ECL),
 MC 14194 (CMOS);
8 bits: SN 74198 (TTL).

10.5 Generation of pseudo-random sequences

Random sequences are often needed to examine analog and digital systems. For their generation, a natural noise source can be connected to the input of a Schmitt trigger. A binary output signal is then obtained, the states of which are randomly distributed. Such a signal is shown in Fig. 10.22, where it can be seen that the duration of the logic states 1 or 0 varies randomly, i.e. no regularity can be detected. If such a random sequence is repeated after a certain time, it is called a pseudo-random sequence. There is no way that a logic device can distinguish between a random and a pseudo-random sequence as long as the period is longer than its memory capacity [10.2]. In most cases, this condition can be easily fulfilled.

The great advantage of pseudo-random sequences is that results are reproducible and displays on the oscilloscope are stationary. Moreover,

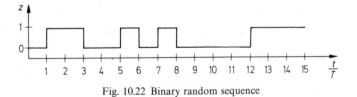

Fig. 10.22 Binary random sequence

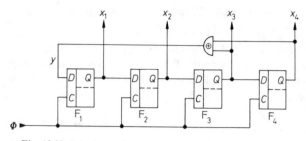

Fig. 10.23 Pseudo-random sequence generator with $n=4$ bits

in the low-frequency range, it is usually easier to generate a pseudo-random sequence than to use a natural noise source.

Pseudo-random sequences are produced by shift registers which use a special kind of feedback logic consisting of exclusive-OR gates.

The largest non-periodic bit sequence which can be generated by a shift register of n stages, is $N=2^n-1$ bits long. A 4-stage shift register can therefore produce a pseudo-random sequence having a maximum period of 15 bits. Figure 10.23 depicts such a circuit. For an explanation of its operation, we assume that the shift register has the state $x_1=1$ and $x_2=x_3=x_4=0$. At the first clock pulse, the information is shifted by one place to the right. Before the clock pulse, $y=x_3 \oplus x_4=0$ and therefore a zero is read in at the first stage. The states after the first clock pulse are then $x_2=1$ and $x_1=x_3=x_4=0$. As y remains zero in this case, another zero is read into the shift register at the second clock pulse. The states are then $x_3=1$ and $x_1=x_2=x_4=0$, but y is 1. This means that, at the next clock pulse, a 1 is read in so that $x_1=x_4=1$ and $x_2=x_3=0$. The remainder of the cycle can be followed in the state table of Fig. 10.24. After the fifteenth clock pulse the initial state is reached. The cycle can obviously be started from any other initial state, with the exception of the state 0000 which would latch the circuit. This state could, however, occur because of disturbances or at switch-on and must be suppressed. This is the purpose of the additional logic circuit given in Fig. 10.25. If the state 0000 arises, the output of the NOR gate

$\varPhi$	0	1	2	3	4	5	6	7	8	9	10	11	12	13	14	15
x_1	1	0	0	1	1	0	1	0	1	1	1	1	0	0	0	1
x_2	0	1	0	0	1	1	0	1	0	1	1	1	1	0	0	0
x_3	0	0	1	0	0	1	1	0	1	0	1	1	1	1	0	0
x_4	0	0	0	1	0	0	1	1	0	1	0	1	1	1	1	0
y	0	0	1	1	0	1	0	1	1	1	1	0	0	0	1	0

Fig. 10.24 State table of the pseudo-random sequence generator with 4 bits

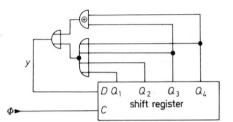

Fig. 10.25 Anti-latch circuit for the pseudo-random sequence generator

changes to 1. This bit is read into the shift register via the OR gate. As the state 0000 does not arise during normal operation, the additional circuit will not interfere with the function of the circuit.

It is immaterial at which output the pseudo-random sequence is monitored, as the same sequence appears, somewhat delayed, at each output.

For longer periods, correspondingly larger shift registers must be used. A shift register with 10 stages produces a period equivalent to 1023 clock pulses, one with 20 stages 1048575. To ensure that the maximum period of $N = 2^n - 1$ is attained, the feedback logic must be connected to particular outputs. The last output is always used. The other outputs to be combined by the feedback logic are determined by the length of the shift register. The logic itself consists entirely of

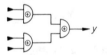

Fig. 10.26 Exclusive-OR operation with four inputs

exclusive-OR gates which are connected as in Fig. 10.26, [10.3]. Sufficient gates must be combined to provide the number of inputs needed. Figure 10.27 shows for 3- to 20-stage shift registers, how many inputs are necessary and to which stages they must be connected. For reasons

n	3	4	5	6	7	8	9	10	11	12	13	14	15	16	17	18	19	20
	3	4	5	6	7	8	9	10	11	12	13	14	15	16	17	18	19	20
	2	3	3	5	4	7	5	7	9	11	10	13	14	14	14	11	18	17
						5					8	6	8		13		17	
						3					6	4	4		11		14	

Fig. 10.27 Table of the terminals required for feedback

of simplicity, we have given only one solution for each case. For each given solution involving the input terminals x_i, a symmetrical solution with the inputs x_{n-i} exists, whereby the last terminal x_n is common to both solutions. Hence, instead of the terminals $i = 3, 5, 7, 8$, the terminals $i = 1, 3, 5, 8$ can be employed. Other combinations are often possible which also give the maximum sequence length [10.4]. The determination of the necessary input terminals is rather difficult and will thus be omitted.

As the state $0000...$ is suppressed in all circuits while all other combinations occur, a 1 is found at any output for $\frac{1}{2} \cdot 2^n$ clock pulses and a 0 for $\frac{1}{2} \cdot 2^n - 1$ clock pulses. The probability that a 1 is obtained with the next clock pulse approaches 50% more closely the longer the shift register.

The state of an output can remain constant for intervals of $1, 2, 3 ... n$ clock pulses. Each of these intervals turns up at least once, its frequency decreasing, however, with increasing length of the interval.

Many applications require the digital noise to be converted into analog noise. For this purpose it is sufficient to connect to an output a lowpass filter, the cutoff frequency of which is small compared to the clock frequency. The more logic 1's that occur consecutively, the higher the voltage. A considerably larger noise bandwidth is obtained if the entire data in the shift register at any time are applied to a digital-to-analog converter. To generate Gaussian noise, the individual output bits must be given appropriate weightings [10.5].

10.6   Processing of asynchronous signals

Sequential logic circuits can be realized either in asynchronous or in synchronous, i.e. clocked, operation mode. Asynchronous operation normally requires simpler circuits but gives rise to a number of problems: it must be ensured that the spurious transitions (hazards) which may temporarily appear because of the difference in delay times, are not decoded as valid states. For a synchronous system, the conditions are far more simple. Any transition within the system can only arise at the edge of a clock pulse. The clock pulse therefore indicates when the system is stationary. It is sensible to construct the system so that all changes appear uniformly at one edge of the clock pulse. If, for instance, all circuits are triggered at the trailing edge, the system is certain to be at steady-state whenever the clock pulse is 1.

As a rule, external data fed to the system are not synchronized with its clock. In order that they may be processed synchronously, they must be adapted by special circuits, some of which are described below.

10.6.1 Bounce elimination for mechanical switches

If a mechanical switch is opened or closed, vibrations usually generate a pulse chain. A counter then registers an undefined number of pulses instead of the one intended. One way of avoiding this is the use of mercury-wetted contacts, although this is rather expensive. A simple method of electronic debouncing by means of an RS flip-flop is presented in Fig. 10.28. When the switch U is in its lower position (break contact) $\bar{R}=0$ and $\bar{S}=1$, i.e. $x=0$. When the switch is operated, a pulse train initially occurs at the $\bar{R}$ input because the break contact is opened. Since $\bar{R}=\bar{S}=1$, this being the storing condition, the output remains unchanged. After the complete opening of the break contact a pulse train is generated by the opposite make contact. With the very first pulse, $\bar{R}=1$ and $\bar{S}=0$, which makes the flip-flop change state so that $x=1$. This state is stored during the bouncing that follows. The flip-flop changes to its former state only when the lower break contact is touched again. The pulse diagram of Fig. 10.29 illustrates this behaviour.

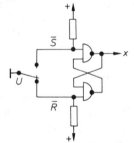

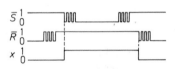

Fig. 10.28 Bounce eliminator
for a switch

Fig. 10.29
Variables as functions of time

10.6.2 Synchronization of pulses

The simplest method of synchronizing pulses employs D-type flip-flops. The external unsynchronized signal x is as in Fig. 10.30 applied to the D-input, and the system clock Φ to the C-input. In this manner, the state of the input variable x is monitored and transferred to the output only at the edge of the clock pulse. Because the input state can change at either $\Phi=0$ or $\Phi=1$, single-edge triggered flip-flops must be used.

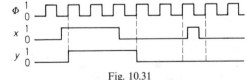

Fig. 10.30
Synchronization circuit

Fig. 10.31
Variables as functions of time

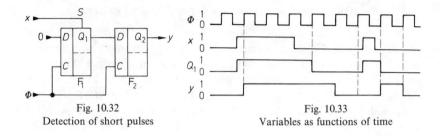

Fig. 10.32	Fig. 10.33
Detection of short pulses	Variables as functions of time

Figure 10.31 shows an example of a pulse diagram with positive-edge triggering. Any pulse too short to be registered by the leading edge of a clock pulse is ignored. This case is also shown in Fig. 10.31. If such short pulses are not to be lost, they must be read into a temporary store before being transferred to the D flip-flop. The D flip-flop F_1 in Fig. 10.32 serves this purpose. It is set asynchronously via the S-input when x becomes 1. With the next positive-going edge, $y = 1$. If, at this moment, x has already returned to zero, flip-flop F_1 is reset by the same edge. A short pulse x is thus prolonged until the next clock edge occurs. This property may also be seen in the example in Fig. 10.33.

10.6.3   Synchronous one-shot

It is possible, with the circuit in Fig. 10.34, to generate a pulse which is in synchronism with the clock. The pulse length equals a period of the clock, and is independent of the length of the trigger signal x. Figure 10.35 illustrates the operation.

If x changes from 0 to 1, $Q_1 = 1$ and $\bar{Q}_2 = 1$ at the positive-going edge of the next clock pulse, i.e. $y = 1$. At the following leading edge, $\bar{Q}_2$ becomes 0 and y becomes 0 again. This state remains unchanged until x has been zero for at least one clock period and has returned to 1. Short trigger pulses which are not registered by the leading edge of a clock pulse are lost, as with the synchronizing circuit in Fig. 10.30. If

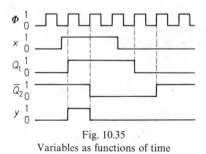

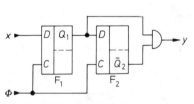

Fig. 10.34 Generation of a single,	Fig. 10.35
but clock-synchronous pulse	Variables as functions of time

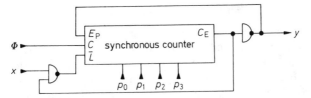

Fig. 10.36 Synchronous one-shot. Suitable IC type e.g. SN 74161

they too are to be considered, an additional flip-flop must store the pulses until they are transferred to the main flip-flop, this process being the same as that for the circuit in Fig. 10.32.

A synchronous one-shot for ON-times longer than one clock period can be realized quite simply by using a synchronous counter, as is shown in Fig. 10.36. If the trigger variable x is at 1, the counter is loaded in the parallel-in mode at the next clock pulse. The following clock pulses are used to count to the maximum output state Z_{max}. At this number, the carry output $C_E = 1$. The counter is then blocked via the count-enable input E_P; the output variable y is 0. The ordinary enable input E cannot be employed for this purpose as it not only affects the flip-flops but also the output C_E directly and this would result in an unwanted oscillation.

A new cycle is started by parallel read-in. Immediately after loading, C_E becomes zero and y is then 1. The feedback from C_E to the NAND gate at the x-input prevents a new loading process unless the counter has reached the state Z_{max}. By this time, x should have returned to 0; if not, the counter is loaded again, i.e. is operating as a modulo-$(M + 1)$ counter.

The pulse sequence is represented in Fig. 10.37 for an ON-time of 9 clock cycles. If a 4-bit straight binary counter is employed, it must, for this particular ON-time, be loaded with $P = 7$. The first clock pulse is taken up for the read-in process and the remaining 8 pulses for counting up to 15.

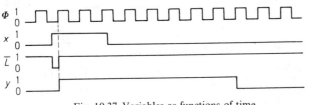

Fig. 10.37 Variables as functions of time

10.6.4 Synchronous edge detector

A synchronous edge detector gives an output signal in synchronism with the clock pulse whenever the input variable x has changed. For the implementation of such an arrangement, we consider the one-shot circuit in Fig. 10.34. It produces an output pulse whenever x changes from 0 to 1. In order that a pulse is also obtained at the transition from 1 to 0, the AND gate must be replaced by an exclusive-OR gate, this resulting in the circuit in Fig. 10.38. Its performance is illustrated by the pulse diagrams in Fig. 10.39.

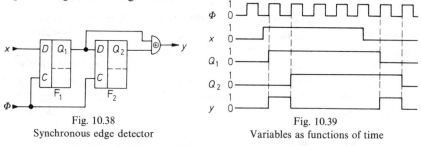

Fig. 10.38
Synchronous edge detector

Fig. 10.39
Variables as functions of time

10.6.5 Synchronous clock switch

The problem often arises of how to switch the clock on and off without interrupting the clock pulse generator. In principle, an AND gate could be used for this purpose, but this would result in the first and the last pulse being of undefined length if the switching signal is not clock-synchronized. This effect can be avoided by employing a positive-edge triggered D-type flip-flop for the synchronization, as is shown in Fig. 10.40. If $E=1$, at the next leading pulse edge $Q=1$ and therefore $\Phi'=1$. The first pulse of the switched clock Φ' always has the full length because of the edge-triggering property. The leading pulse edge cannot be used to switch off as directly after the transition, $Q=0$ which would result in a very short output pulse. The flip-flop is therefore cleared asynchronously via the reset input when E and Φ are 0, achieved by the OR gate at the $\bar{R}$-input. As is obvious from Fig. 10.41, only full-length clock pulses can reach the output of the AND gate.

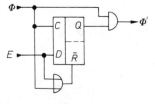

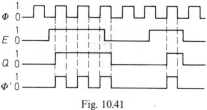

Fig. 10.40
Synchronous clock switch

Fig. 10.41
Variables as functions of time

10.7 Systematic design of sequential circuits

10.7.1 State diagram

To enable the systematic design of sequential circuits, it is necessary to obtain a clear description of the problem in hand. This is the purpose of the state diagram, an example of which is given in Fig. 10.42.

Each state S_Z of the system is illustrated by a circle. The index Z is called the state vector and is represented by the state variables z_i. These are most suitably expressed in straight binary code.

The transition from one state to another is shown by an arrow. The symbol on the arrow indicates under which condition a transition is to occur. For the example in Fig. 10.42, state $S(t_k) = S_1$ is followed by state $S(t_{k+1}) = S_2$ if $x_1 = 1$. For $x_1 = 0$ however, $S(t_{k+1}) = S_0$. An unmarked arrow stands for an unconditional transition.

For a *synchronous* sequential circuit there is an additional condition that a transition will only occur at the next clock pulse edge and not immediately the transition condition is fulfilled. As this restriction applies to all transitions in the system, it is usually not entered in the state diagram, but mentioned in the description. We deal below only with synchronous sequential circuits as their design is less problematic.

If the system is in the state S_Z and no transition condition is fulfilled which might lead out of this state, the system remains in the state S_Z. This obvious fact can sometimes be emphasized by entering an arrow

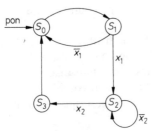

Fig. 10.42
Example of a state diagram.

State S_0: Initial state
State S_1: Branching state
State S_2: Wait state
State S_3: Temporary state

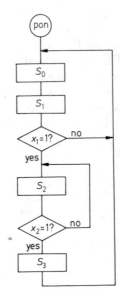

Fig. 10.43 Equivalent flow diagram

which starts and ends at S_Z (wait state). Such a case is illustrated at state S_2 in Fig. 10.42.

After switch-on of the power supply, a sequential circuit must be set to a defined initial state. This is the "pon" condition ("power on"). Its signal is produced by a special logic unit and is 1 for a short time after switch-on of the supply and is otherwise zero.

The operation of a sequential circuit can be described not by the state diagram only, but also by a flow chart, as is shown for the same example in Fig. 10.43. This representation suggests the implementation of a sequential circuit by a micro-computer, as discussed in the next chapter.

The basic arrangement of a sequential circuit has already been shown in Fig. 10.1. To characterize the system state, a memory for the state variables in the form of edge-triggered flip-flops is needed. The combinatorial part of the circuit can be realized with gates or with a ROM.

10.7.2   Design example for a programmable counter

We shall demonstrate the design process for a counter, the counting cycle of which is either $0, 1, 2, 3$ or $0, 1, 2$ depending on whether the control variable x is 1 or 0. The appropriate state diagram is given in Fig. 10.44. As the system can assume 4 stable states, we require two flip-flops for storage of the state vector Z which consists of two variables, z_0 and z_1. Since these variables immediately indicate the state of the counter, they simultaneously serve as output variables. In addition, a carry y should be produced when the counter state is 3 for $x = 1$, or 2 for $x = 0$.

We obtain the circuit in Fig. 10.45 with the truth table given in Fig. 10.46. The left-hand side of the table shows all possible bit combinations of the input- and state-variables. The state diagram in Fig. 10.44 shows which is the next system state for each combination. They are shown at the right-hand side of the table. The respective values of the carry bit y are also entered.

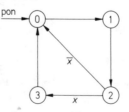

Fig. 10.44 State diagram of a counter with programmable counting cycle.

$$\text{Counting cycle} = \begin{cases} 3 & \text{if } x = 0 \\ 4 & \text{if } x = 1 \end{cases}$$

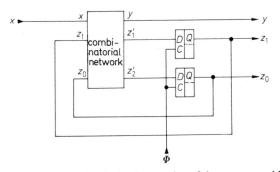

Fig. 10.45 Sequential circuit for the implementation of the programmable counter

	$Z(t_k)$			$Z(t_{k+1})$		
x	z_1	z_0	z_1'	z_0'	y
0	0	0	0	1	0
0	0	1	1	0	0
0	1	0	0	0	1
0	1	1	0	0	0
1	0	0	0	1	0
1	0	1	1	0	0
1	1	0	1	1	0
1	1	1	0	0	1
ROM address			ROM contents		

Fig. 10.46 Truth table for the state diagram in Fig. 10.44

If a ROM is used to realize the combinatorial network, the truth table of Fig. 10.46 can be directly employed to program the memory. The state- and input-variables then serve as address variables. The new value Z' of the state vector Z, and the output variable y are stored at the appropriate addresses. Hence, for the implementation of our example, we require a ROM for 8 words of 3 bits. The smallest PROM can store 32 words of 8 bits (e.g. SN 74S 288) so that only one tenth of its memory capacity is used.

The truth table in Fig. 10.46 supplies the following Boolean functions:

$$z_1' = z_0 \bar{z}_1 + x \bar{z}_0 z_1,$$
$$z_0' = \bar{z}_0 \bar{z}_1 + x \bar{z}_0,$$
$$y = \bar{x} \bar{z}_0 z_1 + x z_0 z_1.$$

Figure 10.47 shows the realization of this combinatorial network by gates. It can be seen that the number of integrated circuits involved is

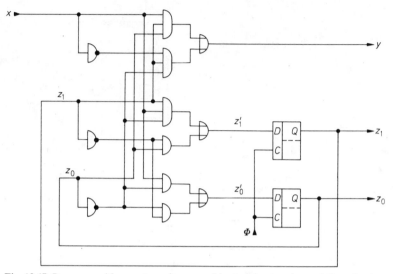

Fig. 10.47 Programmable counter using a combinatorial network consisting of gates

many times greater than when using a ROM. The application of a ROM also has another decisive advantage, flexibility: the ROM need only be reprogrammed and a circuit with different properties is obtained without further changes.

The use of gates for a sequential circuit is thus recommended only in certain simple cases, for instance in the standard counters described in the previous sections.

When constructing complex sequential networks, a limit is soon reached even for the ROM solution, where the required memory capacity rises excessively. The following section therefore describes some strategies by which this problem can, to a great extent, be overcome.

10.7.3   Reduction of memory requirement

As can be seen in the basic circuit of Fig. 10.1, the combinatorial circuit contained in the sequential logic circuit has $(n+l)$ inputs and $(n+m)$ outputs, where n is the number of state variables, l the number of input variables x (qualifiers) and m the number of output variables y. When implementing it using a ROM, a memory of $2^{(n+l)}$ words at $(n+m)$ bits $=(n+m)\,2^{(n+l)}$ bits is needed. It is possible to assign an output vector Y to each combination of state and input variables. In practice, however, the values of most of the output variables are already fully defined by the state variables, and only a few are dependent on only some of the qualifiers.

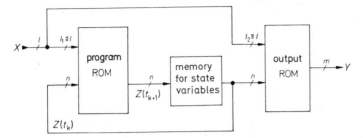

Fig. 10.48 Reduction of the required total memory capacity by replacing a large ROM by two small ones

This fact allows the ROM to be split into two ROMs as in Fig. 10.48. One is the program ROM containing the system states only and no output states. The output states are formed by an output ROM from the state variables and from a small part of the input variables. Therefore l_2 is usually small compared to l. There may also be cases for which an input variable affects only the decoding at the output but not the sequence of the states. For the two-ROM solution in Fig. 10.48, such qualifiers may then be connected directly to the output ROM and omitted for the program ROM. Therefore, $l_1 < l$ is also possible.

As only those qualifiers required for the sequential and output control are connected to the two ROMs, a considerably smaller memory is needed. The most unfavourable case is that where all the l qualifiers are used for both ROMs. The necessary memory capacity of both ROMs together is then just as large as that of the single ROM in Fig. 10.1.

Although there is no saving in memory capacity, the solution employing two ROMs (Fig. 10.48) is advantageous even in this case, since the system can be more easily adapted to different operating conditions. There are many cases for which the state sequence is identical and which differ only in the output instructions. For an adaptation, only the output ROM need be exchanged while the program ROM remains unchanged.

Input multiplexer

A second property of practical sequential logic circuits can also be used for the reduction of required memory space. The number l of qualifiers is often so large that the number of address variables of a ROM is greatly exceeded. On the other hand, only relatively few of the 2^l possible combinations are decoded. It is therefore reasonable not to use the qualifiers directly as address variables but to employ a multi-

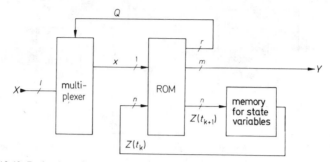

Fig. 10.49 Reduction of required memory capacity using a multiplexer at the input

plexer to read the variables relevant for each state of the system. This results in the block diagram of Fig. 10.49.

Apart from the state variables, only the output x of the multiplexer is connected to the address inputs of the ROM. The multiplexer is controlled by the straight binary number Q taken from some additional outputs of the ROM. The qualifier selected by this number is denoted by x_Q.

If, at a single transition, several qualifiers are to be identified, the determination must be carried out serially as only one variable can be selected at a time. For this purpose, the appropriate state is divided into several sub-states for which only one qualifier need be determined. There is therefore a larger total number of system states, and they can be represented with the help of a few additional state variables. This extension is small, however, compared to the saving in memory locations by using the multiplex qualifier identification.

This fact is demonstrated by the following typical example. A sequential logic circuit having the state diagram in Fig. 10.50 is to be constructed. It has four states and six qualifiers. The realization based on the principle in Fig. 10.1 would require a ROM with 8 inputs and a memory capacity of $2^8 = 256$ words. Assuming that two output variables are to be generated, and taking the two state variables into account, a word length of 4 bits, i.e. a total memory of 1024 bits is needed.

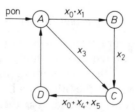

Fig. 10.50 State diagram for the example

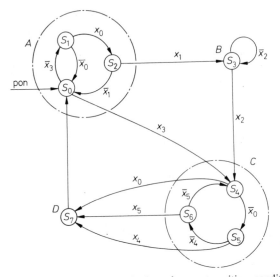

Fig. 10.51 State diagram, modified to obtain only one transition condition for each change of state

For the implementation using an input multiplexer, the states A and C must be each divided into three sub-states for which only one of the qualifiers shown in Fig. 10.50 is determined at a time. In this manner, the modified state diagram in Fig. 10.51 is obtained. There is now a total of eight states, and they are identified by S_0 to S_7. A transition from the macrostate A to the macrostate B occurs only when x_0 and x_1 are 1, in accordance with the original state diagram in Fig. 10.50. The appropriate combination for an OR operation is seen at the macrostate C.

The representation of the eight states requires three state variables. The ROM of Fig. 10.49 must also possess three outputs for the control of the 8-input multiplexer, as well as two y-outputs. This amounts to a word length of 8 bits. Apart from the three state variables, only the output of the multiplexer serves as an address variable so that a memory capacity of 2^4 words at 8 bits each$=128$ bits is needed. This is only about one tenth of that required by the standard solution.

The compilation of the truth table is not difficult. The state table in Fig. 10.52 can be directly deduced from the state diagram in Fig. 10.51. It indicates which state $Z(t_{k+1})$ follows the state $Z(t_k)$, according to whether x equals 1 or 0. The binary number Q identifies the qualifier x_Q selected at the state $S_{Z(t_k)}$. The numbers $Z(t_k)$, $Z(t_{k+1})$ and Q need now only be written as straight binary numbers to obtain the programming table in Fig. 10.53. In the "contents" column we have entered only the 6 bits required for the sequential control. Extra bits for the output can be added at will.

$Z(t_k)$	x	$Z(t_{k+1})$	Q	z_2	z_1	z_0	x	z_2'	z_1'	z_0'	a_2	a_1	a_0
0	0	1	3	0	0	0	0	0	0	1	0	1	1
0	1	4	3	0	0	0	1	1	0	0	0	1	1
1	0	0	0	0	0	1	0	0	0	0	0	0	0
1	1	2	0	0	0	1	1	0	1	0	0	0	0
2	0	0	1	0	1	0	0	0	0	0	0	0	1
2	1	3	1	0	1	0	1	0	1	1	0	0	1
3	0	3	2	0	1	1	0	0	1	1	0	1	0
3	1	4	2	0	1	1	1	1	0	0	0	1	0
4	0	5	0	1	0	0	0	1	0	1	0	0	0
4	1	7	0	1	0	0	1	1	1	1	0	0	0
5	0	6	4	1	0	1	0	1	1	0	1	0	0
5	1	7	4	1	0	1	1	1	1	1	1	0	0
6	0	4	5	1	1	0	0	1	0	0	1	0	1
6	1	7	5	1	1	0	1	1	1	1	1	0	1
7	0	0	×	1	1	1	0	0	0	0	0	0	0
7	1	0	×	1	1	1	1	0	0	0	0	0	0

Table header spanning: "address" spans z_2, z_1, z_0, x; "contents" spans z_2', z_1', z_0', a_2, a_1, a_0.

Fig. 10.52 State table
(× = don't-care condition)

Fig. 10.53
PROM programming table

11   Microprocessors

The previous chapter illustrates how ROMs enable the construction of sequential logic circuits which can be easily modified simply by changing the contents of the ROM. Such an arrangement can carry out a set of instructions whereby conditional and unconditional jumps can be executed. A microprocessor enables, in addition, the processing of subroutines, i.e. the execution of indirect jumps. It also contains an arithmetic logic unit (ALU) for the execution of arithmetic and logic operations, as well as a series of temporary registers.

11.1   Basic structure of a microcomputer

A microprocessor cannot operate on its own. It requires an external memory in which the instructions to be executed (i.e. the program) are stored. This memory can be a RAM (random access memory) into which the instructions are read after switch-on. If the program is not to be changed, it can also be stored in a ROM and is then nonvolatile. For the storage of variables, some additional RAM space is required, the size of which is determined by the problem in hand. The input and output of data is accomplished by special interface circuits.

The entire arrangement is known as a *microcomputer*, and is shown schematically in Fig. 11.1. The *microprocessor* is the central processing unit (CPU) of the microcomputer.

The block diagram in Fig. 11.1 gives no indication of the performance and efficiency of the CPU nor of the capacity of the memory; it is simply the basic block diagram of a computer. Computers can be crudely grouped into the following three families:

Large computers: over 128 k words of 24...64 bits.
Minicomputers: 8 k...64 k words of 12...16 bits.
Microcomputers: 0.5 k...64 k words of 4...16 bits.

The breakthrough of microcomputers occurred with the arrival of monolithic microprocessors. Steeply falling prices are the reason that not only their application as simple universal computers has become more attractive, but also — to a much larger extent — their use in all kinds of other equipment. In the latter case they are given a fixed program and are able to take over relatively complex arithmetic and/or

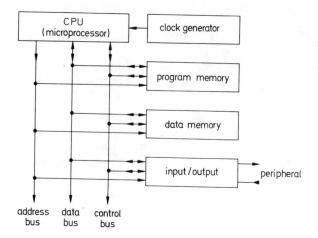

Fig. 11.1 Block diagram of a microcomputer

control tasks. For many applications, a standard hardware unit then suffices so that the real effort of development is more and more directed towards programming (software).

This trend has been accelerated by the introduction of single-chip microcomputers. Such large-scale integrated circuits contain not only the processor but also the clock generator, the input/output ports and a small memory consisting partly of a RAM and partly of an erasable PROM. Such a microcomputer is therefore functional without any additional units. It can be extended by external memories and further input/output ports.

11.2   Operation of a microprocessor

This section deals with the operation of a microprocessor and the organization of its instructions. A general description is extremely difficult; we shall therefore concentrate on a well-known type having a clear structure, e.g. the microprocessor MC 6800. Most other micro-processors have a very similar construction, except for the bit-slice processors which, however, will not be discussed here.

11.2.1   Block diagram

The block diagram of a microprocessor is represented in Fig. 11.2. For the addresses, a word length of 16 bits is normally used with which a maximum of $2^{16} = 64 \text{ k} = 65536$ words can be addressed. Often however, only a small part of these are actually assigned to memory

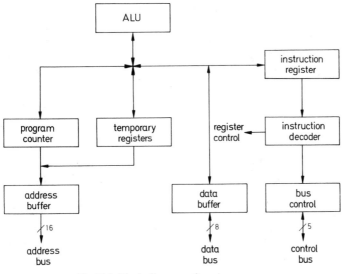

Fig. 11.2 Block diagram of a microprocessor

locations. The word length of the data is usually 8 bits, for some processors 4 or 16 bits.

The Boolean and arithmetic operations are usually carried out with the data stored in the temporary registers. At the beginning of the program, the program counter is set to the program start address which is then transferred to the registers via the address bus. With a read instruction at the control bus, the contents of the appropriate register is transferred by the data bus and stored in the instruction register. The instruction decoder then initiates the operations necessary for the execution of the instruction. As will be seen later, this requires a variable number of clock periods (machine cycles). When the instruction has been executed, the instruction decoder sets the program

flag register	8 bit
accumulator A	8 bit
accumulator B	8 bit
index register	16 bit
stack pointer	16 bit
program counter	16 bit

Fig. 11.3 List of the registers in the MC 6800 microprocessor which are
accessible to the user

counter to the address of the next instruction. This operation will be discussed in more detail for a few examples in the next section.

Figure 11.3 shows the user-accessible registers of the micropro-cessor MC 6800. Most arithmetic operations are carried out with the help of the accumulators A and B. The index register stores frequently used addresses, and the stack pointer serves to organize the subroutine techniques. The flag register (condition code register) contains flags used for the accomplishment of conditional jumps.

11.2.2 Form of an instruction

As already mentioned, the microprocessor MC 6800 operates with an address word length of 16 bits ($= 2$ bytes) and a data word length of 8 bits ($= 1$ byte). For the user, bit strings of this length are difficult to read, and for this reason a shortened notation is used. The bits are collected in groups of 4, each represented by a digit which may thus assume 16 different values. The resulting code is therefore called hexadecimal or hex code. For the digits 0 to 9, the normal decimal symbols are used. The digits "ten" to "fifteen" are represented by the capitals A to F, so that the assignment shown in Fig. 11.4 is obtained.

Since the base 16 of the hex code is a power of 2, there are two different possibilities of converting a many-digit hex coded number to the corresponding decimal number. The relationship

$$Z_{hex} = z_{N-1} \cdot 16^{N-1} + z_{N-2} \cdot 16^{N-2} + \cdots + z_1 \cdot 16 + z_0$$

indicates one method of conversion. The other way would be to convert each digit separately to a straight binary number and join them together. This method results in the corresponding straight binary number which can then be processed using the methods previously described. The following example illustrates the two ways of con-

straight binary	hex	decimal	straight binary	hex	decimal
0000	0	0	1000	8	8
0001	1	1	1001	9	9
0010	2	2	1010	A	10
0011	3	3	1011	B	11
0100	4	4	1100	C	12
0101	5	5	1101	D	13
0110	6	6	1110	E	14
0111	7	7	1111	F	15

Fig. 11.4 Comparison of straight binary, hex and decimal notation

version:

$$A148_{hex} = 10 \cdot 16^3 + 1 \cdot 16^2 + 4 \cdot 16 + 8 = 41\,288_{dec},$$
$$A148_{hex} = \underbrace{1\,0\,1\,0}\ \underbrace{0\,0\,0\,1}\ \underbrace{0\,1\,0\,0}\ \underbrace{1\,0\,0\,0}_2 = 41\,288_{dec}.$$

The 16-bit straight binary address words may thus be represented in the shortened notation by 4-digit hex numbers, and the 8-bit data words by 2-digit hex numbers.

The different instructions which a microprocessor can carry out are represented in machine language (operation code or op code) by 8-bit words, i.e. 2-digit hex numbers. In addition, mnemonics are employed to make programming easier. The instruction "load the accumulator A", for example, is shortened to the mnemonic "LDA A". However, this symbol cannot be interpreted by the microprocessor and must therefore be translated into op code. For this purpose, a translating table or a special translating program (assembler) is used.

The instruction LDA A is not yet complete, because the microprocessor must be told *what* the accumulator is to be loaded with, i.e. to which *operand* the instruction is to be applied. There are several methods of indicating this.

1)   Extended addressing

The two bytes which follow the instruction in the program contain the complete 16-bit address of the memory location, the contents of which are to be loaded into the accumulator A. The result is the following structure:

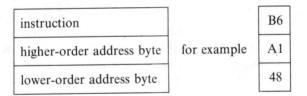

instruction		B6
higher-order address byte	for example	A1
lower-order address byte		48

The op code of the LDA A (ext) instruction for the MC 6800 is $B6_{hex} = 1011\ 0110_2$. In the above example, we have entered the address

$$A148_{hex} = 1010\ 0001\ 0100\ 1000_2.$$

2)   Direct addressing

Only a one-byte address is specified in the program, the microprocessor automatically assuming the higher-order byte to be zero. The corresponding op code for the LDA A (dir.) instruction is "96". This addressing mode allows the user to directly address the location 0000 to $00FF_{hex} = 0$ to 255_{dec} (base page), thereby reducing the execution time. It is therefore sensible to use the base page for the storage of frequently needed variables and constants. The structure of the instruction in the direct addressing mode is as follows:

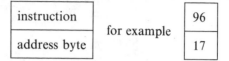

instruction		96
address byte	for example	17

In the example, the contents of the register

$$17_{hex} = 0000\ 0000\ 0001\ 0111_2$$

is read into the accumulator A.

3)   Indexed addressing

With this mode of addressing, the contents of a memory location is read, the address of which is held in the index register. In addition, it is possible to specify an index "offset" by including an 8-bit number in the instruction. The result is a simple method of addressing locations above any desired 16-bit index address, the instruction then reading:

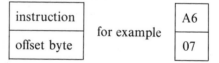

instruction		A6
offset byte	for example	07

The op code for the LDA A (inx.) instruction is $A6_{hex} = 1010\ 0110_2$. If we assume that the address $A148_{hex}$ is stored in the index register, the contents of the memory location $A148_{hex} + 0007_{hex} = A14F_{hex}$, for our example, are read into accumulator A. The specification of the offset is read by the microprocessor as being a positive 8-bit number. A negative offset is not defined. The largest possible offset is therefore

$$1111\ 1111_2 = FF_{hex} = 255_{dec}.$$

It will be seen in Section 11.3.3 (jump and branch instructions) that there is another way of specifying an offset, although this specification is used only to define a program branch and is interpreted as being a signed 7-bit number.

4)   Immediate addressing

In this mode, the operand is included in the instruction as the second byte:

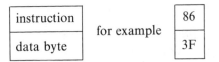

The op code for the LDA A (imm.) instruction is 86_{hex}. In the example, the operand $3F_{hex} = 00111111_2$ is read into accumulator A. In contrast to the example, two data bytes are needed for the immediate loading of the index register and stack pointer, as their word length is 16 bits.

11.2.3   Execution of an instruction

The execution of an instruction usually requires several machine cycles. Referring to the example LDA A (ext.) $= B6_{hex}$, we shall explain the manipulation carried out by the microprocessor: if the program counter calls for the address M at which the instruction is stored, the register answers on the data bus with the op-coded instruction, in our example $B6_{hex}$. The microprocessor decodes the instruction and finds that it must fetch the next two bytes from the program memory in order to obtain the address of the operand. To achieve this, it applies the address $(M+1)$ to the address bus and reads the appropriate byte into an auxiliary register. At the next cycle, it issues the address $M+2$ and reads the appropriate byte into a second auxiliary register. During the fourth cycle, the microprocessor applies the two stored bytes (i.e. the 16-bit address of the operand) in parallel to the address bus and reads the byte which now appears on the data bus into the accumulator A.

The execution of the instruction thus requires the four cycles set out in Fig. 11.5. Similar considerations show that the instruction LDA A (dir.) requires three cycles and the instruction LDA A (imm.) two cycles.

The number of cycles is a direct measure of the execution time of an instruction. For the MC 6800, the cycle time is equal to the clock period. For the maximum clock rate of 1 MHz, a cycle time of 1 μs is therefore attainable so that the operation LDA A (ext.) is executed within 4 μs.

cycle	address bus	data bus
1	address M of instruction	op-coded instruction
2	address $M+1$	higher-order byte of operand address
3	address $M+2$	lower-order byte of operand address
4	address of operand	operand

Fig. 11.5 Activity at address and data bus when executing the operation LDA A (ext.)

11.3 Instruction set

In this section, a review of the instructions for the MC 6800 is given. This microprocessor can carry out 72 different instructions, most of which may be applied in different addressing modes. Considering the manifold addressing modes, a set of 197 instructions is obtained.

11.3.1 Register operations

In Fig. 11.6, we have listed the operations available for the exchange of data between individual registers. Under the heading "addressing modes" we have included a further column containing the inherent addressing operations. These are operations for which the operand is self evident (e.g. clear accumulator A: CLR A).

The symbols under the heading "short description of operation" are as follows:

A: contents of accumulator A,
B: contents of accumulator B,
$[M]$:   contents of the memory location having the address M,
X: contents of the index register X,
X_H: higher-order byte of X,
X_L: lower-order byte of X,
C: carry bit in the flag register.

operations	mne-monic	addressing modes					short description of operation
		ext.	dir.	inx.	imm.	inher.	
load accumulator	LDA A	B6	96	A6	86		$[M] \to A$
	LDA B	F6	D6	E6	C6		$[M] \to B$
store accumulator	STA A	B7	97	A7			$A \to M$
	STA B	F7	D7	E7			$B \to M$
transfer accumulator	TAB					16	$A \to B$
	TBA					17	$B \to A$
clear	CLR	7F		6F			$00 \to M$
	CLR A					4F	$00 \to A$
	CLR B					5F	$00 \to B$
load index register	LDX	FE	DE	EE	CE		$[M] \to X_H$ $[M+1] \to X_L$
store index register	STX	FF	DF	EF			$X_H \to M,$ $X_L \to M+1$

Fig. 11.6 Register operations of the microprocessor MC 6800

11.3.2 Arithmetic and Boolean operations

The instructions for arithmetic and Boolean operations are listed in Fig. 11.7. Logic operations are carried out simultaneously for each bit of the data words and each result is transferred to the corresponding bit of the result. The AND operation, for example, results

$$
\begin{aligned}
\text{for } A: \quad & 1001\ 1101 \\
\text{and } B: \quad & 0110\ 1011 \\
\text{in } A \cdot B: \quad & 0000\ 1001.
\end{aligned}
$$

The instruction set for arithmetic operations is very limited for most microprocessors. Apart from the conversion to 2's-complement, only addition and subtraction are available. When using the operation DA A (decimal adjust), the addition can also be applied to BCD numbers. In this case, the same corrections are carried out after the addition as are discussed in Section 9.5.4. More complex arithmetic operations in the user program must be composed of the basic operations. Only very recent microprocessors (such as type TMS 9900 from Texas Instruments) also have operations available for multiplication and division.

To demonstrate the application of the instruction set, we shall write a program for the addition of two 16-bit numbers: the first term of the sum is assumed to be stored in the two registers 0001 and 0002, i.e. the higher-order byte in 0001 and the lower-order byte in 0002. The second term of the sum is stored similarly in the registers 0003 and 0004. The result is to be transferred to 0005 and 0006.

In the first step, the two lower-order bytes of the straight binary numbers are added, i.e. the contents of the registers 0002 and 0004. As no carry from a previous number is to be taken into account, the operation ADD A is used. The result is stored in register 0006. In the second step, the higher-order bytes are added using the operation ADC A. This ensures that the carry of the previous addition is included, after it has been called by the ALU from the flag register. The result is stored in register 0005. The entire program list is shown in Fig. 11.8.

With the same program, two 4-digit BCD numbers can be added if the NOP instructions (inserted as dummy operations to keep space for other operations to be entered later) are replaced by the BCD correction DA A.

11.3.3 Jump and branch instructions

Flag register

One particular strength of the microprocessor is that it can carry out manifold logic jump and branch instructions. To enable this, various flags are held in the flag register (condition code register) and

operations	mnemonics	addressing modes					short description
		ext.	dir.	inx.	imm.	inher.	
add	ADD A	BB	9B	AB	8B		A plus $[M]$ $\rightarrow$ A
	ADD B	FB	DB	EB	CB		B plus $[M]$ $\rightarrow$ B
	ABA					1B	A plus B $\rightarrow$ A
add with carry	ADC A	B9	99	A9	89		A plus $[M]$ plus C $\rightarrow$ A
	ADC B	F9	D9	E9	C9		B plus $[M]$ plus C $\rightarrow$ B
decimal adjust A	DAA					19	BCD correction
subtract	SUB A	B0	90	A0	80		A minus $[M]$ $\rightarrow$ A
	SUB B	F0	D0	E0	C0		B minus $[M]$ $\rightarrow$ B
	SBA					10	A minus B $\rightarrow$ A
subtract with carry	SBC A	B2	92	A2	82		A minus $[M]$ minus C $\rightarrow$ A
	SBC B	F2	D2	E2	C2		B minus $[M]$ minus C $\rightarrow$ B
2's-complement	NEG	70		60			$[M]^{(2)} \rightarrow M$
	NEG A					40	$A^{(2)}$ $\rightarrow$ A
	NEG B					50	$B^{(2)}$ $\rightarrow$ B
increment by 1	INC	7C		6C			$[M]$ plus 1 $\rightarrow M$
	INC A					4C	A plus 1 $\rightarrow$ A
	INC B					5C	B plus 1 $\rightarrow$ B
	INX					08	X plus 1 $\rightarrow$ X
decrement by 1	DEC	7A		6A			$[M]$ minus 1 $\rightarrow M$
	DEC A					4A	A minus 1 $\rightarrow$ A
	DEC B					5A	B minus 1 $\rightarrow$ B
	DEX					09	X minus 1 $\rightarrow$ X

Fig. 11.7 Arithmetic and logic operations of the microprocessor MC 6800

operations	mnemonics	addressing modes					short description
		ext.	dir.	inx.	imm.	inher.	
1's-complement	COM	73					$[M]^{(1)} \rightarrow M$
	COM A					43	$A^{(1)} \rightarrow A$
	COM B					53	$B^{(1)} \rightarrow B$
AND	AND A	B4	94	A4	84		$A \cdot [M] \rightarrow A$
	AND B	F4	D4	E4	C4		$B \cdot [M] \rightarrow B$
OR	ORA A	BA	9A	AA	8A		$A + [M] \rightarrow A$
	ORA B	FA	DA	EA	CA		$B + [M] \rightarrow B$
exclusive-OR	EOR A	B8	98	A8	88		$A \oplus [M] \rightarrow A$
	EOR B	F8	D8	E8	C8		$B \oplus [M] \rightarrow B$
rotate left	ROL	79		69			$[M]$ 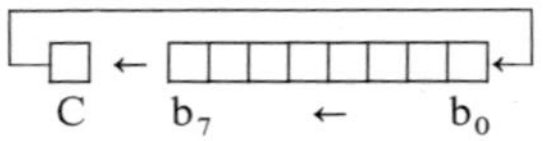
	ROL A					49	A
	ROL B					59	B
shift left, arithmetic	ASL	78		68			$[M]$ 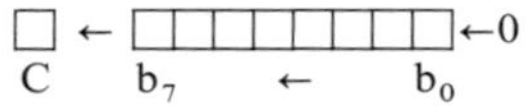
	ASL A					48	A
	ASL B					58	B
rotate right	ROR	76		66			$[M]$ 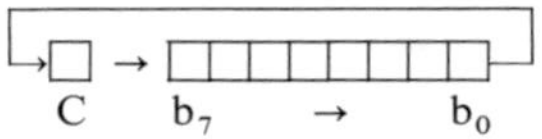
	ROR A					46	A
	ROR B					56	B
shift right, arithmetic	ASR	77		67			$[M]$ 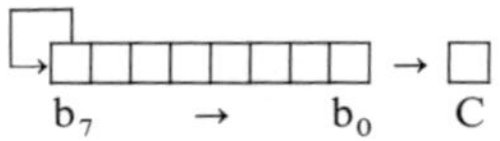
	ASR A					47	A
	ASR B					57	B
shift right, logic	LSR	74		64			$[M]$ 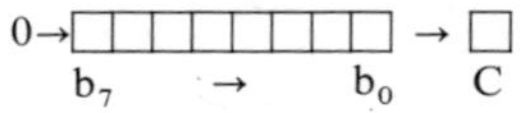
	LSR A					44	A
	LSR B					54	B
no operation	NOP					01	advances program counter only

Fig. 11.7 (continued)

address (hex.)	operation in hex code	mnemonic	comment
FC00 FC01	96 02	LDA A (dir.) 02	
FC02 FC03	9B 04	ADD A (dir.) 04	addition of the two lower-order bytes
FC04	01	· NOP	
FC05 FC06	97 06	STA A (dir.) 06	
FC07 FC08	96 01	LDA A (dir.) 01	
FC09 FC0A	99 03	ADC A (dir.) 03	addition of the two higher-order bytes
FC0B	01	NOP	
FC0C FC0D	97 05	STA A (dir.) 05	
FC0E	3E	WAI	

Fig. 11.8 Program for the addition of two 16-bit numbers

read when required. The flag register is an 8-bit register in which the two most significant bits are always 1. The individual flags are arranged in the following sequence

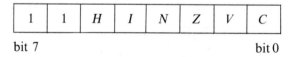

| 1 | 1 | H | I | N | Z | V | C |

bit 7   bit 0

where
C: carry flag, carry from bit 7,
V: overflow flag for 2's-complement representation,
Z: zero flag,
N: sign flag (negative flag) for 2's-complement representation,
I: interrupt flag,
H: half carry flag, half carry from bit 3.

The individual flags are set or reset at all register and arithmetic operations. For example, when a number is loaded into the accumulator, bit 7 of which is 1, flag N is set to 1; if the number is represented in 2's-complement, it will then be interpreted as being negative. If, during an addition or subtraction, the condition for overflow in the 2's-complement representation (see Section 9.5.6) is recognized, the overflow flag V is set to 1. The zero flag is set if the bits 0 to 7 of an operand are zero.

operations	mne-monics	addressing modes					short description
		ext.	dir.	inx.	imm.	inher.	
compare	CMP A	B1	91	A1	81		A minus $[M]$
	CMP B	F1	D1	E1	C1		B minus $[M]$
	CBA					11	A minus B
	CPX	BC	9C	AC	8C		X_H minus $[M]$, X_L minus $[M+1]$
bit test	BIT A	B5	95	A5	85		$A \cdot [M]$
	BIT B	F5	D5	E5	C5		$B \cdot [M]$
test, zero or minus	TST	7D		6D			$[M]$ – 00
	TST A					4D	A – 00
	TST B					5D	B – 00
set carry flag	SEC					0D	$1 \rightarrow C$
clear carry flag	CLC					0C	$0 \rightarrow C$
set overflow flag	SEV					0B	$1 \rightarrow V$
clear overflow flag	CLV					0A	$0 \rightarrow V$
set interrupt mask	SEI					0F	$1 \rightarrow I$
clear interrupt mask	CLI					0E	$0 \rightarrow I$

Fig. 11.9 Operations of the microprocessor MC 6800, manipulating the flag register

There is a series of operations the result of which is given only in the form of flag states. If, for instance, it is of interest to know whether the number in the A register is larger than the number in register B, the difference $(A - B)$ may be calculated using the operation SBA and afterwards the sign flag N is tested. If it is set, then $A < B$. The value of the difference is now stored in register A. Should it be of no interest, the operation CBA may be used instead of SBA. Although the difference $(A - B)$ is still computed and the flag register is set, the value of the difference is not stored and, after the operation, the initial operands are still available in the A and B registers.

A series of further instructions at which no results are stored except the flags, is compiled in Fig. 11.9.

Unconditional jumps

An unconditional jump is carried out without reading the flag register. One must differentiate between absolute addressing (jump) and relative addressing (branch). For a *jump*, the address is specified to which the program counter is to be set. Two different methods may be used, extended addressing or indexed addressing. The address of the next instruction to be executed is then specified as described in Section 11.2.2. This results, for example, in the following program sequence:

a) extended addressing

address	hex code	mnemonics
⋮	⋮	⋮
0107	7E ⎫	
0108	01 ⎬	JMP (ext.) 018F
0109	8F ⎭	
⑁	⋮	⋮
018F		next instruction to be executed
⋮		⋮

b) indexed addressing

address	hex code	mnemonics
⋮	⋮	⋮
0107	6E ⎫	
0108	1A ⎬	JMP (inx.) 1A
⑁	⋮	⋮
$X+1A$		next instruction to be executed
⋮		⋮

For a *branch*, instead of specifying the absolute address of the next instruction to be executed, an offset is given by which the program counter is to be advanced. This has the advantage that the program need not be changed if it is loaded at a different starting address. The specification of the offset is in the form of a 7-bit number with sign, represented in 2's-complement notation. The range of branching is therefore limited to $-128\ldots+127$ program steps. The program sequence may, for example, be as follows:

address	hex code	mnemonics
⋮	⋮	⋮
0107	20 ⎫	
0108	0E ⎬	BRA 0E
⑁	⋮	⋮
$0109+0E$		next instruction to be
$=0117$		executed
⋮		⋮

operations	mnemonics	rel.	ext.	inx.	inher.	description of the branch condition
jump	JMP		7E	6E		
branch always	BRA	20				
branch if $\neq 0$	BNE	26				$Z=0$
branch if $=0$	BEQ	27				$Z=1$
branch if $\geqq 0$ (condition when using unsigned arithmetic)	BCC	24				$C=0$
branch if $\leqq 0$ (condition when using unsigned arithmetic)	BLS	23				$C+Z=1$
branch if >0	BHI	22				$C+Z=0$
branch if <0	BCS	25				$C=1$
branch if $V=0$	BVC	28				$V=0$
branch if $V=1$	BVS	29				$V=1$
branch if $\geqq 0$ (condition when using 2's-complement arithmetic)	BGE	2C				$N\oplus V=0$
branch if $\leqq 0$ (condition when using 2's-complement arithmetic)	BLE	2F				$Z+(N\oplus V)=1$
branch if >0	BGT	2E				$Z+(N\oplus V)=0$
branch if <0	BLT	2D				$N\oplus V=1$
branch if $b_7=0$	BPL	2A				$N=0$
branch if $b_7=1$	BMI	2B				$N=1$
branch to subroutine	BSR	8D				
jump to subroutine	JSR		BD	AD		
return from subroutine	RTS				39	
software interrupt	SWI				3F	
return from interrupt routine	RTI				3B	
wait for interrupt	WAI				3E	

Fig. 11.10 Jump and branch instructions of the microprocessor MC 6800

The offset is counted from the next instruction following the branch instruction. Offset 00 therefore results in a normal program sequence without branch.

Conditional branch

A conditional branch is carried out only if the appropriate condition of the flag register is true. The branch instructions involved employ relative addressing exclusively. If the condition is not true, the program continues without jump and executes the instruction following the branch instruction. Figure 11.10 lists the jump and branch instructions. For operations referring to 2's-complement arithmetic, the sign bit is correctly interpreted even for an overflow, since the overflow flag is also taken into account. The description column indicates the Boolean operations carried out for each particular branch instruction and also gives information on how to test for particular bit combinations.

The application of conditional branch instructions is best explained by an example: the number sequence 0, 1, 2, 3 ... is to be stored in the registers 0200 to $(M-1)$. The higher-order byte of the address M is assumed to be kept in register 0000, the lower-order byte in register 0001.

The program is listed in Fig. 11.11. Firstly the 16-bit number 0200_{hex} is read in the index register and the accumulator A is cleared. At entry into the loop, the contents of accumulator A is stored in the indexed

address hex	operation in hex code	mnemonic	comment
FC00	CE	LDX (imm.) 0200	
FC01	02		
FC02	00		
FC03	4F	CLR A	
→FC04	A7	STA A (inx.) 00	entry into loop
FC05	00		
FC06	4C	INC A	
FC07	08	INX	
FC08	9C	CPX (dir.) 00	
FC09	00		
FC0A	26	BNE (rel.) -08_{dec}	return if $X < M$
FC0B	F8		
FC0C	3E	WAI	

Fig. 11.11 Program for leading the memory with 0, 1, 2 ..., starting at the location 0200

addressing mode. Subsequently, accumulator A and the index register are incremented. If the resulting address is smaller than M, the program counter jumps back to the loop entry. In this manner, the next value of the number sequence is stored in the next higher register, and so on. When $X = M$, the program counter no longer returns and the program stops at the instruction WAI.

Subroutines

The jump or branch to a subroutine (BSR, JSR) is an unconditional jump having the additional feature that the address of the next instruction is stored as a return address in a special register. This enables the user to switch from different parts of the main program to frequently used subroutines. With the instruction RTS (return from subroutine), the program counter is set back to the return address stored at the time.

It is possible to jump from one subroutine to another (nesting). As the return from the first subroutine may only be carried out after the return from the second, the second return address must also be stored. The same applies to all other nested subroutines. The first return jump must be to the address stored last, the second to the one before last and so on. This sequence is organized by a special 16-bit register within the CPU, the so-called *stack pointer*.

For the storage of the return addresses, a range of RAM locations is defined which is not otherwise used. It is known as the *stack*. Its size may be chosen according to the number of nested routines included in the program. After switch-on of the microprocessor, the *highest* address of the stack must be read into the stack pointer of the CPU, using the LDS instruction shown in Fig. 11.12.

If a jump to a subroutine is carried out on the instruction BSR or JSR, the return address (lower-order byte) is automatically stored at the address indicated by the stack pointer. The higher-order byte is stored in the next lower location. The contents of the stack pointer are then decreased by 2 so that it points to the next lower free address of the RAM. If a jump to a second subroutine is incurred, the appropriate return address is stored at that location. With the instruction RTS (return from subroutine), the number in the stack pointer is first increased by 2. This is followed by a return jump to the address stored at the stack location which is currently specified by the stack pointer. In this manner, the return addresses are worked through, as required, in the reverse order of their occurrence.

The stack may also be used for economical temporary storage of the accumulator contents by employing the inherent instructions PSH A or PSH B. The contents are then stored in the memory location

operations	mne-monics	addressing modes					short description
		ext.	dir.	inx.	imm.	inher.	
push accumulator	PSH A					36	$A \to M_{SP}$, SP minus $1 \to SP$
	PSH B					37	$B \to M_{SP}$, SP minus $1 \to SP$
pull accumulator	PUL A					32	SP plus $1 \to SP$, $[M_{SP}] \to A$
	PUL B					33	SP plus $1 \to SP$, $[M_{SP}] \to B$
load stack pointer	LDS	BE	9E	AE	8E		$[M] \to SP_H$, $[M+1] \to SP_L$
store stack pointer	STS	BF	9F	AF			$SP_H \to M$, $SP_L \to M+1$
increment stack pointer	INS					31	SP plus $1 \to SP$
decrement stack pointer	DES					34	SP minus $1 \to SP$
stack pntr. $\to$ index reg.	TSX					30	SP plus $1 \to X$
index reg. $\to$ stack pntr.	TXS					35	X minus $1 \to SP$

Fig. 11.12 Stack operations of the microprocessor MC 6800

specified by the stack pointer. The contents of the pointer is subsequently reduced by 1, since the data word, unlike the address word, is only 8 bits long.

The data are retrieved by the instructions PUL A or PUL B. It is obvious that data stored in this manner must always be retrieved within the *same* subroutine level, as otherwise return addresses and data are mixed up.

Interrupt

An interrupt routine is a special type of subroutine. It may be distinguished from the ordinary subroutine in that the jump from the current program is not initiated by a jump instruction somewhere within the program, but by an arbitrary external control signal. This signal must be applied to the interrupt input *IRQ* (interrupt request) of the CPU.

The start address of the interrupt routine is stored in special locations independent of the program. For the MC 6800, the memory locations FFF8 (higher-order byte) and FFF9 (lower-order byte) are reserved for this purpose.

Since the interrupt request may occur at any stage of the program, steps must be taken to ensure that the program can continue without error after the return from the routine. The original data must therefore

address	stack		
A049	return address	Low	} 1st subroutine
A048	return address	High	
A047	accumulator A		PSH A
A046	accumulator B		PSH B
A045	return address	Low	} 2nd subroutine
A044	return address	High	
A043	return address	Low	
A042	return address	High	
A041	index register	Low	
A040	index register	High	} interrupt
A03F	accumulator A		
A03E	accumulator B		
A03D	flag register		
stack pointer → A03C			

Fig. 11.13 Example of the stack contents after an interrupt

be available in the temporary registers of the CPU. For this reason, at an interrupt, the contents of the accumulators A and B, of the index register and the flag register are automatically stored in the stack. At the instruction RTI (return from interrupt) they are re-read into the CPU.

To demonstrate this process, we have listed in Fig. 11.13 the contents of the stack after an interrupt. We assumed that the program before the interrupt was at the second subroutine level and that during the execution of the first (not yet finished) subroutine, the contents of the accumulators A and B were stored in the stack.

After termination of the interrupt routine, the interrupted program returns to the second subroutine, and later from there to the first. At this level, the effect of the two PSH operations must be cancelled by the corresponding PUL instructions before returning to the main program. At the return, the stack pointer indicates again the highest address of the stack (our example: A049).

Interrupt mask

The interrupt flag I in the flag register enables the user to block the interrupt input IRQ. A jump to the interrupt routine is executed only if the signal IRQ is handed on to the CPU and the flag I cleared. For this reason, the flag is known as the interrupt mask. It may be set or cleared with the instructions SEI or CLI, respectively, which are specified in Fig. 11.9. At a jump to an interrupt routine, the flag is

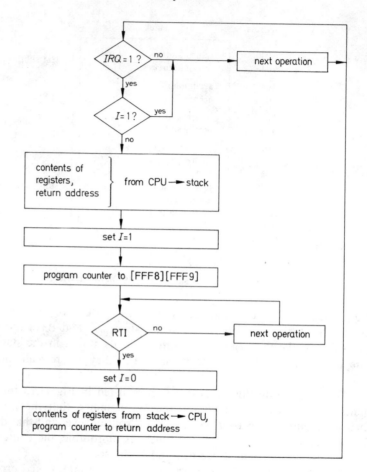

Fig. 11.14 Flow diagram for the execution of the interrupt IRQ and the return RTI

automatically set to prevent this interrupt routine being requested again before it has been completed. Figure 11.14 shows the program sequence following an interrupt request, in the form of a flow diagram.

The control input NMI (non-maskable interrupt) enables the user to switch to a second interrupt routine the starting address of which is stored at FFFC (higher-order byte) and FFFD (lower-order byte). In this kind of interrupt, the interrupt mask bit is not tested and both kinds of interrupt routines can therefore be nested.

A third kind of interrupt routine is started by the SWI instruction (software interrupt). Its start address is not stored in the program either but at the locations FFFA and FFFB. It has an advantage over a normal jump instruction in that the contents of the temporary registers are stored in the stack without the need for additional instructions. The return is initiated by the RTI instruction. The interrupt mask bit is not tested.

Restart

The *RESET* input of the CPU offers an additional possibility for interrupting the current program. It is used for starting the computer. When the CPU recognizes the control signal *RESET*, it loads the *restart* address stored permanently in the memory locations FFFE and FFFF (i.e. hardwired or by switches or by a ROM) into the program counter.

A list of the different start addresses is shown in Fig. 11.15.

address	memory contents		condition	initiation
FFFF	starting address	low	reset	*RESET* input
FFFE	starting address	high		
FFFD	starting address	low	non-maskable	*NMI* input
FFFC	starting address	high	interrupt	
FFFB	starting address	low	software	SWI instruction
FFFA	starting address	high	interrupt	
FFF9	starting address	low	interrupt	*IRQ* input
FFF8	starting address	high	request	

Fig. 11.15 Definition of memory locations for the start addresses of the interrupt routines

11.4   Development aids

As mentioned before, microcomputers are normally employed not as freely programmable computers but as fixed-programmed control and arithmetic units. Their programs are then stored in PROMs.

It has been shown in the previous section that a program may be written directly in hex code by using a programming table. The finished program can be read into a PROM by a PROM programmer, and the PROM inserted in a microcomputer arrangement as in Fig. 11.1. However, in most cases it will be found that the program is not functional because it still contains errors. As the arrangement does not allow experimental changes of individual instructions, the debugging is very difficult and time consuming.

In this section, some ways are discussed allowing the development and testing of programs before they are read into the PROM.

11.4.1   Prototyping computer

If a program is still in the development stage and is to be altered, it should be stored in a RAM rather than a PROM. For the development of a program, a prototyping (or evaluation) microcomputer is used, the configuration of which corresponds to the fixed-programmed original but where RAMs are employed for program storage.

A typical memory organization for a fixed-programmed (or target) computer is illustrated in Fig. 11.16. The program ROMs are situated at the upper end of the memory range so that restart and interrupt addresses can be stored at the locations indicated in Fig. 11.15. The RAMs for data storage are best installed at the bottom of the address range to enable efficient utilization of the direct addressing mode.

Since, when being loaded, the stack grows towards lower addresses, the first instruction in the user program must set the stack pointer to the highest RAM address available. The number of stack locations required depends on the intended number of subroutine and interrupt levels.

The memory organization of the corresponding prototyping computer is shown in Fig. 11.17. The PROMs for program storage are here replaced by RAMs, except for the two uppermost register addresses

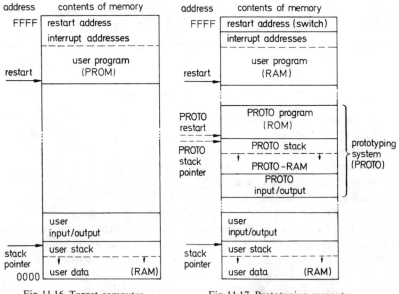

Fig. 11.16 Target computer,
memory assignment

Fig. 11.17 Prototyping computer,
memory assignment

FFFE and FFFF, where the restart address is stored permanently. Switches may be used for this purpose, the positions of which are tested via the data bus when the addresses FFFE and FFFF are called. In most applications of the target computer, far less than the 64 k memory locations theoretically available will actually be implemented. The middle field of the address range in Fig. 11.16 will thus be free. In the prototyping computer, this space is therefore available for the installation of a prototyping system. It consists of a ROM containing the prototyping operating system programs, a RAM for temporary storage and an input/output interface. The main elements of the prototyping operating system (PROTO) program are the routines for input and output.

Input routine:

Scanning of a hexadecimal keyboard or reading of a symbol from a teletypewriter, and loading the corresponding binary number in the accumulator.

Output routine:

Read-out of the accumulator contents in hexadecimal format or converted to teletypewriter symbols.

These two subroutines are combined in different ways to make up the operating programs. These can be called by specific keys.

Display memory:

The address *M* of the desired memory location is entered in the form of a four-digit hex number, and the corresponding contents are obtained on a display as a two-digit hex number, as in Fig. 11.18. The displayed contents can then be changed by entering new numbers.

Set memory:

The contents displayed last are stored in the memory location called. Subsequently, the next higher memory location is automatically read and displayed.

With these two instructions of the operating system program (PROTO program), the user program can be written into the desired RAM range. The switches for encoding the restart address are then

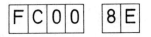

Fig. 11.18 Display after calling the memory location FC00 and writing 8E into it

changed to indicate the start address of the user program instead of the PROTO program. The user program can now be started by a reset signal.

Most PROTO programs have additional PUNCH and LOAD instructions with which the user program can be punched on or read from the paper tape of a teletypewriter. If no teletypewriter is available, the program can be stored on magnetic tape using a *modem.*

The punched paper tape can be directly employed as input for a PROM programmer. A particularly elegant solution, however, uses the prototyping computer directly to transfer the user program into the PROM, simply by executing an additional routine of the PROTO program.

A series of prototyping computers of different manufacturers is available, the ROMs of which contain such a convenient kind of program (e.g. EVK 300 from AMI using the microprocessor MC 6800).

11.4.2   Development systems

The previous section has shown that writing and testing a micro-computer program with a prototyping computer requires very little effort. The precondition for this is that the programmer translates the user program into hex code by employing a programming table. This work may be taken over, however, by the computer if an additional program (assembler) is used to translate the mnemonic into hex code. Since the translation is carried out by the prototyping computer, this program is termed a *resident assembler.* For its storage in the computer, the memory capacity must be considerably larger than for the application discussed in the previous section.

An assembler offers the advantage that, apart from the immediate translation of the instructions, it is also able to compute the address of jump/branch instructions. The appropriate address need only be specified by a symbol. However, a translation of the complete program requires several runs and for each run, a new part of the assembler must be loaded. Even when using a fast paper-tape reader, the work is considerable and a translation "by hand" is still preferable.

Fast and comfortable use of an assembler requires a disk memory (e.g. floppy disk) with an accompanying disk operating system. The sequence of operation can then be automated. A schematic representation of such a *development system* is shown in Fig. 11.19.

The great financial investment (at least $8000. −) is only justified if several programmers constantly need to edit large programs. One must also remember that such a system ties the user down for a long time to one specific type of microprocessor.

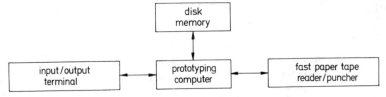

Fig. 11.19 Arrangement of a development system

If a larger computer system is available, there is a further possibility for programming; the *cross-assembler*. Compilation in this case is carried out not in the prototyping computer, but in a third machine. In addition to the cross-assembler, a simulation program is often used to simulate the instruction set of the microprocessor. This enables the compiled program to be tested before it is punched on tape.

In this manner, the tasks of a development system can be achieved solely by software although such programs are not inexpensive (c. $2500.−). A great disadvantage is that the input/output hardware cannot be tested as well. The correct operation of the developed program cannot, therefore, be checked together with the hardware of the apparatus in which the microcomputer is to be used.

11.5 Microcomputer hardware

This section deals with the individual components of a microcomputer in more detail.

11.5.1 CPU

The pin assignment of the microprocessor MC 6800 is shown in Fig. 11.20. All inputs and outputs are TTL compatible except for the clock inputs $\Phi 1$ and $\Phi 2$ which require somewhat larger voltage levels. The effects of most of the inputs and outputs on the operation have been described in the previous sections and are listed below.

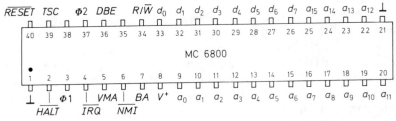

Fig. 11.20 Pin assignment of the microprocessor MC 6800

$\Phi1$ } Non-overlapping clock pulses at $200\,kHz\ldots1\,MHz$.
$\Phi2$ } A suitable clock oscillator, for instance, is the type XC6875.

HALT: When the input is in the low state, the processor activity is stopped. All three-state outputs are floated. For single-instruction operation, the HALT input must be high for one clock pulse of $\Phi1$.

BA: Bus available. The BA signal being in the high state indicates that the microprocessor is in the HALT or WAIT state and that the three-state outputs are floated.

$a_0\ldots a_{15}$:   Address bus (three-state).

TSC: Three-state control for the address outputs. The signal is normally kept in the H-state.

VMA: Valid memory address. The H-state indicates that there is a valid address on the address bus.

$d_0\ldots d_7$:   Bi-directional data bus (three-state).

DBE: Data bus enable; three-state control for the data lines; normally connected to $\Phi2$.

R/$\overline{W}$:   Read/Write. This three-state output indicates to the RAM whether the CPU is in a read (H) or write (L) state.

IRQ }
NMI }   Interrupt inputs.
RESET }

Details of the most common microprocessors are given in Fig. 11.21. When specifying the number of instructions we do not discriminate between the different addressing modes or between the applications of an instruction to different registers. The operations LDAA and LDAB are therefore treated as being identical and count as one instruction.

At present, the trend in development is to design all microprocessors for a single supply voltage of $+5\,V$ and to permit simpler clock waveforms.

A next step in development is the integration of the clock generator and a small RAM on the microprocessor chip (e.g. MC6802). For the 8748, an additional program memory in the form of an UV-erasable PROM (EPROM) and even an input/output interface is integrated on the same chip. A comparison with Fig. 11.1 shows that a single-chip microprocessor evolves in this way into a single-chip microcomputer. The data and address lines are externally accessible so that the single-chip microcomputer can be extended by external memory locations to

type	4040	8080	Z 80	6800	9900
word length, data [bits]	4	8	8	8	16
word length, arithmetic [bits]	4	8	8/16	8	8/16
available address range [kbytes]	12	64	64	64	64
stack	7 levels	external	external	external	external
number of instructions	60	78	158	72	69
relative branch	no	no	yes	yes	yes
clock phases	2	2	1	2	4
max. clock frequency [MHz]	0.75	2	2.5	1	3
average duration of instructions [µs]	22	4	3.2	4	8
supply voltage [V]	−10, +5	−5, +5, +12	+5	+5	−5, +5, +12
technology	PMOS	NMOS	NMOS	NMOS	NMOS
internal memories	dynamic	dynamic	static	dynamic	dynamic
number of pins	24	40	40	40	64
manufacturer	Intel Siemens	Intel Siemens	Zilog Mostek	Motorola AMI	Texas Instr.
start of production	1973	1973	1976	1974	1976
special features			refresh control for dynamic memories		built-in multi- plication/ division

Fig. 11.21 Specifications of the most common types of microprocessors

the largest possible capacity of 64 kbyte. A list of some of these next-generation microprocessors is given in Fig. 11.22.

The arithmetic capabilities of microprocessors are limited to 8-bit or, at best, 16-bit addition and subtraction, with the exception of the TMS 9900 family which has additional instructions for multiplication and division. All other mathematical functions must be implemented by software. Such programs may be rather bulky. A floating-point arithmetic for the four basic operations, using only a simple instruction set, requires a program memory of the order of 1 kbyte. The computing

type	based on type	supply voltage	clock	RAM	EPROM	begin of production
6802	6800	5 V	internal	128 byte	−	1977
8035	8080	5 V	internal	64 byte	−	1977
8748	8080	5 V	internal	64 byte	1 kbyte	1977
9940	9900	5 V	internal	128 byte	2 kbyte	1978

Fig. 11.22 Development trends of microprocessors

times are in the ms-range. The programming of such a routine is usually not necessary, however, since most manufacturers of micropro- cessors also offer program lists of the more important mathematic routines.

The computation time can be reduced by a factor of about 100, by employing a special hardware arithmetic circuit which is addressed in the same way as an input/output port. The arithmetic unit Am 9511 from AMD for instance, along with the four fundamental arithmetic operations, has routines for the computation of transcendental func- tions.

11.5.2 Connecting program- and data-memories

The table in Fig. 11.21 indicates that most microprocessors have a 16-bit address bus, i.e. $2^{16} = 64$ k words can be addressed. Often, however, only part of the available memory range is used although, as a rule, a single memory IC is not sufficient. Several memory ICs must then be connected to the address and the data bus and operated in a multiplex manner by employing their three-state outputs. The less significant address lines are connected simultaneously to each IC of the memory. The remaining more significant address bits are decoded by comparators and used to select individual ICs by their chip-select inputs CS.

In small systems, the address decoding can be simplified by testing only those address bits which are needed to differentiate between memory locations *actually existing* in hardware. This method is known as *partial decoding*. It must be noted, however, that an extension of the memory capacity is then much more difficult.

Read/Write memories

The list in Fig. 11.23 shows a selection of RAMs particularly suited for use in microcomputers. Since the logic voltage levels of memories and microprocessors are TTL compatible, the designer is free to combine microprocessors and memories from different manufacturers. We have limited the selection to static RAMs as they are most simple to operate. Dynamic RAMs are of economic advantage only for systems with very large read/write memories, as additional facilities for refreshing the memories are required. In this respect, the micropro- cessor Z80 offers the useful feature of a refresh controller.

Static RAMs are available for word lengths of 1, 4 and 8 bits. For an extension of word length, the address and chip-select pins of an appropriate number of ICs are connected in parallel. If the memory types listed in Fig. 11.23 are employed, the following advantageous

type	MCM 6810	2112	IM 6561	2102	IM 6508	TMS 4045	TMS 4044
manufacturer	Motorola	Intel	Intersil	Intel	Intersil	Texas Inst.	Texas Inst.
memory capacity [bits]	1 k	1 k	1 k	1 k	1 k	4 k	4 k
organization	128 × 8	256 × 4	256 × 4	1024 × 1	1024 × 1	1024 × 4	4096 × 1
supply voltage [V]	5	5	5	5	5	5	5
power dissipation (active) [mW]	650	150	10	150	10	450	450
power dissipation (standby) [mW]	650	150	<0.001	150	<0.001	450	450
access time [ns]	500...1000	450...1000	250...350	450...1000	250...350	200	200
technology	NMOS	NMOS	CMOS	NMOS	CMOS	NMOS	NMOS
number of pins	24	16	18	16	16	18	18
equivalent type		AM 9112	HM 6561	AM 9102	HM 6508	2114	2147
equiv. type manufacturer		AMD	Harris	AMD	Harris	Intel	Intel

Fig. 11.23 List of suitable static RAMs in MOS technology and their data

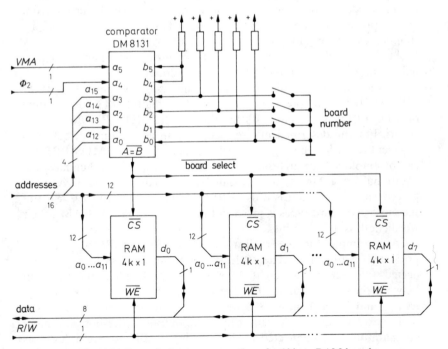

Fig. 11.24 Example for the construction of a 4 kbyte RAM board

configurations for read/write memory blocks are obtained:

$$128 \, \text{byte}: \quad 1 \times \quad 128 \times 8 \, \text{bits,}$$
$$256 \, \text{byte}: \quad 2 \times \quad 256 \times 4 \, \text{bits,}$$
$$1 \, \text{kbyte}: \quad 2 \times 1024 \times 4 \, \text{bits,}$$
$$4 \, \text{kbyte}: \quad 8 \times 4096 \times 1 \, \text{bits.}$$

An example for the construction of a 4 kbyte RAM board is presented in Fig. 11.24. The lower 12 address bits are applied simultaneously to the eight 4 kbit RAMs. To distinguish between individual boards, the four most significant bits are decoded with, for instance, the comparator DM 8131 from National. It can compare two words at 6 bits each and can thus simultaneously decode the condition $VMA \cdot \Phi_2 = 1$ necessary for the operation of RAMs. The condition prevents invalid "transient" addresses initiating unwanted write operations. The board numbers 0000 ... 1111 can be either hardwired on each board or adjusted by means of switches.

Nonvolatile memories

CMOS-RAMs have the special feature that, in the standby state, i.e. for constant input signals, they only draw a very small current from the supply, i.e. in the µA region. In this state they can therefore be supplied over a very long period from a battery. A nonvolatile memory is obtained which can be easily and quickly programmed and in which, unlike the EPROM, individual bits can be erased.

A genuine nonvolatile memory is the user-programmable ROM (PROM). It is available in TTL and MOS technology. The access time of TTL PROMs is below 100 ns, that of the MOS PROMs below 500 ns. For the described microprocessors, an access time of 500 ns is sufficient, even for the highest clock frequency, so that MOS PROMs can be employed throughout. Apart from the low fan-in, they have the additional advantage that they can be erased by UV light. A more precise characterization is thus "EPROM" (erasable PROM). A list of the most important types is given in Fig. 11.25. Unlike for RAMs, their wordlength is always 8 bits.

An example for the realization of an 8 kbyte ROM board is presented in Fig. 11.26. The 10 least significant address lines are connected to each of the eight 1 kbyte EPROMs. The differentiation between the eight memory ICs is effected by the three address bits a_{10} to a_{12} and by a 1-out-of-8 decoder. It is suitable to choose a demultiplexer for decoding. This allows the output of the board number comparator to be connected to the data input of the demultiplexer, as shown in Fig. 11.26. A 1 therefore appears at *all* outputs if the board is

type		1702 A	2704	2708	2758	2716
memory capacity	[bits]	2 k	4 k	8 k	8 k	16 k
organization		256 × 8	512 × 8	1 k × 8	1 k × 8	2 k × 8
supply voltages	[V]	+5, −9	+12, +5, −5	+12, +5, −5	+5	+5
additional programming voltages	[V]	+12, −35, −48	+26	+26	+25	+25
power dissipation	[mW]	700	730	730	300	300
access time	[ns]	1000	450	450	450	450
technology		PMOS	NMOS	NMOS	NMOS	NMOS
number of pins		24	24	24	24	24
manufacturer		Intel National	Intel	Intel Texas Inst. Intersil	Intel	Intel

Fig. 11.25 List of suitable EPROMs and their data

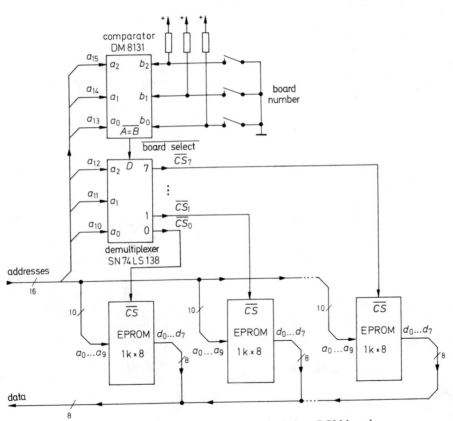

Fig. 11.26 Example for the construction of a 8 kbyte ROM board

not addressed by the three most significant address bits. As the chip-select inputs are inverted (active low), none of the memory ICs would be activated in this case.

11.5.3   Bus system

The address outputs of the microprocessor MC 6800 can accept a current of 1.6 mA when in the L-state, and supply 100 μA in the H-state. This is sufficient to drive one standard-TTL input, 5 low-power-Schottky inputs or 10 MOS inputs.

As is shown in the previous section, the address inputs of all memories are parallel-connected to the address lines. Such a system is called a *bus*. Since the addresses are always supplied by the micropro-cessor and received by the memories, the address bus is unidirectional. The data bus on the other hand, is bidirectional as the data are transmitted and received (i.e. transceived) by the microprocessor on the same lines.

If a larger number of memories is to be connected to the micropro-cessor, buffers for the address and data bus are needed. For the bidirec-tional data bus, bidirectional buffers must be used. They consist of two amplifiers in anti-parallel connection, with three-state outputs which are alternately switched on by means of the output-enable terminal. This pin thus acts as a direction reversal switch. It is controlled by the

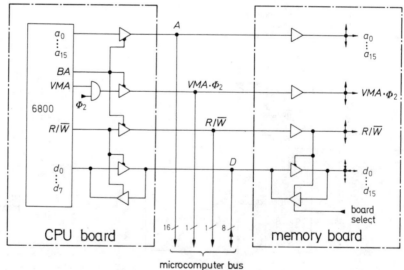

Fig. 11.27 Use of buffers in larger systems

read/write signal $R/\overline{W}$, see Fig. 11.27. When using a buffer, up to 100 MOS inputs can be driven.

Most microprocessors also have three-state gates for the address outputs which are floated (i.e. have a high impedance) in the HALT or WAIT state. This enables the user to apply addresses externally, bypassing the CPU, and thereby exchanging memory contents in the quickest possible manner (direct memory access, DMA). If this method is to be employed, the address buffers on the CPU board must also be three-state buffers. Their output enable pin is controlled by the BA signal of the microprocessor. The following three-state buffers in low-power-Schottky technology are well suited for microprocessor applications:

Unidirectional:

6 bits: 8 T 97 (Signetics), SN 74 LS 367 (Texas Instr.);
8 bits: 74 LS 541 (Fairchild);
          SN 74 LS 241 A (Texas Instr.).

Bidirectional:

4 bits: 8 T 28 (Signetics), 8216 (Intel);
8 bits: SN 74 LS 245 (Texas Instr.).

11.6   Input/output circuits

Exchange of data with the microcomputer requires interfaces which can be called by the CPU. The simplest solution for an input circuit is the connection of three-state buffers to the data bus. They are activated by the address decoder. When calling the desired address, the external data appear at the data bus and are read in the microprocessor. This procedure is identical to the calling of a memory location. The input operation thus differs from a memory operation only in the choice of address. By calling one address, it is possible for an 8-bit data bus to test 8 external data bits simultaneously. Such an input circuit arrangement is shown in Fig. 11.28. For an octal (eight-line) three-state buffer, the IC type SN 74 LS 241 A may be used.

An output port can be constructed in a similar manner. Since the data must not change until new data are to be transferred, flip-flops are used for temporary storage, as is shown in Fig. 11.29 (e.g. SN 74 LS 373). When the microprocessor signals that the address marked by the address decoder is true, C becomes 1. This causes the data to be propagated from the data bus to the output and to remain unchanged until the next output operation is carried out.

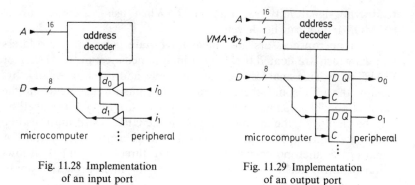

Fig. 11.28 Implementation
of an input port

Fig. 11.29 Implementation
of an output port

11.6.1   Integrated universal interfaces

For more universal applications, integrated birectional interfaces are available, e.g. the PIA (peripheral interface adapter) MC 6820. It contains two peripheral interfaces of 8 bits each. Each of the peripheral data lines can be made to act as an input or a buffered output. For this purpose, a data direction register of 8 bits is assigned to each of the

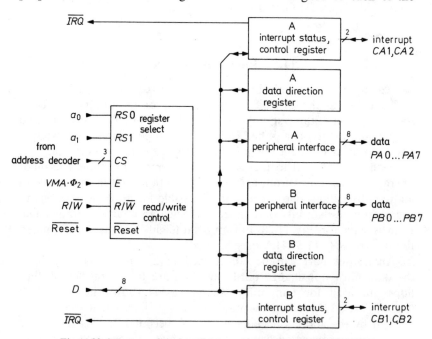

Fig. 11.30 Structure of the parallel input/output circuit MC 6820 (PIA)
Similar types: 8255 from Intel and Z80-PIO from Zilog

interfaces, and also a control register having two additional inputs to allow for the processing of interrupt requests.

The PIA thus contains six 8-bit registers altogether, the arrangement of which is illustrated in Fig. 11.30. They are called by four addresses (two address bits), where each interface and the appropriate direction register have a common address. Discrimination between these two registers is effected by one bit in the control register.

For addressing, the two least significant address lines are directly connected to the PIA (register select, RS0 and RS1). The more significant bits must be applied to a separate address decoder controlling one of the chip-select inputs CS. For small systems, the address decoding can be achieved directly by the AND connected chip-select inputs.

11.6.2 Special modules for serial input/output operations

The interfaces described can also be employed for bit-serial data exchange by using only one output. The data word to be transmitted must then be shifted by one bit place after each output operation. This is effected by software. When receiving, the data word is collated by bit-wise shifting and adding. It is obvious that serial input/output operations using an universal interface require an extensive program and a long computing time.

The parallel-to-serial or serial-to-parallel conversion is better carried out by special hardware. The heart of such a circuit is a shift register having parallel-loading inputs, as described in Section 10.4.3. An additional sequential controller is required which ensures that, at read-out, the 8 bits are propagated at the correct bit rate. The data transfer can be either synchronous or asynchronous.

With synchronous data transfer, the data are propagated down one line and the clock pulses down a second line. Each bit of the data word is assigned to one clock pulse, so that the word can be reconstructed by the receiver in an unambiguous manner.

With asynchronous data transfer, only the data bits are transmitted. In this case, the clock pulses must be generated within the receiver and adjusted to a defined transmission rate. The following bit rates are in use (unit: 1 bit/s = 1 baud):

50	150	1200	4800
75	300	1800	9600
110	600	2400	19200

To attain sufficiently accurate synchronization between the clocks of transmitter and receiver, the clock frequency is derived from a quartz

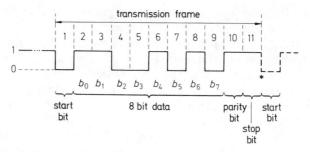

Fig. 11.31 Example of a transmission frame. * Earliest possible instant for the next start bit

time base. The programmable bit-rate generators 34702 from Fairchild or the BR 1941 from Western Digital may be employed for this purpose. They use quartz-crystal frequencies of 2.4576 MHz and 5.0688 MHz, respectively. Adjustable frequency division is available to produce 16-times the bit rates mentioned above.

To enable checking of synchronization, the bit sequence is divided into individual groups (transmission frames). They are marked by additional start and stop bits. For further checking, the data bits may be supplemented by a parity bit, whereby either odd or even parity is stipulated. For even parity, the parity bit is set so that the number of ones in the transmitted data word plus the parity bit is always even, and for odd parity is always odd. This results in bit sequences, an example of which is shown in Fig. 11.31. The following transmission frames are in use:

1) 1 start bit, 8 data bits, 1 parity bit, 1 stop bit;
2) 1 start bit, 7 data bits, 1 parity bit, 2 stop bits.

The parity check may be either for even or for odd parity or left out altogether.

It is now obvious that the control of serial data transmission requires considerable circuitry. It is therefore useful to employ special integrated circuits such as the ACIA MC 6850 (asynchronous communications interface adapter), the block diagram of which is represented in Fig. 11.32. The circuit contains four registers:

a transmit data register for the parallel-to-serial conversion,
a receive data register for the serial-to-parallel conversion,
a control register defining the operation mode,
a status register indicating the operational state.

The four registers are selected by the address bit a_0 and the read/write signal. This is possible as the receive-data and the status register

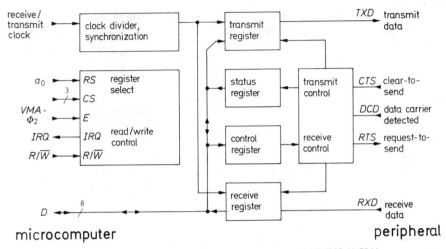

receive/
transmit → | clock divider,
clock | synchronization

a_0 → | RS | register
  → /3 → | CS | select

VMA·
Φ_2 → | E

IRQ ← | IRQ | read/write
  | control

$R/\overline{W}$ → | $R/\overline{W}$

D ← /8 →

transmit
register

status
register

control
register

receive
register

transmit
control

receive
control

TXD   transmit
      data

CTS   clear-to-
      send

DCD   data carrier
      detected

RTS   request-to-
      send

RXD   receive
      data

microcomputer   **peripheral**

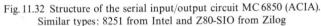

Fig. 11.32 Structure of the serial input/output circuit MC 6850 (ACIA).
Similar types: 8251 from Intel and Z80-SIO from Zilog

are read-only registers whereas the transmit-data and the control register are write-only registers.

The transmission frame and the parity condition can be chosen using the 8-bit control register. It can also define which condition is to initiate an interrupt request. The frequency division ratio for the clock can also be programmed ($n = 1, 16, 64$). When selecting clock rates of 16 or 64 times the bit rate, automatic synchronization with the start bit of received data is accomplished.

If a data word is to be transmitted, bit number 1 of the status register must be tested. It becomes 1 as soon as the previous data word has been completely transmitted, i.e. when the transmit data register is empty. If this condition is recognized, the next data word can be written into the transmit data register. It is subsequently independently transmitted by the ACIA.

If the computer is waiting for input, bit number 0 of the status register must be tested. It changes to 1 as soon as the receive data register is full, i.e. when a data word has been received completely. By appropriately setting the control register, this condition may be used to initiate an interrupt.

Further bits in the status register indicate whether or not the parity condition is fulfilled, or if a loss of information has been incurred by overrun of the receive data register. This may happen if the computer does not call the previous word in time.

Three control lines CTS, DCD and RTS are available for organizing the exchange of data. The signal RTS (request to send) can be set to either 0 or 1, by a bit in the control register. It may be used in the program to start or stop a paper-tape reader or puncher. Both the inputs *CTS* (clear to send) and *DCD* (data carrier detected) are intended to provide automatic control of a modem. If *CTS* is zero, the ACIA will not transmit data; if *CDC* is zero, it will not receive.

The name "data carrier detected" refers to data transfer by frequency modulation in the audio range, using a modem (modulator/demodulator). If the modem signals to the computer that the carrier frequency is detected, the computer interprets this as an indication that valid data are available. In this sense, this signal can also be used for other peripherals. When the automatic peripheral control is not used, the inputs *CTS* and *DCD* must be kept at 1 so as not to block the ACIA.

11.6.3   ASCII code

Data communication to and from a teletypewriter or a CRT (cathode ray tube) terminal is a particularly important application of serial data transmission. A special 7-bit code was standardized for this purpose known as the ISO-7-bit code or the ASCII code (American standard code for information interchange). The coding list of individual symbols is shown in Fig. 11.33.

When transmitting these symbols in a serial manner, bit number 0 is transmitted first and bit number 7 is added as the parity bit. If the transmission format "7 bits + parity" is selected for the ACIA, the most significant bit, though monitored for the parity check, is always written on the data bus as being zero. In this manner, the 8-bit word of a symbol becomes independent of the mode of parity checking. The shortened hexadecimal notation of the 8-bit word is also shown in Fig. 11.33.

It can be seen that the digits 0 to 9 are assigned to the hex numbers 30 to 39. The straight binary numbers assigned to an ASCII number may thus be evaluated simply by subtracting 30_{hex}.

The columns 0 and 1 in Fig. 11.33 contain special instructions concerning the teletypewriter operation. The most important instructions can be called up by special keys, e.g.

CR, carriage return $\cong 0D_{hex}$,
LF, line feed $\cong 0A_{hex}$.

The remainder are called up by the control (CTRL) key in connection with the corresponding symbol in column 4 or 5. The symbol BEL (bell) $\cong 07_{hex}$, for instance, is obtained by simultaneous keying of CTRL and G.

	hex equiv.	0	1	2	3	4	5	6	7
	$b_6 b_5 b_4$	000	001	010	011	100	101	110	111
hex equiv.	$b_3 b_2 b_1 b_0$								
0	0 0 0 0	NUL	DLE	SP	0	@	P	`	p
1	0 0 0 1	SOH	DC1	!	1	A	Q	a	q
2	0 0 1 0	STX	DC2	"	2	B	R	b	r
3	0 0 1 1	ETX	DC3	#	3	C	S	c	s
4	0 1 0 0	EOT	DC4	$	4	D	T	d	t
5	0 1 0 1	ENQ	NAK	%	5	E	U	e	u
6	0 1 1 0	ACK	SYN	&	6	F	V	f	v
7	0 1 1 1	BEL	ETB	'	7	G	W	g	w
8	1 0 0 0	BS	CAN	(	8	H	X	h	x
9	1 0 0 1	HT	EM	)	9	I	Y	i	y
A	1 0 1 0	LF	SUB	*	:	J	Z	j	z
B	1 0 1 1	VT	ESC	+	;	K	[	k	{
C	1 1 0 0	FF	FS	,	<	L	\	l	¦
D	1 1 0 1	CR	GS	–	=	M	]	m	}
E	1 1 1 0	SO	RS	.	>	N	↑	n	~
F	1 1 1 1	SI	US	/	?	O	←	o	DEL

Fig. 11.33 ASCII code

Transmission of teletypewriter signals

The level of signals on the transmission lines to and from a teletype-writer is defined by special standards. The following methods of transmission are available: the current loop interface and the voltage interface. In the current loop method, a 1 is characterized by an open circuit, whereas a 0 is determined by a current of 20 mA. For a voltage interface (CCITT recommendation V.24, US standard RS 232 C, DIN standard 66020), a 1 is represented by a voltage between -3 V and -25 V, a 0 by a voltage between $+3$ V and $+25$ V. For transmission of the ASCII symbol "S", the voltage waveform in Fig. 11.34 is then obtained.

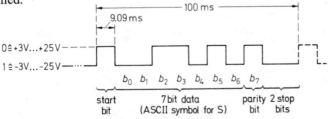

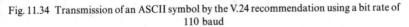

Fig. 11.34 Transmission of an ASCII symbol by the V.24 recommendation using a bit rate of 110 baud

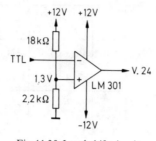

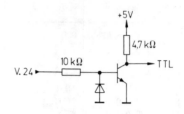

Fig. 11.35 Level-shift circuit,   Fig. 11.36 Level-shift circuit,
        from TTL to V.24   from V.24 to TTL

The bipolar voltage levels cannot be generated by TTL circuits and special level shift circuitry is required. Simple solutions are shown in Figs. 11.35 and 11.36. Such level translators are available as monolithic ICs, e.g.

$4 \times$ TTL-to-V.24:   MC 1488 L,
$4 \times$ V.24-to-TTL: MC 1489 L.

If a computer is to be connected to a terminal, the standard requires that the computer is the control unit and the terminal is the transmission unit. Often, however, terminals possess a V.24 connection primarily for the operation of a modem, i.e. the terminal itself is used as the control unit. Therefore, the connectors of the computer interface cannot simply be connected to the corresponding socket of the terminal, as output would then be connected to output, and input to input. If the automatic peripheral control lines (CTS, DCD) are not used, a transfer of data is still possible by having the DCD and CTS inputs permanently at 1 and cross-connecting the two data lines, as shown in Fig. 11.37. The free ACIA output RTS may be employed as

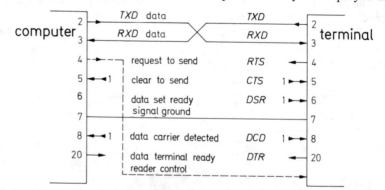

Fig. 11.37 Connection of a computer to a terminal if both have connectors designed for modem use

reader control. This signal serves to switch on and off the paper-tape reader in a teletypewriter.

The numbers entered in Fig. 11.37 define the connector pins as set out in the V.24 standard. The arrows indicate that, contrary to the standard, *both* units are designed as controlling units. The levels for the control signals are defined in the V.24 standard as having a polarity opposite to that of the data. A 1 is thus represented by a voltage between $+3\,V$ and $+25\,V$. This voltage level is available at the connector pin 20 (data terminal ready, DTR).

11.6.4   Direct memory access (DMA)

The *direct memory access* (DMA) is a special way of exchanging data with the computer. In this mode of operation, the CPU is disconnected from the address and the data bus. For the MC 6800, for instance, this can be done by a hardware $HALT$ or a software $WAIT$ instruction. At these instructions, all three-state outputs of the microprocessor are floated (high impedance).

This allows the user to externally apply in succession different addresses and to exchange data using the $R/\overline{W}$ control. The rate is limited only by the access time of the memories. In this manner, transmission rates of over 1 Mbyte/s are attainable.

Since it is often necessary to transfer whole fields of memory locations, a programmable counter is needed with which the desired range can be assigned, one location after the other. Such DMA controllers are available as integrated circuits, e.g. the MC 6844 from Motorola or the 8257 from Intel.

11.7   Minimum systems

This section deals in more detail with the implementation of a minimum microcomputer system. Such systems consist of only a few large-scale ICs but replace a multitude of SSI and MSI circuits.

11.7.1   Memory location assignment

Figure 11.16 has already shown the basic pattern of memory utilization in a microcomputer. Figure 11.38 illustrates the allocation of addresses for the case when a 1 kbyte PROM is employed for program storage and a 128 byte RAM for data storage. In addition, a PIA is provided for input/output operations. Hence, a total of

$$1024 + 128 + 4 = 1156$$

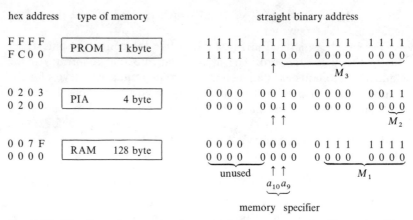

Fig. 11.38 Allocation of memories for a minimum system and address decoding. M_1, M_2, M_3 denote address inputs of the appropriate memory. The highest and the lowest address is entered for each memory group

memory locations are to be used. To distinguish between them, eleven address bits are necessary, decoded as shown in Fig. 11.38: the address bits a_0 to a_9, designated by M_3, are directly applied to the PROM. Bit a_{10} specifies the range of program storage and is applied to the chip-select input of the PROM.

The locations of the input/output registers are designated by the 01 combination at the address lines a_{10} and a_9. The internal registers of the PIA are selected by the address bits a_0 and a_1.

The RAM locations are identified by the combination 00 at the address lines a_{10} and a_9. The 128 locations are distinguished by the address bit a_0 to a_6 (M_1). By the additional decoding of bit a_7 and a_8, the RAM range can be extended to 512 bytes without having to change the memory specifier decoder.

With the partial decoding technique used, the address bits a_{11} to a_{15} are not tested, and therefore each location may be called by 32 different 16-bit addresses. The highest digit of the 4-digit hexadecimal number may thus have any value. However, it is recommended to choose the addresses in the program as if they were fully decoded. This ensures that the program is operable even in a prototyping computer which decodes the address completely.

11.7.2   Construction of hardware

The implementation based on the memory location assignment in Fig. 11.38, is shown in Fig. 11.39. To minimize the number of ICs,

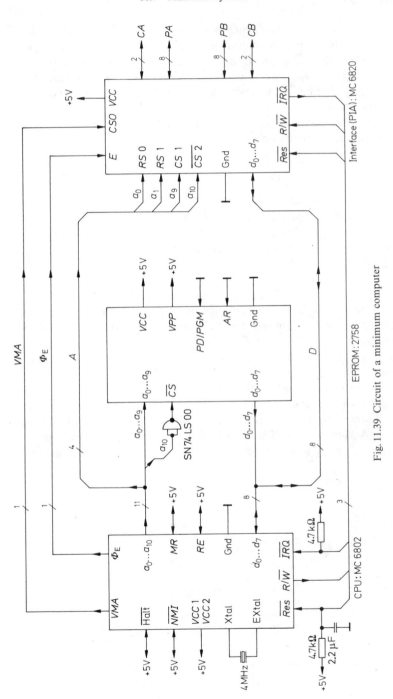

Fig. 11.39 Circuit of a minimum computer

the microprocessor MC 6802 is chosen as the CPU, since it already contains the clock generator and the 128-byte RAM. The RAM addresses are fully decoded to enable an extension to 64 kbyte.

The quartz frequency must be four times the clock frequency since a frequency divider is incorporated. This allows the use of inexpensive quartz crystals.

The RC lowpass filter at the reset input automatically starts the program after switch-on of the supply.

11.7.3 Scanning of keyboards

Microcomputer minimum systems are always devised for a specific application. The peripheral control is then usually effected by a number of key switches and only rarely by a terminal. The following section therefore deals with a method of scanning a keyboard without involving additional hardware.

To minimize the number of connections, the keys are arranged in a matrix, as in Fig. 11.40, and scanned using the multiplex method. For a 4×4 matrix, eight connections are thus required. One side of the PIA MC 6820 can be employed for this purpose, if the pins PA0 to PA3 are operated as outputs and connected to the horizontal lines (rows) as shown in Fig. 11.40. The pins PA4 to PA7 are operated as inputs and connected to the vertical lines (columns).

If no key is pressed, the inputs are open-circuit. In this case, the input potential must be defined, and pull-up (to the supply voltage) or pull-down (to ground) resistors are therefore required at the inputs. The A side of the PIA already contains pull-up resistors so that it can be directly used for key scanning. An open input then assumes H-

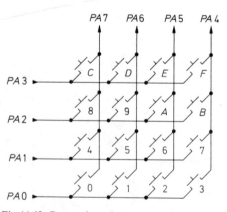

Fig. 11.40 Connection of a keyboard to the PIA

key	columns (input)	rows (output)
	PA: 7654	PA: 3210
0	1000	0001
1	0100	0001
2	0010	0001
3	0001	0001
4	1000	0010
5	0100	0010
6	0010	0010
7	0001	0010
8	1000	0100
9	0100	0100
A	0010	0100
B	0001	0100
C	1000	1000
D	0100	1000
E	0010	1000
F	0001	1000

Fig. 11.41 Input and output signals of the PIA

state. To ensure that a closed key defines an L-state, the PIA outputs are employed to apply an L-voltage to each row in succession and simultaneously to apply an H-voltage to all other rows.

For example, if key "5" is pressed, an L-state appears at a PIA input only if the output PA1 is at L. It can also be seen from Fig. 11.40 that the L-voltage then appears at input PA6.

A list of the key assignment to the logic states at the inputs and outputs is shown in Fig. 11.41, using negative true logic. The key scanning program can be deduced directly: the rows are tested one after the other for a 1 and the number of testing steps (beginning at zero) is counted. When a 1 is found, the state of the counter is read, this being equal to the hexadecimal-coded number of the key. If no 1 has been found by the time the counter has reached the state F_{hex}, no key has been pressed.

The corresponding flow chart is represented in Fig. 11.42. Input and output takes place via the accumulator A, and counting of the scanning steps is accomplished by the lower 4 bits of accumulator B. A 1 in the highest bit of B indicates that no pressed key has been found.

Advancing the logic 1 from one row to the next is achieved by left-shifting within accumulator A. When checking the lines, the contents of A are temporarily stored in the stack and subsequently, an input

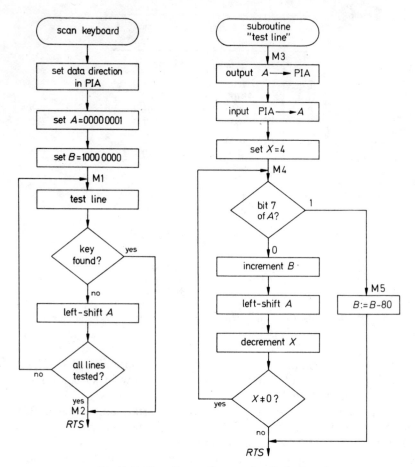

Fig. 11.42 Flow diagram for scanning the keyboard

operation from the PIA to A is made. The most significant bit of A is then tested, and if it is zero, the A-contents are shifted by one bit to the left and a new test of the most significant bit is carried out, etc. After 4 unsuccessful tries, the step to the next row is made. The index register counts the number of scanning steps within a row.

When a pressed key has been recognized, the MSB of B is cleared (subtraction of 80_{hex}) and the scanning is stopped. For the example of key "5" being pressed, accumulator B then contains the number 05_{hex}.

If no key is pressed, a transition occurs in accumulator B from $8F_{hex}$ to 90_{hex}, after the last testing step. This state serves as a criterion for the unsuccessful run of the key scanning program. It may be utilized in the user program to initiate a new scanning sequence.

address	op code	labels	mnemonics	comments
FC00	8E 00 7F		LDS #H, 007F	*main program*
FC03	BD FF C0		JSR → M0	
⋮	⋮ ⋮ ⋮		⋮ ⋮	
FFDF				*scan keyboard*
FFC0	7F 02 01	M0	CLR 0201	
FFC3	86 0F		LDA A #B, 0000 1111	
FFC5	B7 02 00		STA A 0200	definition of the data
FFC8	86 04		LDA A #H, 04	direction in the PIA
FFCA	B7 02 01		STA A 0201	
FFCD	86 01		LDA A #B, 0000 0001	
FFCF	C6 80		LDA B #B, 1000 0000	
FFD1	36	M1	PSH A	
FFD2	8D 0A		BSR → M3	branch to "test line"
FFD4	32		PUL A	
FFD5	5D		TST B	
FFD6	2A 05		BPL → M2	key found?
FFD8	48		ASL A	
FFD9	C1 90		CMP B #H, 90	
FFDB	2D F4		BLT → M1	all keys tested?
FFDD	39	M2	RTS	
FFDE	43	M3	COM A	*test line*
FFDF	B7 02 00		STA A 0200	
FFE2	B6 02 00		LDA A 0200	
FFE5	43		COM A	complementation for
FFE6	CE 00 04		LDX #4	negative true logic
FFE9	4D	M4	TST A	
FFEA	2B 06		BMI → M5	bit 7 of A = 1?
FFEC	5C		INC B	
FFED	48		ASL A	
FFEE	09		DEX	
FFEF	26 F8		BNE → M4	entire line tested?
FFF1	39		RTS	
FFF2	C0 80	M5	SUB B #H, 80	clear bit 7 of B
FFF4	39		RTS	to indicate that key
				has been found
FFFE	FC			restart address
FFFF	00			

Fig. 11.43 Complete program for scanning the keyboard

The complete listing of the program is shown in Fig. 11.43. In contrast to the program examples of Section 11.3, we have chosen a shorter notation which is used in a similar form by most assembler programs. The bytes belonging to one instruction are written on the same line, its length being dependent on the address mode. The address indicated by the instruction refers to the first byte.

It is usual in the mnemonic description to write the operand directly after the instruction, e.g. LDAA 0200. If there is no indication to the contrary, the number will be interpreted as a hex address, i.e. as an extended address if it consists of four digits, or as a direct address if it contains only two digits. In the immediate addressing mode, either the two-digit hex number may be inserted or the expanded bit combination, or the corresponding decimal number or even the corresponding ASCII symbol. The notation normally used is represented below:

adressing mode: representation of operand:

extended	□ □ □ □	(address hex number)
direct	□ □	(address hex number)
indexed	□ □, X	(offset hex number)
relative	label, e.g. M1	

immediate:

# H,	□ □	(data hex number)
# B,	□...□ (8)	(data binary number)
#	□ □	(data decimal equiv.)
#	ʹ□	(data ASCII equiv.)

In order that the keyboard scanning routine may be called from anywhere within the user program, it is written as a subroutine, i.e. the program is terminated by RTS. The call for this subroutine from the main program is accomplished by the instruction JSR (jump to subroutine). All subroutines are best stored in the upper part of the program memory, to keep the lower part coherent and available for the main program. The highest locations FFF8...FFFF are reserved for the interrupt and restart addresses.

The main program begins at the lowest available program memory location, i.e. at FC00, according to Fig. 11.38. It is started via the reset pin of the CPU, so this address must be stored at FFFE/FFFF. With the first instruction of the main program, the stack pointer is set to the highest available RAM address because the stack always grows towards lower addresses. According to Fig. 11.38, this is the address 007F.

The program can then be run without the help of a prototyping operating system.

12 Digital filters

In Chapter 3, several transfer functions are discussed and their realization by active filters is described. The processed signals are voltages which in turn are continuous functions of time. The circuits are made up of resistors, capacitors and amplifiers.

Recently, the trend has been towards signal processing by digital rather than analog circuits. The advantages are high accuracy and consistency in the results and a lower sensitivity to disturbances. The high number of digital components required is a disadvantage, however, but in view of the increase in integration of digital circuitry, is becoming less important.

Sequences of discrete numbers are processed instead of continuous signals, and the circuit elements are memories and arithmetic circuits. When changing over from analog to digital filters, two questions arise: (1) Is it possible to transform a continuous input voltage into a sequence of discrete numbers without loss of information? (2) How is this sequence of numbers to be processed to obtain a desired transfer function? These two questions are dealt with in the sections below.

12.1 Sampling theorem

12.1.1 Theory

A continuous input signal can be converted to a series of discrete values, by using a sample-and-hold circuit for sampling the signal at equidistant instants $t_\mu = \mu T_s$, where $f_s = 1/T_s$ is the sampling rate. It is obvious from Fig. 12.1 that a staircase function arises, and that the approximation to the continuous input function is better the smaller the sampling time T_s. Depending on the accuracy required, an appropriately high sampling rate will be necessary, a demand which cannot normally be met by the circuitry available. Shannon's sampling theorem, however, states that it is possible to sample the input function at a relatively low frequency and, despite this, recover the original signal by appropriate filtering. This involves the precondition that the input function $U_1(t)$ is *band limited*, i.e. that its spectrum $|F_1(jf)|$ above a frequency f_{max} is, at least approximately, zero. If this condition is not fulfilled, it is often possible to enforce it by using a lowpass filter at the input without unduly changing the signal.

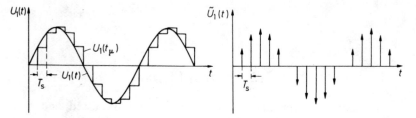

Fig. 12.1 Example of an input signal $U_1(t)$ Fig. 12.2 Representation of the input sig-
and the sampled values $U_1(t_\mu)$ nal by a Dirac impulse sequence

In order to obtain a simpler mathematical description, the staircase function in Fig. 12.1 is replaced by a series of Dirac impulse functions, as illustrated in Fig. 12.2:

$$\tilde{U}_1(t) = \sum_{\mu=0}^{\infty} U_1(t_\mu) \, T_s \delta(t - t_\mu). \tag{12.1}$$

Their impulse area $U_1(t_\mu) \cdot T_s$ is represented by an arrow. The arrow must not be mistaken for the height of the impulse, as a Dirac function is, by definition, an impulse with infinite height but zero width although its area has a finite value. This area is often misleadingly known as the impulse amplitude. The characteristics of the impulse function are shown by Fig. 12.3, where the Dirac impulse function is approximated by a square wave pulse r_ε; the limit of the approximation is

$$U_1(t_\mu) \, T_s \delta(t - t_\mu) = \lim_{\varepsilon \to 0} U_1(t_\mu) \, r_\varepsilon(t - t_\mu). \tag{12.2}$$

For an investigation of the information contained in the impulse function sequence represented by Eq. (12.1), we consider its spectrum. By applying the Fourier transformation to Eq. (12.1), we obtain

$$\tilde{F}_1(j f) = T_s \sum_{\mu=0}^{\infty} U_1(\mu T_s) \, e^{-2\pi j \mu f / f_s}. \tag{12.3}$$

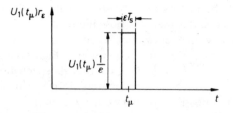

Fig. 12.3 Approximation of a Dirac impulse by a finite voltage pulse

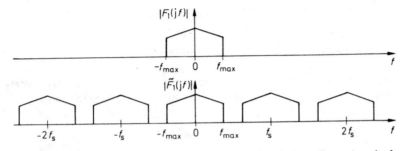

Fig. 12.4 Spectrum of the input voltage before sampling (upper diagram) and after sampling (lower diagram)

It can be seen that this spectrum is a periodic function, the period being identical to the sampling frequency f_s. When Fourier-analyzing this periodic function, it can be shown that the spectrum $|\tilde{F}_1(jf)|$ is, for $-f_{max} \leq f \leq f_{max}$, identical to the spectrum $|F_1(jf)|$ of the original waveform [12.1]. It thus still contains all the information although only a few values of the function were sampled.

There is only one restriction, and this is explained with the help of Fig. 12.4. The original spectrum reappears unchanged only if the sampling rate is chosen such that consecutive bands do not overlap. According to Fig. 12.4, this is the case for

$$\boxed{f_s \geq 2f_{max}}, \tag{12.4}$$

this condition being known as the sampling theorem.

Recovery of the analog signal

From Fig. 12.4, the condition for recovery of the analog signal can be directly deduced. The frequencies above f_{max} in the spectrum must be cut off by a lowpass filter. The lowpass filter must therefore be designed so as to have zero attenuation at f_{max} and infinite attenuation at $(f_s - f_{max})$ and above. To summarize: the original waveform can be recovered from the sampled values of a continuous band-limited function of time as long as the condition $f_s = 2f_{max}$ is fulfilled. To ensure this, the sampled values must be transformed into a sequence of Dirac impulse functions which is in turn applied to an ideal lowpass filter with $f_c = f_{max}$.

If the sampling rate is lower than the frequency demanded by the sampling theorem, spectral components arise which have the difference frequency $(f_s - f) < f_{max}$. They are not suppressed by the lowpass filter and are found as a beat in the output signal (aliasing).

12.1.2   Practical aspects

For a practical realization, the problem arises that a real system is unable to generate Dirac impulse functions. The impulses must thus be approximated, as in Fig. 12.3, by a finite amplitude and a finite time interval, thereby abandoning the limit concept of Eq.(12.2). By inserting Eq.(12.2) in (12.1), we obtain, for finite ε, the approximated impulse sequence

$$\tilde{U}_1'(t) = \sum_{\mu=0}^{\infty} U_1(t_\mu)\, r_\varepsilon(t - t_\mu). \qquad (12.5)$$

The Fourier transformation yields the spectrum

$$\tilde{F}_1'(\mathrm{j}f) = \frac{\sin \pi \varepsilon T_s f}{\pi \varepsilon T_s f} \cdot \tilde{F}_1(\mathrm{j}f), \qquad (12.6)$$

which is the same as for the Dirac impulse sequence, except for a superposed weighting function causing an attenuation of the higher frequency components. The case of the staircase function is particularly interesting, as the pulse width εT_s is identical to the sampling interval T_s. The spectrum is then given by

$$\boxed{\tilde{F}_1'(\mathrm{j}f) = \frac{\sin (\pi f / f_s)}{\pi f / f_s} \cdot \tilde{F}_1(\mathrm{j}f)}. \qquad (12.7)$$

The magnitude of the weighting function is represented in Fig. 12.5, along with the symbolic spectrum of the Dirac impulse functions. At half the sampling rate, an attenuation of 0.64 is obtained; at $0.2f_s$ it is only 0.94. The influence of the weighting function on the spectrum below the band limit f_{max} thus remains negligibly small as long as $f_s \approx 5 f_{max}$.

As shown above, the recovery of the original signal requires a lowpass filter which stops all frequencies in the spectrum above f_{max}. A real filter, however, possesses a finite roll-off. In order that these components of the spectrum can still be safely cut off, f_s must again be larger than $2 f_{max}$. A sufficiently high attenuation at the lowest detri-

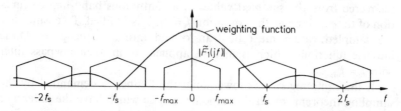

Fig. 12.5 Transition from the spectrum of a Dirac impulse sequence to the spectrum of the staircase function by means of the weighting function $|(\sin \pi f / f_s)/(\pi f / f_s)|$

mental frequency $(f_s - f_{max})$ can then be achieved. It is even advantageous to use a staircase function instead of the Dirac impulse functions, as the resulting weighting function itself has lowpass filter characteristics.

The deformation of the amplitude spectrum in the passband region can be compensated by increasing the gain of the lowpass filter near the band limit. For a sufficient decrease in attenuation above f_{max}, the insertion of a zero in the frequency response, at $f_s - f_{max}$, is recommended.

12.2   Digital transfer function

It is shown in Chapter 3 that analog filters can be realized by integrators, summing amplifiers and coefficient circuits. A digital filter is obtained if the integrators are replaced by delay elements and the summing amplifiers by adders. The delay elements may, for instance, be shift registers through which the sampled values of the input waveform are shifted at the sampling rate f_s. The simplest case is that of unit delay, i.e. the delay by one time interval T_s. An example of such a first-order digital filter is illustrated in Fig. 12.6. To begin with, its performance is to be described in the time domain.

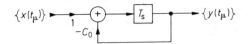

Fig. 12.6   Example of first-order digital filter

12.2.1   Time domain analysis

The number sequence $\{x(t_\mu)\}$ is given and applied as the input signal to the circuit in Fig. 12.6. The resulting output sequence $\{y(t_\mu)\}$ is to be determined. At the time t_μ, the number $x(t_\mu) - C_0 y(t_\mu)$ is at the input of the register. This value appears, delayed by one clock period, at the output of the register. We therefore obtain for the values of the output sequence the relationship

$$y(t_{\mu+1}) = x(t_\mu) - C_0 y(t_\mu).\qquad(12.8)$$

This *difference equation* is analogous to the differential equation of the continuous system. It can be used as a recursion formula for the determination of the output sequence if the initial value $y(t_0)$ is given. As an example, we choose $y(t_0) = 0$ and calculate the step response for $C_0 = -0.75$. The result is shown in Fig. 12.7; it can be seen that the circuit has a lowpass filter characteristic.

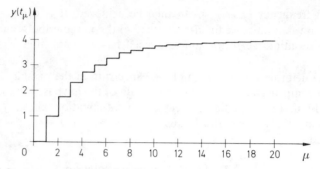

Fig. 12.7 Step response of the digital filter in Fig. 12.6, for $C_0 = -0.75$ and a unity step
at the input

12.2.2 Frequency domain analysis

For an investigation of the frequency response, the sinusoidal
sequence $x(t_\mu) = x_0 \sin \omega t_\mu$ is applied to the input. If the system is linear,
a sinusoidal sequence appears at the output. As for analog filters, the
ratio of the amplitudes is equivalent to the magnitude of the transfer
function for $p = j\omega$. The linearity of a digital filter is indicated by the
linearity of the difference equation. According to Eq. (12.8), the filter in
Fig. 12.6 is therefore linear.

The transfer function may be inferred from the circuit with the
help of complex calculus, as for analog filters. This requires that the
frequency response of a delay element is known. With the harmonic
input sequence

$$x(t_\mu) = x_0 e^{j\omega t_\mu},$$

the harmonic output sequence

$$y(t_\mu) = x_0 e^{j\omega(t_\mu - T_s)} = x_0 e^{j\omega t_\mu} \cdot e^{-j\omega T_s}$$

is obtained, and with $j\omega = p$, the transfer function

$$A(p) = 1 \cdot e^{-pT_s}. \tag{12.9}$$

It is a periodic function, the period being $\omega_s = \dfrac{2\pi}{T_s} = 2\pi f_s$, where f_s is the
sampling, i.e. clock frequency.

To abbreviate,

$$\boxed{z = e^{+pT_s}}, \tag{12.10}$$

which results, together with Eq. (12.9), in the transfer function of a delay
element

$$\boxed{\tilde{A}(z) = z^{-1}}. \tag{12.11}$$

It was mentioned in Chapter 3, that the transfer function $A(p)$ describes the relationship between the output signal and any desired time-dependent input signal if the Laplace transforms according to the equation

$$L\{y(t)\} = A(p) \cdot L\{x(t)\} \tag{12.12}$$

are used. This relationship also holds for a digital system. Using the converted transfer function of Eq. (12.11), the determination for number sequences can be simplified since

$$Z\{y(t_\mu)\} = \tilde{A}(z) \cdot Z\{x(t_\mu)\}, \tag{12.13}$$

where

$$Z\{x(t_\mu)\} = X(z) = \sum_{\mu=0}^{\infty} x(t_\mu) z^{-\mu} \tag{12.14}$$

is the Z-transform of the input sequence. The output sequence is obtained by the corresponding reverse transform [12.2]. Because of this property, $A(z)$ is called the *digital transfer function*.

With the digital transfer function Eq. (12.11) for the delay element, we are able to directly determine the digital transfer function for the digital filter in Fig. 12.6. From

$$Y(z) = [X(z) - C_0 Y(z)] z^{-1},$$

we obtain

$$\tilde{A}(z) = \frac{Y(z)}{X(z)} = \frac{1}{C_0 + z}. \tag{12.15}$$

For the determination of the frequency response $\underline{A}(j\omega)$, we let $z = e^{j\omega T_s}$ and obtain

$$\underline{A}(j\omega) = \frac{1}{C_0 + e^{j\omega T_s}} = \frac{1}{C_0 + \cos\omega T_s + j\sin\omega T_s}. \tag{12.16}$$

This function is periodic in $2\pi f_s$, a property which is common to all digital filters. For the amplitude frequency response we calculate

$$|\underline{A}(j\omega)| = \frac{1}{\sqrt{(C_0 + \cos\omega T_s)^2 + \sin^2\omega T_s}}.$$

This function is shown in Fig. 12.8; it may be seen that the amplitude frequency response for $0 \leq \omega T_s \leq \pi$, i.e. $0 \leq f \leq \frac{1}{2}f_s$ has clear lowpass filter characteristics, as is to be expected from the step response in Fig. 12.7.

The sampling theorem demands that the frequency of a sinusoidal sequence $\{x(t_\mu)\}$ in a discrete delay system clocked with the frequency f_s, is not larger than $\frac{1}{2}f_s$. The amplitude frequency response can therefore not be used in the range $f > \frac{1}{2}f_s$. If it were to be implemented

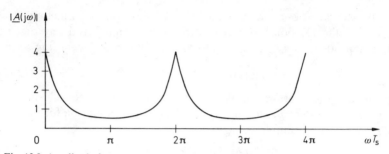

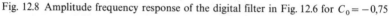

Fig. 12.8 Amplitude frequency response of the digital filter in Fig. 12.6 for $C_0 = -0,75$

for measuring purposes, the clocked delay elements would have to be replaced by continuous delay elements, e.g. delay lines or allpass filters.

12.3   Bilinear transformation

It would be of advantage for the design of digital filters if the optimized transfer function of the analog filters in Chapter 3 could be modelled digitally. However, it has been shown in the previous section that this is not feasible since digital filters differ from analog filters in that they possess a transfer function periodic for $0 \le f \le \infty$. Because the useful frequency range is limited to $0 \le f \le \frac{1}{2}f_s$, the frequency response need be modelled only up to $\frac{1}{2}f_s$ and is repeated periodically in the non-relevant range $f > \frac{1}{2}f_s$. In a similar manner to the lowpass-bandpass transformation, the amplitude frequency response of the analog filter may be modified for this purpose by a transformation of the frequency axis. This is done in such a manner that the frequency range $0 \le f \le \infty$ is mapped into the range $0 \le f \le \frac{1}{2}f_s$ and continued periodically at higher frequencies. Therefore,

$$f = \frac{f_s}{\pi} \tan \frac{\pi f'}{f_s}. \tag{12.17}$$

For $f \to \infty$, $f' \to \frac{1}{2}f_s$ as required. For $f' \ll f_s$, the frequency becomes $f \approx f'$. Hence, the compression of the frequency axis is less significant the higher the clock frequency f_s compared to the frequency range of interest.

The optimized transfer functions in Chapter 3 are always plotted against a normalized frequency $\Omega = f/f_0$, where f_0 is the cutoff frequency or the resonant frequency of the filter. To enable direct calculation in this normalized notation, we introduce a normalized sampling frequency

$$\Omega_s = \frac{f_s}{f_0}. \tag{12.18}$$

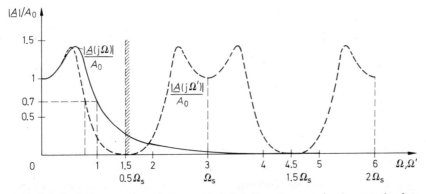

Fig. 12.9 Establishment of a periodic amplitude frequency response, for the example of a second order Chebyshev characteristic with 3 dB ripple. Normalized sampling frequency: $\Omega_s = 3$. Linear representation

Equation (12.17) thereby becomes

$$\Omega = \frac{\Omega_s}{\pi} \tan \frac{\pi \Omega'}{\Omega_s}. \tag{12.19}$$

As an example of the transformation of the frequency axis, we have illustrated the amplitude frequency response of a second order Chebyshev lowpass filter in Fig. 12.9. It can be seen that the typical passband characteristic remains unchanged, but the cutoff frequency is shifted. To avoid this effect, we shift (in the logarithmic representation) the frequency response curve before transformation so that the cutoff frequencies coincide after transformation. With Eq. (12.19), it follows that

$$\Omega = l \tan \frac{\pi \Omega'}{\Omega_s}, \tag{12.20}$$

where

$$l = \cot \frac{\pi}{\Omega_s}. \tag{12.21}$$

This results in Ω' being 1 for $\Omega = 1$. The frequency response curve so transformed is represented in Fig. 12.10. We interpret the previously introduced quantity Ω' as the new frequency variable and denote the transformed frequency response by $\underline{A}'(j\Omega)$. It can be seen that this is a good approximation of the analog filter curve.

The above statements indicate that the transformed frequency response $\underline{A}'(j\omega)$ has a form easily implemented by a digital filter. For the calculation of the digital transfer function $\tilde{A}(z)$, the transformation equation for the complex frequency variable P is needed. With $P = j\Omega$,

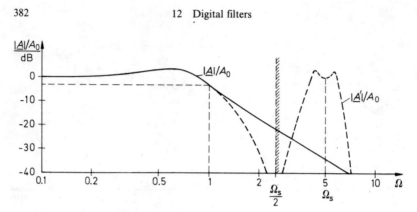

Fig. 12.10 Matching of the cutoff frequency, for the example of a second order Chebyshev characteristic with 3 dB ripple. Normalized sampling frequency: $\Omega_s = 5$. Logarithmic representation

one obtains from Eq. (12.20)

$$P = l \tanh \frac{\pi P'}{\Omega_s} = l \frac{e^{2\pi P'/\Omega_s} - 1}{e^{2\pi P'/\Omega_s} + 1}.$$

With $\Omega_s = \dfrac{f_s}{f_0} = \dfrac{1}{T_s f_0}$, $2\pi P' f_0 = p'$ and $e^{p' T_s} = z$, the result is given as

$$\boxed{P = l\frac{z-1}{z+1} \quad \text{where} \quad l = \cot\frac{\pi}{\Omega_s}} \qquad (12.22)$$

This relationship is called the bilinear transformation.

To summarize, an analog filter function is transformed to a digital filter function in the following way: in the analog transfer function $A(P)$, the normalized complex frequency variable P is replaced by $l(z-1)/(z+1)$, whereby the transfer function $\tilde{A}(z)$ is obtained which can be implemented by a digital filter. The amplitude frequency response is then very similar in shape to that of the analog filter. The characteristic is compressed along the Ω axis so that the value $|A(j\infty)|$ appears at the frequency $\frac{1}{2}\Omega_s$. The deviations caused thereby are smaller, the larger Ω_s compared to the frequency range of interest $0 < \Omega < \Omega_{max}$.

The phase frequency response, on the other hand, is changed to a far greater extent. Statements concerning analog filters may therefore not be directly carried over to digital applications. For this reason, it would not be wise to approximate the amplitude frequency response, for instance, of a Bessel filter because the linearity in the phase delay is lost. It is more appropriate to solve such approximation problems in the z domain [12.2].

When realizing digital filters, it is easiest to cascade first- and second-order filter blocks, just as has been described for analog filters. The recalculation of the filter coefficients is therefore given explicitly for this case. From the *analog* transfer function

$$A(P) = \frac{d_0 + d_1 P + d_2 P^2}{c_0 + c_1 P + c_2 P^2},$$ (12.23)

the *digital* transfer function

$$\tilde{A}(z) = \frac{D_0 + D_1 z + D_2 z^2}{C_0 + C_1 z + C_2 z^2}$$ (12.24)

is obtained by applying the bilinear transformation. The equations for the recalculation of the first-order coefficients ($c_2 = d_2 = 0$) are then:

$$D_0 = \frac{d_0 - d_1 l}{c_0 + c_1 l}; \quad C_0 = \frac{c_0 - c_1 l}{c_0 + c_1 l};$$

$$D_1 = \frac{d_0 + d_1 l}{c_0 + c_1 l}; \quad C_1 = 1;$$ (12.25)

$$D_2 = 0; \quad C_2 = 0.$$

Correspondingly, for a second-order filter block ($c_2 \neq 0$):

$$D_0 = \frac{d_0 - d_1 l + d_2 l^2}{c_0 + c_1 l + c_2 l^2}; \quad C_0 = \frac{c_0 - c_1 l + c_2 l^2}{c_0 + c_1 l + c_2 l^2};$$

$$D_1 = \frac{2(d_0 - d_2 l^2)}{c_0 + c_1 l + c_2 l^2}; \quad C_1 = \frac{2(c_0 - c_2 l^2)}{c_0 + c_1 l + c_2 l^2};$$ (12.26)

$$D_2 = \frac{d_0 + d_1 l + d_2 l^2}{c_0 + c_1 l + c_2 l^2}; \quad C_2 = 1.$$

From Eq. (12.24) with Eq. (12.10), the magnitude of the transfer function is obtained as

$$|\underline{A}'(j\Omega)|$$

$$= \sqrt{\frac{\left[D_0 + D_1 \cos\dfrac{2\pi\Omega}{\Omega_s} + D_2 \cos\dfrac{4\pi\Omega}{\Omega_s}\right]^2 + \left[D_1 \sin\dfrac{2\pi\Omega}{\Omega_s} + D_2 \sin\dfrac{4\pi\Omega}{\Omega_s}\right]^2}{\left[C_0 + C_1 \cos\dfrac{2\pi\Omega}{\Omega_s} + C_2 \cos\dfrac{4\pi\Omega}{\Omega_s}\right]^2 + \left[C_1 \sin\dfrac{2\pi\Omega}{\Omega_s} + C_2 \sin\dfrac{4\pi\Omega}{\Omega_s}\right]^2}}.$$

(12.27)

The phase shift is given by

$$
\varphi = \tan^{-1} \frac{D_1 \sin \dfrac{2\pi\Omega}{\Omega_s} + D_2 \sin \dfrac{4\pi\Omega}{\Omega_s}}{D_0 + D_1 \cos \dfrac{2\pi\Omega}{\Omega_s} + D_2 \cos \dfrac{4\pi\Omega}{\Omega_s}}
$$

$$
- \tan^{-1} \frac{C_1 \sin \dfrac{2\pi\Omega}{\Omega_s} + C_2 \sin \dfrac{4\pi\Omega}{\Omega_s}}{C_0 + C_1 \cos \dfrac{2\pi\Omega}{\Omega_s} + C_2 \cos \dfrac{4\pi\Omega}{\Omega_s}} .
$$

It can be seen that both functions are periodic in Ω_s. If the digital transfer function $\tilde{A}(z)$ in Eq. (12.24) has been obtained from an analog transfer function, the amplitude and phase frequency response can obviously be determined much more simply directly from Eq. (12.23) by transforming the Ω axis according to Eq. (12.20), as has already been shown in Fig. 12.10.

12.4   Construction of digital filters

As has been described in the previous section, a digital filter is used to change an input sequence $\{x(t_\mu)\}$ into an output sequence $\{y(t_\mu)\}$ and thereby implement a desired digital transfer function $\tilde{A}(z)$. If the input signal is continuous, the number sequence $\{x(t_\mu)\}$ must first be produced from it, this giving the block diagram in Fig. 12.11. To fulfill the conditions imposed by the sampling theorem, an (analog) lowpass filter must be used for band limiting. The sample-and-hold circuit that follows, takes samples from the band limited signal, at intervals $T_s = 1/f_s$. These samples are converted to the number sequence $\{x(t_\mu)\}$ by an analog-to-digital converter and applied to the input of the digital filter. The output sequence $\{y(t_\mu)\}$ may be processed digitally, or reconverted to a continuous output signal by means of a digital-to-analog converter and a lowpass filter. The aspects discussed in Section 12.1 must then be taken into account.

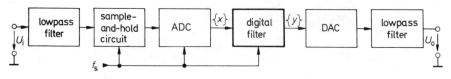

Fig. 12.11 Filtering a continuous signal by a digital filter

12.4.1 Structure of first order digital filters

Figure 3.44 illustrates an analog filter in which all coefficients of a general second order transfer function can be adjusted according to Eq. (3.42) or (12.23). Since Eq. (12.24) for the digital transfer function* $\tilde{A}(z)$ possesses the same structure, it can also be implemented by the same arrangement, if the integrators are replaced by delay elements. In the case of a first order filter, only a single delay element is required, this leading to the circuit in Fig. 12.12. Its transfer function $\tilde{A}(z)$ can be directly determined using the method described in Section 12.2.2: with the help of Eq. (12.11) for the delay element, we obtain the Z-transformed output sequence

$$Y(z) = D_1 X(z) + z^{-1}[D_0 X(z) - C_0 Y(z)]$$

and therefore the digital transfer function

$$\tilde{A}(z) = \frac{Y(z)}{X(z)} = \frac{D_0 + D_1 z}{C_0 + z}. \qquad (12.28)$$

When evaluating $\tilde{A}(z)$ from the specific analog transfer functions, some particular properties of the coefficients become apparent. As for the analog filters they can be used as an indication of the type of filter. By application of Eq. (12.25) the following relationships are found:

Lowpass filter:

$$A(P) = \frac{d_0}{c_0 + c_1 P} \quad \Rightarrow \quad \tilde{A}(z) = D_0 \frac{1 + z}{C_0 + z}. \qquad (12.29)$$

Highpass filter:

$$A(P) = \frac{d_1 P}{c_0 + c_1 P} \quad \Rightarrow \quad \tilde{A}(z) = D_0 \frac{1 - z}{C_0 + z}. \qquad (12.30)$$

Hence, a lowpass filter is characterized by $D_1 = D_0$, a highpass filter by $D_1 = -D_0$. It is now clear why the circuit in Fig. 12.6 is not a genuine lowpass filter but only has a similar characteristic. Coefficient D_1 in Eq. (12.28) is not equal to $+D_0$; in the corresponding analog filter, this would be equivalent to the coefficient d_1 not being zero. The damping at high frequencies thus remains finite.

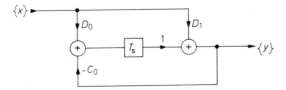

Fig. 12.12 First order digital filter

12.4.2 Structure of second order digital filters

If the circuit in Fig. 12.12 is extended by a further delay element, a second order filter block is obtained, represented in Fig. 12.13. There are also other possible implementations [12.2], which are, however, not dealt with here. Higher-order filters can be realized by adding further delay elements. Cascading first- and second-order filter blocks is the clearest solution. The factorized form of the transfer function, needed for this approach, is best obtained by recalculating the coefficients of Fig. 3.14 using Eq. (12.26).

The transfer function $\tilde{A}(z)$ of the filter block in Fig. 12.13 is found in the same way as that of the first order filter. From Fig. 12.13 it follows directly that,

$$Y = D_2 X + z^{-1}[D_1 X - C_1 Y + z^{-1}(D_0 X - C_0 Y)].$$

Hence,

$$\tilde{A}(z) = \frac{Y(z)}{X(z)} = \frac{D_0 + D_1 z + D_2 z^2}{C_0 + C_1 z + z^2}. \tag{12.31}$$

The circuit can thus be used to implement any desired second order transfer function.

As for the first order filters, the coefficients indicate the special properties of the filter. By applying the recalculation equation (12.26) to the different types of filter, the following relationships are found:

Lowpass filter:

$$A(P) = \frac{d_0}{c_0 + c_1 P + c_2 P^2} \quad \Rightarrow \quad \tilde{A}(z) = D_0 \frac{1 + 2z + z^2}{C_0 + C_1 z + z^2}. \tag{12.32}$$

Highpass filter:

$$A(P) = \frac{d_2 P^2}{c_0 + c_1 P + c_2 P^2} \quad \Rightarrow \quad \tilde{A}(z) = D_0 \frac{1 - 2z + z^2}{C_0 + C_1 z + z^2}. \tag{12.33}$$

Bandpass filter:

$$A(P) = \frac{d_1 P}{c_0 + c_1 P + c_2 P^2} \quad \Rightarrow \quad \tilde{A}(z) = D_0 \frac{1 - z^2}{C_0 + C_1 z + z^2}. \tag{12.34}$$

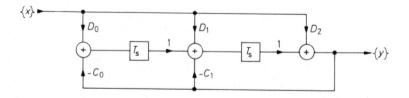

Fig. 12.13 Second order digital filter

The design process of the circuit is best explained by a worked example. A Chebyshev second order lowpass filter having 0.5 dB ripple and a 3 dB cutoff frequency of $f_c = 100\,\text{Hz}$, is required. The bandwidth of the analog signal is to be 3.4 kHz and the signal sampled at the rate $f_s = 10\,\text{kHz}$. The normalized sampling frequency is therefore

$$\Omega_s = \frac{f_s}{f_c} = \frac{10\,\text{kHz}}{100\,\text{Hz}} = 100.$$

With Eq. (12.21), the normalizing factor is given as

$$l = \cot\frac{\pi}{\Omega_s} = 31.82.$$

From the table in Fig. 3.14, the continuous transfer function

$$A(P) = \frac{1}{1 + 1.3614P + 1.3827P^2}$$

is obtained. Comparing the coefficients with those of Eq. (12.23) yields the values

$$d_0 = 1, \quad d_1 = d_2 = 0, \quad c_0 = 1, \quad c_1 = 1.3614, \quad c_2 = 1.3827.$$

With these and Eq. (12.26)

$$D_0 = 6.923 \cdot 10^{-4}, \quad C_0 = 0.9400, \quad C_1 = -1.937.$$

Using Eq. (12.32), the digital transfer function is then given by

$$\tilde{A}(z) = 6.923 \cdot 10^{-4} \cdot \frac{1 + 2z + z^2}{0.9400 - 1.937z + z^2},$$

which can be implemented by the circuit in Fig. 12.13.

The ratio of the sampling frequency to the cutoff frequency is defined by the circuit parameters chosen and has the value 100. Hence, the cutoff frequency is proportional to the sampling frequency. It may therefore be easily controlled by the sampling frequency. This property is characteristic of all digital filters.

In a second example, we design a bandpass filter. As before, the sampling rate is to be 10 kHz. The resonant frequency is assumed to be $f_r = 1\,\text{kHz}$ so that

$$\Omega_s = \frac{f_s}{f_r} = 10.$$

For a Q-factor of 10, the continuous transfer function according to Eq. (3.24) is represented by

$$A(P) = \frac{0.1P}{1 + 0.1P + P^2}.$$

With $l = \cot\dfrac{\pi}{\Omega_s} = 3.078$ and Eq. (12.26), the digital transfer function is determined as

$$\tilde{A}(z) = -2.855 \cdot 10^{-2} \cdot \frac{1 - z^2}{0.9429 - 1.572\,z + z^2}.$$

Correspondingly, for a Q-factor of 100,

$$\tilde{A}(z) = -2.930 \cdot 10^{-3} \cdot \frac{1 - z^2}{0.9941 - 1.613\,z + z^2}.$$

In the case where $Q = 10$ and $\Omega_s = 100$, the transfer function changes to

$$\tilde{A}(z) = -3.130 \cdot 10^{-3} \cdot \frac{1 - z^2}{0.9937 - 1.990\,z + z^2}.$$

It may be seen that with increasing Q-factor and increasing normalized sampling frequency, the coefficient D_0 decreases continually, whereas C_0 approaches 1 and C_1 approaches -2. The information about the filter type is then held in the very small deviations from 1 or -2, respectively. This requires an increased accuracy of the coefficients, i.e. the word length within the filter must be correspondingly extended. In order to keep the circuitry reasonably small, the sampling frequency ought not to be larger than absolutely necessary.

12.4.3 Practical aspects

When designing a digital filter, the main outlay goes into the implementation of the coefficient circuits which multiply the signal by the appropriate coefficient. The internal word length of the filter must be equal to the word length of the input or output signal plus the word length of the coefficients so that relevant information is not lost at multiplication. The filter characteristic would otherwise be amplitude-dependent, resulting in distortions.

The accuracy required of the multiplier for the denominator coefficients increases the closer C_0 comes to 1 and C_1 to -2. In this case, simplification is attained by the following conversion:

$$C_0\,y = (1 - C_0')\,y = y - C_0'\,y,$$

where $C_0' = 1 - C_0$ is the deviation of C_0 from 1. This coefficient needs considerably fewer significant bits than C_0. For most applications, a resolution of 4 bits is sufficient for first order filters and of 8 bits for second order filters. The price of the subtractor is small compared to the saving in multiplier bits.

Coefficient C_1 is split accordingly.

$$C_1 y = (-2 + C_1') y = -2y + C_1' y$$

where $C_1' = 2 + C_1$.

Simple example for the hardware implementation of a digital filter

The implementation of a first order highpass filter is demonstrated as an example. Its cutoff frequency is to be $100\,\text{Hz}$, the bandwidth of the input signal is $3.4\,\text{kHz}$. We choose the sampling rate as $f_s = 10\ \text{kHz}$, i.e. $\Omega_s = 100$. From

$$A(P) = \frac{P}{1+P},$$

we obtain with Eq. (12.25)

$$\tilde{A}(z) = -0.9695 \frac{1-z}{-0.9391+z} = -0.9695 \frac{1-z}{(-1+\underbrace{0.0609})+z}.$$
$$C_0'$$

For a calculation in the straight binary system, C_0' is represented as a straight binary fraction:

$$(0.0609)_{\text{dec}} = (0.000011111001\ldots)_2.$$

This number is close to $2^{-4} = 0.0625$. The number of components required for the multiplication $C_0' y$ could be considerably reduced if C_0' were *exactly* equal to an integral power of 2, in our case 2^{-4}. The multiplication could then be carried out by simply shifting y by 4 bits.

Considering Eq. (12.25) it can be seen that this requirement may be met by a slight change in the normalizing factor l, i.e. by a change in Ω_s. From $C_0 = -1 + 2^{-4}$

$$l = \frac{2 - 2^{-4}}{2^{-4}} = 31 \quad \text{and} \quad \Omega_s = 97.423,$$

whereby the cutoff frequency for a sampling frequency of $10\,\text{kHz}$ is increased from $100\,\text{Hz}$ to $102.6\,\text{Hz}$. A further simplification results from this modification in that the coefficient D_0 also assumes an integral value:

$$D_0 = -1 + 2^{-5} = -0.9687\ldots.$$

To simplify even further, we set this factor to -1 as it influences only the total gain, i.e. the gain of the filter for high frequencies now approaches the value $(1 + l)/l = 1.0323$. As a result, the block diagram in

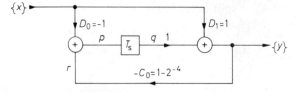

Fig. 12.14 Digital first order highpass filter.

Cutoff frequency: $f_c = \dfrac{f_s}{\Omega_s} = \dfrac{f_s}{97.42}$

Gain: $A_\infty = 1 + \dfrac{1}{l} = 1.032$

Fig. 12.14 is obtained, having the digital transfer function

$$\tilde{A}(z) = \frac{-1+z}{(-1+2^{-4})+z}.$$

The actual circuit is represented in Fig. 12.15 for an input word length of 4 bits. In order that positive and negative numbers can be processed, the 2's-complement representation discussed in Section 9.2.6 is chosen. The MSB is therefore the sign bit. Since the multiplication can be carried out simply by shifting, adders only are needed. We have employed 4-bit arithmetic units of the type Am 25 LS 2517. By appropriate control signals, they can be made to act also as subtractors. In this manner, the evaluation of the 2's-complement for the coefficients $D_0 = -1$ and $-C_0 = 1 - 2^{-4}$ may be incorporated in the adder operation.

The two arithmetic units, IC8 and IC9, evaluate the expression

$$r = -C_0 y = y - 2^{-4} y.$$

The multiplication of y by 2^{-4} is achieved by connecting y to the subtraction inputs in such a way that it is shifted by 4 bits. This effectively increases the word length from 4 bits to 8 bits.

The sign bit v_y must be connected to *all* the inputs made available by the shift so that correct multiplication of y by 2^{-4} is achieved for positive as well as negative values of y. An example clarifies the procedure:

$$+48 \qquad 0\ 011\ 0000$$

$$\cdot\, 2^{-4} = 0\ 000\ 0011 \ = 3;$$

$$-48 \qquad 1\ 101\ 0000$$

$$\cdot\, 2^{-4} = 1\ 111\ 1101 \ = -3.$$

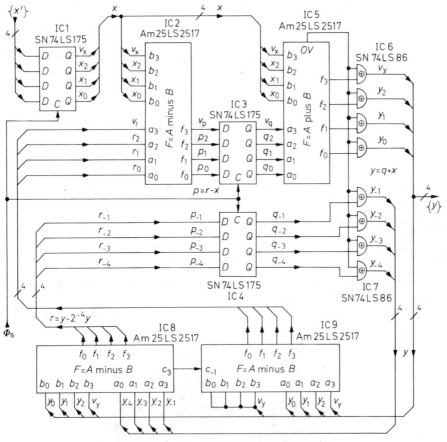

Fig. 12.15 Detailed circuit of the digital highpass filter

The arithmetic unit IC2 corresponds to the first adder in Fig. 12.14, IC5 to the second. The delay by one clock pulse is achieved by ICs 3 and 4, each containing four edge-triggered D-type flip-flops. The flip-flops in IC1 serve to synchronize the input signal.

The exclusive-OR gates of IC6 and 7 protect the circuit from latching. As is shown in Section 9.5.6, exceeding the positive number range would result in a jump from $+127$ to -128 since the MSB is read as the sign. As a result of the undesired change in the sign, the filter may become unstable and may not return to the normal operating mode. This effect exactly corresponds to the latch-up known in analog circuitry. It may, for example, be avoided by setting the numbers at the adder output to $+127$ for a positive overflow and to -128 for a negative overflow, but this would require separate decoding facilities.

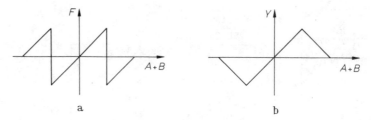

Fig. 12.16 (a) and (b) Characteristic of the arithmetic units. (a) Without limiting logic circuitry, (b) with limiting logic circuitry

A distinction between these two cases is not necessary if the outputs are complemented as soon as the permissible range is exceeded. This leads to the characteristic in Fig. 12.16 b. It is implemented, as is shown in Fig. 12.15, by connecting exclusive-OR gates to the outputs f_i of those arithmetic units where the permissible output range may be exceeded. The gates effect the complementing when $OV = 1$. Compared to the standard arithmetic units SN 74181, the types Am 25 LS 2517 have the advantage of a directly available overflow variable OV which need not be externally evaluated.

The operation of the digital highpass filter can be visualized by its step response shown in Fig. 12.17.

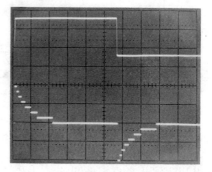

Fig. 12.17 Step response of the digital highpass filter in Fig. 12.15, for maximum input signal

Sequential signal processing

It can be seen from the circuit design in Fig. 12.15 that the arrangement involves considerable outlay although we have chosen the most basic example with all possible simplifications. The question therefore arises whether the choice of a digital filter is justified, bearing in mind the relatively simple analog solution. One such case would be the processing of extremely low signal frequencies, which can be carried out without difficulties by a digital system if a correspondingly low sampling frequency is chosen. An analog system for such an application involves extremely long time constants, the realization of which may be impossible.

For low sampling frequencies, digital filters can be considerable simplified by using only a single arithmetic unit to carry out all the operations *consecutively*. Intermediate results must be stored until the next clock pulse arrives. A microcomputer is well suited for the execution of these operations. As an example, the block diagram in Fig. 12.14 is transformed to the corresponding flow diagram in Fig. 12.18. In order that the next value for x can actually be read-in at each loop, the period T_s of the filter clock must be larger than the computing time within the loop. With standard types (8080, 6800), values around $100\,\mu s$ may be attained for a first order filter having an 8-bit word length. The maximum sampling frequency in this case is about 10kHz. For larger word lengths and higher filter orders it is reduced correspondingly while the circuit size is not significantly increased.

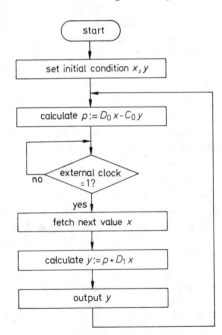

Fig. 12.18 Flow diagram for programming a first order digital filter on a microcomputer

13 Data transmission and display

13.1 Transmission lines

We have assumed so far that digital signals are not affected when transmitted from one integrated circuit to the next. However, the effect of the interconnecting lines, particularly on the steep signal edges, may be significant. As a general rule, the simple connecting wire is no longer adequate if the time delay along that wire is in the same order of magnitude as the rise time of the signal. Such connections should therefore have a maximum length of about 100 mm for each nanosecond of rise time. If this limit is exceeded, serious pulse deformations, reflections and oscillations are incurred. These disturbances can be avoided by the use of lines having a defined characteristic impedance (coaxial line, microstrip line). The line is then terminated by its characteristic impedance which usually has a value between 50 Ω and 300 Ω.

Microstrips can be realized by placing all the connecting conductors on the underside of the printed circuit board and entirely metallizing the component side of the board, leaving only very small eyes for isolation of the component connections. In this manner, all conductors on the underside of the board become microstrip lines. If the board has a dielectric permittivity of $\varepsilon_r = 5$ and a thickness of $d = 1.2$ mm, a characteristic impedance of 75 Ω is obtained for a conductor width of $w = 1$ mm [13.1].

For connections between two printed circuit boards, coaxial lines can be used but have the disadvantage that they are difficult to incorporate in multiway connectors. It is considerably easier to transmit the signal via two single insulated, but twisted wire (twisted-pair line) which can be connected to two neighbouring pins of standard multiway connectors. If the wires are twisted approximately 100 times per meter, a characteristic impedance of about 110 Ω is obtained [13.1].

The simplest method of transmitting data via a twisted-pair line is shown in Fig. 13.1. Since the characteristic impedance required for the termination is low, the transmitting gate must be able to produce a correspondingly high output current. Such gates are known as IC line drivers (buffers). At the receiving end it is useful to employ a Schmitt trigger gate to re-shape the signal edges.

The unsymmetrical signal transmission in Fig. 13.1 is relatively sensitive to external disturbances such as noise or spikes induced in the

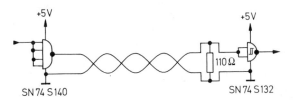

Fig. 13.1 Data transmission via an unsymmetrically operated twisted-pair line

ground line. In larger systems, the *symmetrical driver* (differential line driver, complementary driver) shows in Fig. 13.2 is therefore to be preferred. Two complementary signals are applied to the interconnecting twisted-pair line and are monitored at the receiving end by a comparator. In this mode of operation, the information is contained in the polarity of the voltage difference rather than in its absolute value. A noise voltage therefore only effects a common-mode voltage which remains ineffective due to the differential operation of the comparator.

When generating the complementary signal, it must be ensured that there is no time delay between the two signals. In TTL circuits it is therefore not sufficient to use a simple inverter, and a special circuit having complementary outputs (e.g. Am 26 LS 31 from Advanced Micro Devices) should be employed instead.

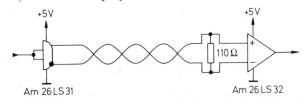

Fig. 13.2 Data transmission via a symmetrically operated twisted-pair line

Such complementary outputs are inherent in ECL gates, and they are thus particularly well suited to symmetrical data transmission. In order that full use can be made of their high switching speed, a simple differential amplifier having an ECL compatible output is used as comparator. It is known as a line receiver, and the resulting circuit is shown in Fig. 13.3.

The comparators in Figs. 13.2 and 13.3 can only evaluate the difference of the two inputs correctly if their common-mode signal range is not exceeded. If larger potential differences arise (e.g. for floating operation of a digital voltmeter), an opto-coupler can be used, as shown in Fig. 13.4. The receiver is a phototransistor which gives a TTL compatible output signal. The transmission rate, however, is limited to values around 100 kbit/s. If the receiver is a photodiode,

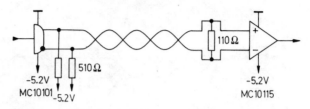

Fig. 13.3 Data transmission in ECL systems via a symmetrically operated
twisted-pair line

considerably higher rates can be attained, but an additional amplifier is
required since the photo-electric current is very small. Such opto-
couplers with an integrated amplifier enable transmission rates of up to
20 Mbit/s (e.g. type 5082-4364 from Hewlett-Packard).

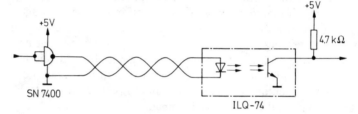

Fig. 13.4 Data transmission in floating circuits. Suitable opto-coupler e.g. ILQ-74 from
Litronix

13.2   Error detection and correction

When transmitting or storing data, it is impossible to prevent the
occurrence of errors. For this reason, transmission methods are employed which are designed to indicate when an error arises. One or
more check bits are transmitted in addition to the data bits; the more
check bits used the more errors can be detected or even corrected.

13.2.1   Parity bit

The simplest method of error detection is the transmission of a
parity bit p. Several data bits are combined to make up a data word
which is then transmitted in a serial or a parallel mode. Either even or
odd parity can be defined. For the even parity check, the parity bit,
added to the data word, is zero if the number of ones in the data word
is even. It is one if the number of ones is odd. The total of transmitted
ones in a data word including the parity bit, is therefore always even.
For odd parity, it is always odd.

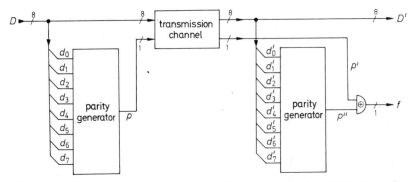

Fig. 13.5 Data transmission with parity check (for the example of an 8-bit data word)

At the receiver end of the transmission line, the parity bit is evaluated from the data bits in the same way as at the transmitting end, and is then compared with the transmitted parity bit. If there is a difference, an error has occurred in transmission. In this manner any single error can be detected. However, correction is not possible as the erroneous bit cannot be located. If *several* bits of the transmitted word have been interfered with, an odd number of errors can be recognized, an even number not.

The block diagram for error detection by a single parity bit is presented in Fig. 13.5. The comparison of the transmitted parity bit with that generated by the receiver is achieved by an exclusive-OR gate. If the two bits are different, the error signal is $f = 1$.

The implementation of the parity generator for even parity is shown in Fig. 13.6. With the exclusive-OR operation, p becomes 1 if the number of ones in the data word is odd. Such parity generators are available as integrated circuits:

8 bits   SN 74180 (TTL);
9 bits: SN 74S 280 (TTL);
12 bits: MC 10160 (ECL), MC 14531 (CMOS).

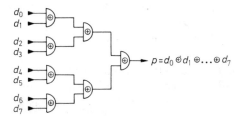

Fig. 13.6 Parity generator for even parity, with 8 inputs

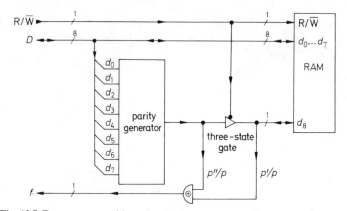

Fig. 13.7 Data memory with parity check (for the example of an 8-bit data word)

As the sequence of the exclusive-OR operations is not significant, the number of inputs can be increased by combining several parity generators by means of additional exclusive-OR operations. A change to even parity check is achieved by inverting the output signal, or by simply having one free input permanently at 1. The external exclusive-OR gate in Fig. 13.5 used for the comparison of the transmitted parity bit p' with that generated by the receiver (p''), can also be included in the parity generator if p' is simply connected to a free input.

Error-detecting encoding is important not only for data transmission but also for data *storage*. The difference between the two applications is that in the latter, transmitter and receiver are the same so that only one parity generator is required. To enable comparison of the generated parity bit with that read out of the memory, the connection for the parity bit is opened up and a three-state gate is inserted, as in Fig. 13.7, which has a high-impedance output during read-out.

13.2.2   Hamming code

The Hamming code improves error detection by having several check bits to not only detect but also locate single-bit errors. If an erroneous bit in a binary code is located, it can be corrected by complementing it.

The number of check bits required for this purpose can be determined very simply: with k check bits, 2^k different binary digits can be identified. With m data bits, a total word length of $(m+k)$ bits is obtained. An additional combination is required to indicate whether the received data word is correct. Therefore, the condition must hold that

$$2^k \geqq m+k+1.$$

number of data bits	m	1...4	5...11	12...26	27...57	58...120	121...247
number of check bits	k	3	4	5	6	7	8

Fig. 13.8 Least number of check bits required for the detection and correction of a single-bit error, depending on the length of the data word

Some solutions of this equation important for practical applications are listed in Fig. 13.8. It can be seen that the relative contribution of the check bits to the total word length is the smaller the larger the word length.

The method for the determination of the check bits is to be illustrated for the example of a 16-bit number. To enable correction of 16 bits, we require 5 check bits according to Fig. 13.8, and thus a total word length of 21 bits. According to Hamming, the individual check bits are evaluated as parity bits for different parts of the data word. In our example, we therefore require five parity generators. Their inputs are assigned to the data bits such that each data bit is applied to at least 2 of the 5 generators. If an incorrectly transmitted data bit exists, a difference arises only between those parity bits which are affected by that particular data bit. With this method, we obtain instead of the parity signal f, a 5-bit error word which can assume 32 different values and allows us to pinpoint the bit in error. It can be seen that the identification us unique for each single-bit error if every binary digit of the data word affects a different combination of check bits. If the receiver detects a difference for only *one* parity bit, then only *that particular* parity bit can be in error, as at least *two* parity bits must be different for an incorrect data bit. If all data and parity bits are transmitted without error, the parity bits evaluated by the receiver are identical to those transmitted.

A practical example for the assignment of the 5 parity bits to the individual data bits is given in Fig. 13.9. The data bit d_0, for instance, affects the parity bits p_0 and p_1; data bit d_1 affects p_0 and p_2, etc. As required, every data bit affects a different combination of check bits. To simplify the circuit, we have distributed the combinations so that each parity generator comprises 8 inputs.

Figure 13.10 shows the combinations for the receiving end. In the same way as for the 1-bit parity method, the expression

$$f_i = p_i' \oplus p_i''$$

must be evaluated, where p_i'' is calculated from the transmitted data bits, using the same procedure as the transmitter, and is compared with the transmitted p_i'. The appropriate circuit is presented in Fig. 13.11.

parity bits	data bits d_i															
	0	1	2	3	4	5	6	7	8	9	10	11	12	13	14	15
p_0	×	×	×	×							×	×	×		×	
p_1	×					×	×	×			×	×	×	×		
p_2		×			×			×	×		×			×	×	×
p_3			×			×		×		×		×		×	×	×
p_4			×	×	×		×		×	×			×			×

Fig. 13.9 Example for the determination of the parity bits according to Hamming, for a word length of 16 bits

The check word $F = (f_4\, f_3\, f_2\, f_1\, f_0)$ is decoded in the ROM. If a data bit error is detected, the straight binary number of the bit in error appears at the outputs $y_0 \ldots y_3$, and the 1-out-of-16 decoder is activated by the output y_4. The exclusive-NOR gate selected in this manner effects correction by complementing the data bit in error.

An error in one of the check bits is indicated by a 1 at output y_5. The straight binary number of the check bit in error appears at the outputs $y_0 \ldots y_3$. Every error recognized is indicated at the output y_7. Of all 32 possible combinations of the check bits, only 22 are used in our example. The remaining 10 combinations occur only if several bits are in error. Such multiple errors are indicated by output y_6, but they cannot be corrected by the above method. The circuits for the correction of multiple errors are of course considerably larger [13.2, 13.3]. The ROM table shown in Fig. 13.12 and used for the implementation of the operations described, follows directly from Fig. 13.10.

As is shown by the circuit in Fig. 13.11, data error detection and correction according to Hamming involves relatively little additional

error word	data bits d_i'																check bits p_i'				
	0	1	2	3	4	5	6	7	8	9	10	11	12	13	14	15	0	1	2	3	4
f_0	×	×	×	×							×	×	×		×		×				
f_1	×					×	×	×			×	×	×	×				×			
f_2		×			×			×	×		×			×	×	×			×		
f_3			×			×		×		×		×		×	×	×				×	
f_4			×	×	×		×		×	×			×			×					×

Fig. 13.10 Determination of the error word

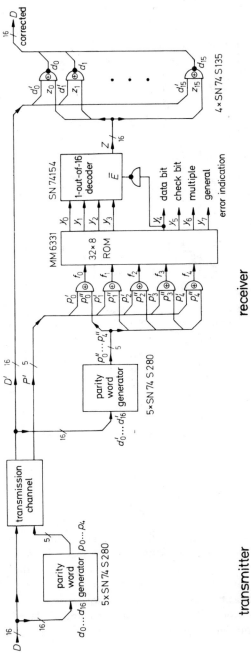

Fig. 13.11 Data transmission with correction of single-bit errors

error code	decoded error	wrong bit	kind of error
f_4 f_3 f_2 f_1 f_0	y_7 y_6 y_5 y_4 y_3 y_2 y_1 y_0		
0 0 0 0 0	0 0 0 0 0 0 0 0		no error
0 0 0 1 1	1 0 0 1 0 0 0 0	d_0	
0 0 1 0 1	1 0 0 1 0 0 0 1	d_1	
1 1 0 0 1	1 0 0 1 0 0 1 0	d_2	
1 0 0 0 1	1 0 0 1 0 0 1 1	d_3	
1 0 1 1 0	1 0 0 1 0 1 0 0	d_4	
0 1 0 1 0	1 0 0 1 0 1 0 1	d_5	
1 0 0 1 0	1 0 0 1 0 1 1 0	d_6	
0 1 1 0 0	1 0 0 1 0 1 1 1	d_7	
1 0 1 0 0	1 0 0 1 1 0 0 0	d_8	data error
1 1 0 0 0	1 0 0 1 1 0 0 1	d_9	
0 0 1 1 1	1 0 0 1 1 0 1 0	d_{10}	
0 1 0 1 1	1 0 0 1 1 0 1 1	d_{11}	
1 0 0 1 1	1 0 0 1 1 1 0 0	d_{12}	
0 1 1 1 0	1 0 0 1 1 1 0 1	d_{13}	
0 1 1 0 1	1 0 0 1 1 1 1 0	d_{14}	
1 1 1 0 0	1 0 0 1 1 1 1 1	d_{15}	
0 0 0 0 1	1 0 1 0 0 0 0 0	p_0	
0 0 0 1 0	1 0 1 0 0 0 0 1	p_1	
0 0 1 0 0	1 0 1 0 0 0 1 0	p_2	check bit error
0 1 0 0 0	1 0 1 0 0 0 1 1	p_3	
1 0 0 0 0	1 0 1 0 0 1 0 0	p_4	
0 0 1 1 0	1 1 0 0 0 0 0 0		
0 1 0 0 1	1 1 0 0 0 0 0 0		
0 1 1 1 1	1 1 0 0 0 0 0 0		
1 0 1 0 1	1 1 0 0 0 0 0·0		
1 0 1 1 1	1 1 0 0 0 0 0 0		multiple error
1 1 0 1 0	1 1 0 0 0 0 0 0		(can not be
1 1 0 1 1	1 1 0 0 0 0 0 0		corrected)
1 1 1 0 1	1 1 0 0 0 0 0 0		
1 1 1 1 0	1 1 0 0 0 0 0 0		
1 1 1 1 1	1 1 0 0 0 0 0 0		

Fig. 13.12 List of error codes and their decoding

hardware, and since the appropriate circuitry is connected in parallel to the signal path, the loss in transmission speed is very small. This method has therefore been widely adopted to increase the reliability of memories. Its particular advantage is that faults in the memory are detected although, because of the correction procedure, they have no effect. Faulty memory ICs can be located and replaced at an early stage so that the dependability of a computer system is considerably improved [13.4].

13.3   Static digital displays

There are many different ways of displaying digital information, for instance with filament lamps, cold-cathode tubes, light-emitting diodes (LEDs) or liquid crystals. The LED display has become most popular since it is TTL compatible and particularly reliable.

13.3.1   Binary display with LED

Light-emitting diodes require a forward current of 5...20 mA for good visibility in daylight. A current of 15 mA can be drawn from a standard TTL gate in the H-state if the LED is directly connected to the output, as in Fig. 13.13. However, the resulting voltage level is too low for further logic processing. Figure 13.14 shows a method for driving a LED by the gate in the L-state. In this mode of operation, the current must be limited by a resistor to the desired value. Light-emitting diodes with built-in current limiters are also available (e.g. type RLC 201 from Litronix).

Another method involves the application of special driver circuits with current-source outputs. Type DM 8859 from National comprises six drivers, the output currents of which can be simultaneously adjusted to a required value by an additional input. This enables adjustment of the diode current according to the surrounding light intensity.

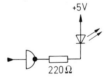

Fig. 13.13 Using a TTL gate to drive an LED with a H-Signal   Fig. 13.14 Using a TTL gate to drive an LED with a L-Signal

13.3.2   Decimal display

When displaying numbers, a decimal display is usually preferred. One method of display, based on the BCD notation, represents each decimal digit in the 1-out-of-10 code. Ten cathodes shaped in the form of decimal digits, are positioned one behind the other in a single cold-cathode tube (Nixie Tube), and can be made to discharge one at a time. There is the disadvantage, however, that a relatively high anode voltage of about 200 V is required.

It is better to arrange several light-emitting diodes in one plane so that all the decimal digits can be synthesized from them. The *7-segment display* is most common. Seven bar-shaped segments *a* to *g* are used

Fig. 13.15 7-segment display

and arranged as in Fig. 13.15. If all seven segments are activated, the figure "8" is displayed. If the segments b and c are emitting, a "1" is seen. The digits are shown in Fig. 13.15. Obviously, the logic operations for activating individual segments are more complicated than for the 1-out-of-10 code, because they must be 1 for more than one digit. Figure 13.16 shows the truth table for the BCD 7-segment decoder. If the table is read from left to right, it can be seen which segments are activated for which decimal digit, for example for the digit "5" the segments a, c, d, f and g, see also Fig. 13.15.

digit	BCD-input				7-segment output						
Z	2^3	2^2	2^1	2^0	a	b	c	d	e	f	g
0	0	0	0	0	1	1	1	1	1	1	0
1	0	0	0	1	0	1	1	0	0	0	0
2	0	0	1	0	1	1	0	1	1	0	1
3	0	0	1	1	1	1	1	1	0	0	1
4	0	1	0	0	0	1	1	0	0	1	1
5	0	1	0	1	1	0	1	1	0	1	1
6	0	1	1	0	1	0	1	1	1	1	1
7	0	1	1	1	1	1	1	0	0	0	0
8	1	0	0	0	1	1	1	1	1	1	1
9	1	0	0	1	1	1	1	1	0	1	1

Fig. 13.16 Truth table of BCD-7-segment decoder

BCD 7-segment decoders are available as integrated circuits. Type SN 74247 has open-collector outputs and is therefore suited to drive common-anode displays, such as that in Fig. 13.17. For the adjustment of the desired diode current, seven external resistors are required.

For some applications it is desirable to control the display intensity electronically. The simplest method is to switch the display on and off periodically and to vary the duty cycle. For this purpose, most BCD 7-segment decoders possess a *blanking* input *BI* with which all output transistors can be turned off.

A further useful characteristic for the representation of many-digit numbers is the automatic blanking of leading zeros. For this purpose,

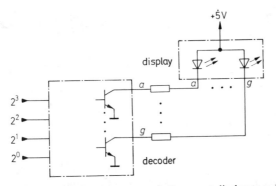

Fig. 13.17 Connection of a common-anode 7-segment display to a decoder

the decoder has a *ripple blank* input *RBI* and a *ripple blank* output *RBO*. If the input *RBI* is in the L-state, the display extinguishes when the decimal digit 0 is applied, and the output *RBO* changes from H to L. If this output is connected to the *RBI* input of the next lower digit, as in Fig. 13.18, a zero is blanked out if the higher digit is also zero. A zero is therefore displayed only if any one of the higher digits is not zero.

BCD 7-segment decoders are sometimes combined with the display unit to comprise a hybrid integrated circuit. Usually a buffer memory for the input variables is also included (e.g. in type TIL 308 from Texas Instruments). If an L is applied to the *strobe* input, the input data are read, if an H is applied, they are stored.

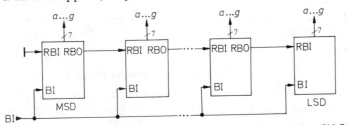
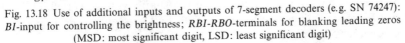

Fig. 13.18 Use of additional inputs and outputs of 7-segment decoders (e.g. SN 74247): *BI*-input for controlling the brightness; *RBI-RBO*-terminals for blanking leading zeros (MSD: most significant digit, LSD: least significant digit)

13.3.3 Hexadecimal display

In hexadecimal notation, the additional digits *ten* to *fifteen* are usually represented by the capital letters A to F. These symbols cannot be displayed by 7-segment units. It can be seen from Fig. 13.15 that the symbol "B" would not be distinguishable from "8" and that there

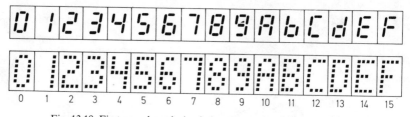

Fig. 13.19 First row: hexadecimal characters on a 7-segment display.
Second row: hexadecimal characters on a 20-dot display

would be no difference between "D" and "0". However, if lower case letters are used for these two symbols, differentiation becomes possible, as is shown in the first row of Fig. 13.19. One decoder having this facility is the type 8 T 75 from Signetics.

Display modules with built-in decoders having 20 dots instead of the 7 segments are also available. The capitals A to F can then be displayed unambiguously as shown in the second row of Fig. 13.19. The types TIL 311 from Texas Instruments and 5082-7359 from Hewlett-Packard are displays of this kind.

13.4   Multiplex displays

The operation of many-digit displays based on the above principle has the disadvantage that each digit requires a separate decoder and that a large number of connecting lines is needed. For this reason, such displays are not usually operated in a parallel but in a sequential mode. The connection of the individual segments in a matrix and the use of multiplexer techniques can considerably reduce the number of lines.

13.4.1   Multi-digit 7-segment displays

If, for example, an 8-digit 7-segment display is to be operated in a parallel mode, and the segment anodes of each digit are connected in parallel, one common anode line for the supply voltage is required and 8×7 cathode lines controlled via 8 BCD 7-segment decoders, i.e. altogether 57 connecting lines.

With the multiplex method, on the other hand, the corresponding cathodes $a, b, ..., g$ of each digit are paralleled. So as not to activate, for instance, all the segments a at the same time, the desired digit is selected by supplying the anode voltage via 8 switches to only one of the 8 digits. This matrix arrangement can be visualized in Fig. 13.20. Only $8 + 7 = 15$ connecting leads and only one decoder are needed for this method.

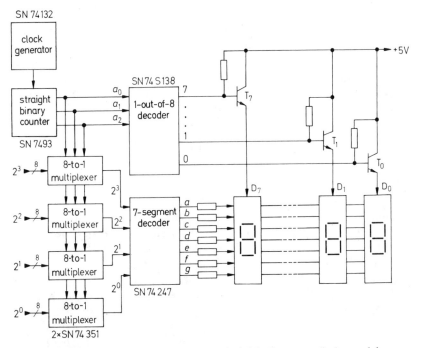

Fig. 13.20 Example for the multiplex control of eight 7-segment display modules

If switching between the individual digits is carried out sufficiently fast, an observer will have the impression that all digits are emitting light simultaneously. At a frequency above 60 Hz, the human eye no longer notices the flicker.

The cyclic switching is performed by the straight binary counter and the 1-out-of-8 decoder. The straight binary number at the counter output is used simultaneously to control the four multiplexers at the input of the 7-segment decoder. This ensures that only the BCD data signal assigned to the currently activated digit, is applied to the decoder. As every digit is switched on for only one of eight clock periods, the resistors at the decoder output must be designed to make the segment current eight times the desired mean value.

The four multiplexers may be omitted if data signal sources with three-state outputs are employed. The eight outputs of the same weighting can then be parallel-connected, the selection being made by means of the chip-select inputs. Suitable data signal sources, for instance, are the octal D latch SN 74 LS 373 or the decade counter Am 25 LS 2568 (Advanced Micro Devices).

Integrated counters that also contain the 7-segment decoder and the multiplexer logic circuit are available. They have seven outputs for segment control and n outputs which specify the individual decimal digits in the 1-out-of-n code. Some types not only contain the segment drivers, but also the digit drivers, for example the 7-digit counter ICM 7208 from Intersil. It is designed for the operation of common-cathode display units.

13.4.2   Dot matrices

If, in addition to the decimal characters, the entire alphabet is to be displayed, the division into 7 segments is no longer sufficient. For such *alphanumeric* displays, 5×7 dot matrices as in Fig. 13.21 are usually employed. As in the multi-digit 7-segment display, they are controlled sequentially, e.g. row by row. A counter and a 1-out-of-8 decoder is used to apply the anode voltage to terminal r_i of each row in succession. The LEDs required in each row are selected by grounding the terminal c_j of the corresponding column via a resistor. For the example of the letter "K" shown in Fig. 13.22, the columns c_1 and c_5 must be selected for the first row, the columns c_1 and c_4 for the second, etc.

These column codes are best generated by ROMs available as mask-programmed *character generators*. The ROM MM 6061 from Monolithic Memories, for instance, holds the codes for all 128 ASCII characters. These symbols are selected by 7 address bits, a_3 to a_9,

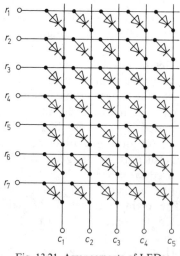

Fig. 13.21 Arrangements of LEDs
in a 5×7 dot matrix

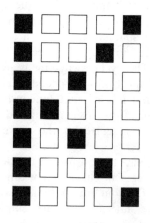

Fig. 13.22 Symbolic representation
of the matrix

row number i	ASCII-"K" $a_9\ a_8\ a_7\ a_6\ a_5\ a_4\ a_3$							i $a_2\ a_1\ a_0$			column code $c_1\ c_2\ c_3\ c_4\ c_5$				
1	1	0	0	1	0	1	1	0 0 1			0	1	1	1	0
2	1	0	0	1	0	1	1	0 1 0			0	1	1	0	1
3	1	0	0	1	0	1	1	0 1 1			0	1	0	1	1
4	1	0	0	1	0	1	1	1 0 0			0	0	1	1	1
5	1	0	0	1	0	1	1	1 0 1			0	1	0	1	1
6	1	0	0	1	0	1	1	1 1 0			0	1	1	0	1
7	1	0	0	1	0	1	1	1 1 1			0	1	1	1	0

Fig. 13.23 Part of the ROM table of the character generator, for the display of the character "K"

according to the table in Fig. 11.33. Instead of the ASCII control signals, which cannot be displayed, other characters, for example a section of Greek letters, are available. The required row number $i = 1 \ldots 7$ is straight-binary coded and applied to the address inputs a_0, a_1, a_2. As an example, the part of the ROM table for character "K" is given in Fig. 13.23 in positive true logic. A zero therefore stands for an L level, i.e. a column selected.

The implementation of the control by means of the described character generator, is represented in Fig. 13.24. An extension to a multi-character display is possible, for instance, by paralleling the

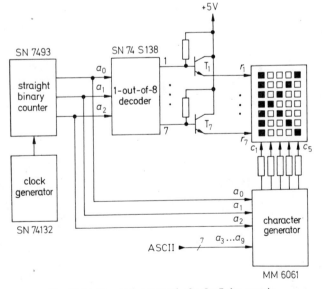

Fig. 13.24 Circuit for control of a 5×7 dot matrix

corresponding rows of all display modules and using a separate character generator for each display unit. However, it is more economical to use a single character generator in multiplex operation for all modules and to temporarily store the column code for each module, row after row.

When employing this method, the 4-character dot-matrix display HDSP-2000 from Hewlett-Packard has a special advantage, namely that the intermediate memories are integrated in the display in the form of a shift register. The transmission of the character code is therefore possible via a single data line.

14   D/A and A/D converters

If a continuous signal is to be processed using digital methods, the analog variable at the input must be converted to the equivalent number. This is the purpose of an analog-to-digital converter (A/D converter, ADC). For the reconversion of the digital number to a proportional analog voltage or current, a digital-to-analog converter is used (D/A converter, DAC). The following sections deal with the most important principles of D/A and A/D conversion.

14.1   Design principles of D/A converters

14.1.1   Summation of weighted currents

A simple circuit for the conversion of a straight binary number to a proportional voltage is given in Fig. 14.1. The resistors are designed so that when the switches are closed, currents corresponding to the appropriate bit weightings flow through them. The switches must be closed whenever the appropriate bit is a logic 1. Because of the feedback of the operational amplifier via resistor R_N, the summing point remains at zero potential. The current components are thus added without influencing each other, and we obtain

$$U_o = -U_{ref}\frac{R_N}{R_0}(8z_3 + 4z_2 + 2z_1 + z_0), \tag{14.1}$$

$$U_o = -U_{ref}\frac{R_N}{R_0}Z \quad \text{with } 0 \leq Z \leq 15. \tag{14.2}$$

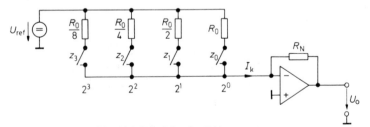

Fig. 14.1 Principle of a D/A converter.

$$U_o = -U_{ref}\cdot\frac{R_N}{R_0}\cdot Z$$

The circuit can be extended for pure straight binary numbers by paralleling further resistors $\frac{1}{16}R_0$, $\frac{1}{32}R_0$ etc., according to the number of bits to be converted. The highest accuracy is required for the resistor of the most significant bit. If the least significant bit is not to be swamped by the errors at the more significant bits, the resistor tolerance for the 2^n-bit (z_n) must be smaller than

$$\frac{\Delta R}{R} = \frac{1}{2^{n+1}}.$$

Hence, the resistor tolerance for the 2^4-bit must be better than 3% and that for the 2^{10}-bit, 0.05%. For straight BCD numbers, an extension is made by using an additional resistor quartet for each decade. The values of the resistors are smaller than those of the previous quartet by a factor of 10.

14.1.2 D/A converters with double-throw switches

A disadvantage of the above D/A converter is that large voltages occur across the switches. The application of electronic switches is therefore rather involved. Moreover, the attainable switching frequency is low because of the required charging and reverse-charging of stray capacitances. These disadvantages can be avoided if double-throw switches are used as in Fig. 14.2, to connect the resistors either to the summing point or to ground. The current through each resistor therefore remains constant. This has a further advantage that, unlike for the previous circuit, the load of the reference voltage source is constant and its internal resistance need not be zero. The input resistance of the network, and thus the load resistance of the reference voltage source, is given by

$$R_i = R_0 \left\| \frac{R_0}{2} \right\| \frac{R_0}{4} \left\| \frac{R_0}{8} = \frac{1}{15}R_0. \right. \tag{14.3}$$

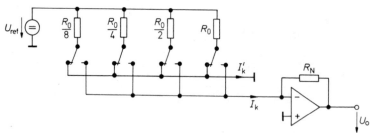

Fig. 14.2 D/A converter with double-throw switches.

$$U_o = -U_{ref} \cdot \frac{R_N}{R_0} \cdot Z$$

14.1.3 Ladder network

When producing integrated D/A converters, the manufacture of accurate resistances of widely differing values is extremely difficult. The weighting of the bits is therefore often effected by successive voltage division using a ladder network as in Fig. 14.3. The basic element of such a ladder network is the loaded voltage divider in Fig. 14.4, which is required to have the following characteristics: if it is loaded by a resistor R_p, its input resistance R_i must also assume the value R_p. At this load, the attenuation $\alpha = U_2/U_1$ along the ladder element is to have a predetermined value. With these two conditions, we obtain

$$R_d = \frac{(1-\alpha)^2}{\alpha} R_q \quad \text{and} \quad R_p = \frac{(1-\alpha)}{\alpha} R_q. \tag{14.4}$$

For the case of the straight binary code, $\alpha = 0.5$. We define $R_q = 2R$, and obtain

$$R_d = R \quad \text{and} \quad R_p = 2R, \tag{14.5}$$

in accordance with Fig. 14.3.

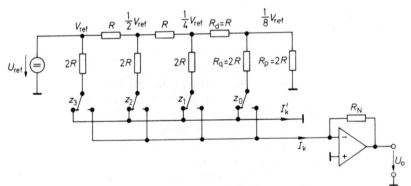

Fig. 14.3 D/A converter with ladder network.

$$U_o = -\frac{U_{ref}}{16} \cdot \frac{R_N}{R} \cdot Z$$

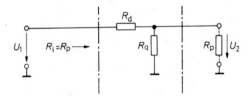

Fig. 14.4 Element of the ladder network

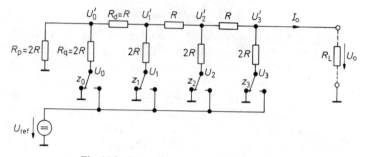

Fig. 14.5 Inversely operated ladder network

$$U_0 = \frac{U_{ref}}{16} \cdot \frac{R_L}{R + R_L} \cdot Z$$

The reference voltage source is loaded by the constant resistance

$$R_i = 2R \| 2R = R.$$

The output voltage of the summing amplifier is

$$U_o = -R_N I_k = -U_{ref} \frac{R_N}{16R}(8z_3 + 4z_2 + 2z_1 + z_0) = -U_{ref} \frac{R_N}{16R} Z. \quad (14.6)$$

Occasionally, the ladder network is operated as in Fig. 14.5 with the input and output interchanged, as then no summing amplifier is required. However, the disadvantage of this method is that a large voltage swing occurs across the switches and that the load of the reference voltage source varies.

To calculate the output voltage we require the relationship between the applied voltages U_i and the corresponding node voltages U_i'. Using the superposition principle, we set all voltages to zero except one voltage U_i and determine its contribution to the output voltage. All individual contributions are then added. When the network is terminated by resistances R_p to the right and to the left, the load to the left and right of each node is also R_p, as stipulated. Hence, with Eq. (14.4), the voltage contributions are found to be

$$\Delta U_i' = \frac{1-\alpha}{1+\alpha} \Delta U_i. \quad (14.7)$$

With $\alpha = 0.5$, we obtain by adding the correctly weighted contributions

$$U_o = \tfrac{1}{3}(U_3 + \tfrac{1}{2}U_2 + \tfrac{1}{4}U_1 + \tfrac{1}{8}U_0), \quad (14.8)$$

$$U_o = \frac{U_{ref}}{24} Z \quad \text{with } 0 \leq Z \leq 15. \quad (14.9)$$

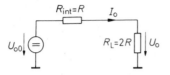

Fig. 14.6 Equivalent circuit for the determination of no-load voltage and short-circuit current

Since the internal resistance of the network is independent of the applied binary number and has the constant value

$$R_{int} = R_p \| R_q = (1-\alpha)R_q = R, \qquad (14.10)$$

the weighting remains unchanged even if the load resistance R_L does not have the value $R_p = 2R$ defined initially. From the equivalent circuit in Fig. 14.6, we can determine the no-load voltage and the short-circuit current. With Eq. (14.9), we obtain

$$U_{oo} = \frac{U_{ref}}{16}Z, \qquad I_{osc} = \frac{U_{ref}}{16R}Z.$$

14.1.4   Ladder network for decade weighting

The ladder network in Fig. 14.3 can be extended to any length if longer straight binary numbers are to be converted. For the conversion of BCD numbers the method is somewhat modified, see Fig. 14.7. Each decade Z_i is converted by a 4-bit D/A converter as in Fig. 14.2 or 14.3, and the individual converters are connected to a ladder network. This effects, from stage to stage, an attenuation of $\alpha = \frac{1}{10}$. In Eq. (14.4), resistance R_q must then be replaced by the input resistance R_i of the D/A converter stages so that the coupling resistors are

$$R_d = 8.1\,R_i \qquad (14.11)$$

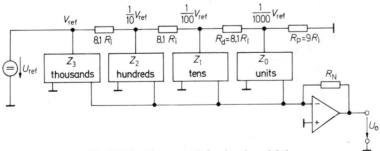

Fig. 14.7 Ladder network for decade weighting

and the terminating resistor is

$$R_p = 9R_i, \tag{14.12}$$

as indicated in Fig. 14.7. In this manner, each input voltage for a D/A converter stage is $\frac{1}{10}$ that of the previous stage. For the example of 4 decades, the output voltage

$$U_o = -\frac{U_{ref}R_N}{16R} \cdot 10^{-3}(10^3 Z_3 + 10^2 Z_2 + 10 Z_1 + Z_0) \tag{14.13}$$

is obtained if, for each decade, a ladder network as in Fig. 14.3 is used.

14.2   Design of D/A converters using electronic switches

For the description of D/A converter operating principles we have assumed mechanical switches to simplify the problem. However, the digital input signal is usually an electrical one and in this case, electronic switches may be used.

14.2.1   D/A converter with CMOS switches

Low-resistance CMOS switches are well suited for D/A conversion as they have no offset voltage. For a low resolution, a circuit can be derived from the principle of Fig. 14.1 by simply replacing the switches by the outputs of standard CMOS gates, as shown in Fig. 14.8. The gate supply voltage is then also the reference voltage. To improve accuracy, the output resistance of the gate in the H-state is taken into account when determining the weighting resistor. It is about $1\,k\Omega$ for a supply voltage of $5\,V$ and about $500\,\Omega$ for $10\,V$.

The implementation of high resolution D/A converters is based on the ladder network in Fig. 14.3 since the change in voltage across the switch is practically zero. The switching from summing point to ground

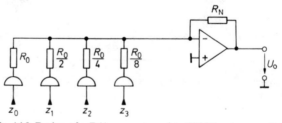

Fig. 14.8 Design of a D/A converter using CMOS gates as switches.

$$U_o = -V^+ \frac{R_N}{R_0} \cdot Z$$

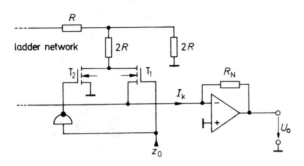

Fig. 14.9 Ladder network with MOS double-throw switches

can be carried out, as in Fig. 14.9, by two enhancement n-channel FETs which are controlled in opposition by an inverter.

If H-state is applied to the digital input z_0, T_1 is ON and T_2 is OFF. The drain potential is zero. An H-level of a few volts is therefore sufficient for safe operation, independent of the magnitude of the reference voltage. If zero potential is applied to the control input z_0, T_1 is turned off. The drain potential remains zero in this case as T_2 is now ON. Therefore FET T_1 remains turned off even if the reference voltage is negative. Such D/A converters are available as monolithic ICs in CMOS technology:

10 bits: AD 7530 from Analog Devices;
12 bits: AD 7531 from Analog Devices.

The settling time is about 0.5 µs.

The reference voltage may have any value within the range $-10\,V$ to $+10\,V$. The relationship

$$U_o \sim Z U_{ref}$$

therefore applies to any analog voltage at the reference input, and for this reason such D/A converters are known as *multiplying* DACs. They can be used to advantage as digitally controlled coefficient circuits, e.g. in active filters.

14.2.2   D/A converter with current switches

The accuracy of the currents in the D/A converters described so far is strongly influenced by the voltage drop across the switches. This effect is avoided in the circuit of Fig. 14.10 where the currents are generated by constant current sources. If a positive voltage is applied to the binary input, the input diode is forward, the other one reverse biased. The constant current thus flows via the binary control input. With a negative input voltage, the diode at the input is OFF and

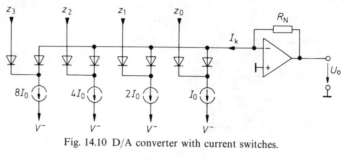

Fig. 14.10 D/A converter with current switches.

$$U_o = R_N I_0 Z$$

the other one is ON. The constant current therefore flows via the
summing point. If the currents are given the weights shown, the output
voltage is always proportional to the straight binary number at the
input.

The summing amplifier is not absolutely essential since the currents are generated by current sources. It may be replaced by a load
resistor R_L, but it should be noted that the permissible change in
voltage of the summing node is usually limited to values below 1 V due
to the construction of the current sources.

The principle of current switching is preferred in D/A converters
integrated in bipolar technology. One implementation is presented
in Fig. 14.11. The weighted currents are generated by the ladder network of Fig. 14.3. Dissimilar base-emitter voltages of the transistors T_1
to T_6 would distort the weighting of the ladder network. So as to obtain
equal base-emitter voltages despite the different collector currents, the
transistors are given emitter areas bearing the weighting factors.

Because of the floating operation of the ladder network, the terminating resistor R_p must not, in this application, be connected to

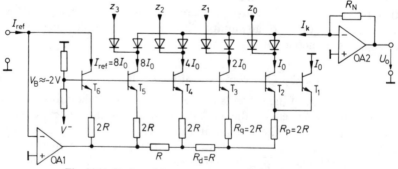

Fig. 14.11 Design of the current sources in bipolar technology.

$$U_o = \tfrac{1}{8} R_N I_{ref} \cdot Z$$

ground but to the emitter of an additional transistor. It is advantageous to use this transistor as an additional current source, i.e. for the conversion of the least significant bit. However, equal currents flow in the terminating resistor R_p and the parallel resistor R_q so that a transistor T_1 must be parallel-connected to T_2 to effect a further halving of the current for the z_0 input.

The input voltage for the ladder network is provided indirectly by the reference transistor T_6 and the operational amplifier OA1. The value of its output voltage is such that the collector current of T_6 assumes the externally adjustable value I_{ref}. Hence, $8I_0 = I_{ref}$, and the output current is given by

$$I_k = \frac{I_{ref}}{8}Z \quad \text{where } 0 \leq Z \leq 15.$$

The following monolithic DACs operate on this principle:

8 bits: MC 3408   from Motorola,
          AM 1408   from Advanced Micro Devices;
10 bits: MC 3410   from Motorola.

The settling time is approximately 250 ns.

Differential amplifier as current switch

Particularly fast current switches can be realized using differential amplifiers. To switch the current from one transistor to the other, a voltage step of less than 1 V is required. The change-over time is short as the transistors are not saturated.

A possible implementation is shown in Fig. 14.12 using standard ECL gates. The differential amplifier T_3/T_3' is composed of the output

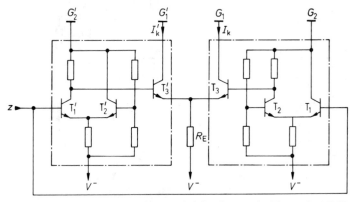

Fig. 14.12 Differential-amplifier current switch implemented with standard ECL gates

emitter followers of two gates situated on different chips. It is useful in this application that the collectors of the output emitter followers of each chip are already connected to a separate ground pin G_1. The current I_k flowing at this terminal is used as the output signal.

If input z is in the H-state, T_3 is conducting and T_3' is turned off. The output current is then given by

$$I_k \approx \frac{|V^-| - 0.9\,\text{V}}{R_E}.\qquad(14.14)$$

With z in the L-state, $I_k = 0$. Two ECL chips make a very fast 4-bit D/A converter if the external emitter resistors R_E are weighted in powers of two. The resulting circuit is given in Fig. 14.13. To prevent the potential at terminal G_1 becoming negative, a positive quiescent potential V_0 is produced by a voltage divider. The output voltage is then given by

$$U_o = V_0 - R_i I_k,\qquad(14.15)$$

where R_i is the internal resistance of the voltage divider. By superposing the four current contributions and using Eq. (14.14), the result is

$$U_o = V_0 - \frac{R_i}{R_0}(|V^-| - 0.9\,\text{V})(8z_3 + 4z_2 + 2z_1 + z_0).$$

The settling time is only a few nanoseconds, the output voltage swing is 2.5 V at 75 Ω.

The accuracy can be considerably increased by replacing the external emitter resistors by constant current sources. The 8-bit D/A converter MC 10318 from Motorola uses this principle.

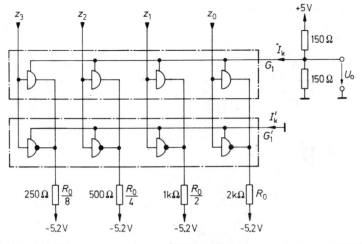

Fig. 14.13 4-bit D/A converter using two ECL ICs. Suitable type of gate: e.g. MC 10101

14.3 D/A converters for special applications

14.3.1 Conversion of signed numbers

In our description of D/A converters we have assumed so far that the binary input information is available in the form of an unsigned number. Signed numbers are usually given in 2's-complement representation (see Section 9.5.6). With 8 bits for example, the range from -128 to $+127$ can be represented. To prepare these numbers for the input of the D/A converter, the range is shifted by adding 128 and is then from 0 to 255. Numbers above 128 are thus to be taken as positive, those below 128 as being negative. The middle number 128 denotes zero in this case. This method of representing signed numbers by positive numbers only is known as offset binary. The addition of 128 can be carried out by simply complementing the sign bit.

To obtain an output voltage having the correct sign, the addition of the offset is cancelled in the analog circuitry by subtracting 128 current units I_{LSB}. This compensating current is generated from the reference voltage by the inverting amplifier OA1 and the resistor R_2 in Fig. 14.14. Hence, the total current

$$I_{tot} = I_k - 128 I_{LSB} = I_k - \tfrac{128}{255} I_{k\,max}. \tag{14.16}$$

Some values of this current are listed in Fig. 14.15. If a D/A converter is used which is able to operate for positive and negative reference voltages, the arrangement in Fig. 14.14 represents a multiplying four-quadrant D/A converter.

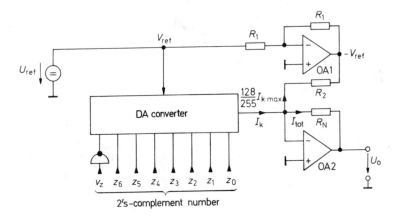

Fig. 14.14 D/A converter for 2's-complement numbers.

$$U_o = R_N \frac{I_{k\,max}}{255} \cdot Z \quad \text{for} \quad -128 \leqq Z \leqq 127$$

decimal	2's-complement	offset binary	analog $I_{tot}/I_{k\,max}$
		z_7 z_6 z_5 z_4 z_3 z_2 z_1 z_0	
127	0 1 1 1 1 1 1 1	1 1 1 1 1 1 1 1	127/255
1	0 0 0 0 0 0 0 1	1 0 0 0 0 0 0 1	1/255
0	0 0 0 0 0 0 0 0	1 0 0 0 0 0 0 0	0
−1	1 1 1 1 1 1 1 1	0 1 1 1 1 1 1 1	−1/255
−127	1 0 0 0 0 0 0 1	0 0 0 0 0 0 0 1	−127/255
−128	1 0 0 0 0 0 0 0	0 0 0 0 0 0 0 0	−128/255

Fig. 14.15 Equivalence of digital and analog values

There is a disadvantage in this method in that zero output voltage is computed as the difference of two large quantities. This results in poor zero stability and this is defined essentially by the difference in the temperature coefficients of the DAC and resistor R_2.

Good stability can be attained if the compensating current is generated within the D/A converter itself. DACs with double-throw switches are well suited for this purpose. The ground terminals of the switches are usually combined at a second output. If this is connected to a further summing amplifier, the output current

$$I_k' = I_{k\,max} - I_k$$

is obtained. This current is increased by one current unit $I_{LSB} = \frac{1}{255} I_{k\,max}$ and inverted by amplifier OA1 in Fig. 14.16. Hence, the total current of the summing amplifier OA2 is given by

$$I_{tot} = I_k - (I_{k\,max} - I_k + \tfrac{1}{255} I_{k\,max}),$$
$$I_{tot} = 2(I_k - \tfrac{128}{255} I_{k\,max}), \tag{14.17}$$

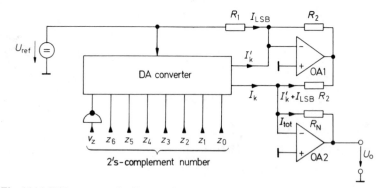

Fig. 14.16 D/A converter for 2's-complement numbers, having improved zero stability.

$$U_o = 2 R_N \frac{I_{k\,max}}{255} \cdot Z \quad \text{for} \quad -128 \leqq Z \leqq 127$$

in analogy to Eq. (14.16). If the influence of the small correction I_{LSB} is disregarded, the absolute values of the resistances R_2 are not important as long as they are the same. This is not so for the previous circuit.

14.3.2   D/A converter as divider

The output voltage of the D/A converters described so far is proportional to the product of reference voltage and binary input number Z. If the reference voltage is taken as an analog input voltage, the D/A converter becomes a multiplier, as has been discussed previously.

Figure 14.17 shows a circuit suitable for the *division* of analog signals by a binary input number. Its operation is related to that of the circuit in Fig. 14.1, where the reference voltage is fed to resistors, the resistance of the parallel connection of which is inversely proportional to the number Z. The current flowing into the summing amplifier is then proportional to Z.

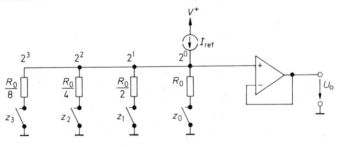

Fig. 14.17 Dividing D/A converter.

$$U_o = \frac{I_{ref} R_0}{Z}$$

In this circuit, we impress on the same parallel connection of resistors a current I_{ref}, and consequently obtain a voltage which is inversely proportional to Z. From Fig. 14.17, we obtain

$$U_o = \frac{I_{ref}}{\dfrac{8z_3}{R_0} + \dfrac{4z_2}{R_0} + \dfrac{2z_1}{R_0} + \dfrac{z_0}{R_0}} = \frac{I_{ref} R_0}{Z}.$$

It is now obvious that the circuit for a dividing DAC is no more complicated than that of a multiplying DAC. A digital or analog division, which always involves large circuitry, especially when high accuracy is required, can often be avoided using this simple circuit.

14.3.3   D/A converter as function generator

The output voltage U_o of the usual D/A converter is proportional to the applied number Z: $U_o = a \cdot Z$. If, instead of the proportional function, any other relationship $U_o = f(Z)$ is to be realized, the function $X = f(Z)$ must be generated by a digital function network (see Section 9.7) and this applied to a D/A converter.

If no stringent requirements are imposed on the accuracy, there is a considerably more simple solution: the binary number Z is used to control an analog multiplexer. One applies constant analog input values, each of which is assigned to the appropriate binary number. For each analog value, a separate switch is needed, and the attainable resolution is therefore limited to about 16 steps.

A possible implementation is shown in Fig. 14.18. In contrast to the usual D/A converter, only one of the switches S_0 to S_7 is closed at any time. The values of the output voltage function are thus given by the expression

$$U_o(Z) = \begin{cases} +U_{\text{ref}} \dfrac{R_N}{R_Z} & \text{for } Z = 0 \ldots 3 \\[2ex] -U_{\text{ref}} \dfrac{R_N}{R_Z} & \text{for } Z = 4 \ldots 7. \end{cases}$$

An important application of this principle is the digital generation of sine waves (e.g. in modems). A simple and widely used method of generating signals of different frequencies, all synchronized to a common time base, is to employ frequency division. However, a serious drawback for the application in analog systems is that the signals obtained are square waves. Sine waves can be produced by filtering the fundamental with a lowpass or bandpass filter, but these filters must always be tuned to the appropriate frequency. The D/A converter described avoids these problems in that it allows the frequency-independent generation of sine waves. According to Fig. 14.19, we require a digital input signal representing a rising and falling sequence of equidis-

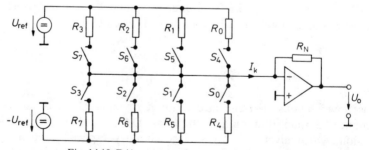

Fig. 14.18 D/A converter for any desired weighting

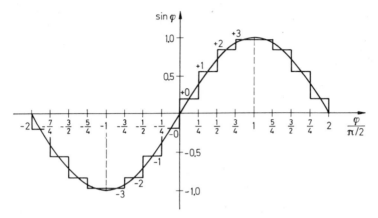

Fig. 14.19 Approximation of a sine wave by 16 steps

tant numbers. This input signal corresponds to the triangular wave-shape for the sine wave generation by an analog function network, as described in Section 1.7.4.

If the sign-magnitude representation is chosen for the binary numbers, a number sequence having the desired properties can be easily generated by a cyclic straight binary counter [14.1]. The most significant bit represents the sign. The second most significant bit effects a change of counting direction for all lower bits by complementing the corresponding outputs with the help of exclusive-OR gates. These bits

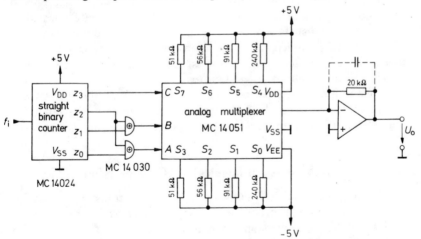

Fig. 14.20 Circuit for the generation of a continuous sine wave.

$$U_o = 2\,\text{V} \sin 2\pi \frac{f_i}{16} t$$

Z	counter outputs				multiplexer inputs			switch closed	step number	output voltage $U_o/\hat{U}_o$
	z_3	z_2	z_1	z_0	C	B	A			
0	0	0	0	0	0	0	0	S_0	$+0$	0.20
1	0	0	0	1	0	0	1	S_1	$+1$	0.56
2	0	0	1	0	0	1	0	S_2	$+2$	0.83
3	0	0	1	1	0	1	1	S_3	$+3$	0.98
4	0	1	0	0	0	1	1	S_3	$+3$	0.98
5	0	1	0	1	0	1	0	S_2	$+2$	0.83
6	0	1	1	0	0	0	1	S_1	$+1$	0.56
7	0	1	1	1	0	0	0	S_0	$+0$	0.20
8	1	0	0	0	1	0	0	S_4	-0	-0.20
9	1	0	0	1	1	0	1	S_5	-1	-0.56
10	1	0	1	0	1	1	0	S_6	-2	-0.83
11	1	0	1	1	1	1	1	S_7	-3	-0.98
12	1	1	0	0	1	1	1	S_7	-3	-0.98
13	1	1	0	1	1	1	0	S_6	-2	-0.83
14	1	1	1	0	1	0	1	S_5	-1	-0.56
15	1	1	1	1	1	0	0	S_4	-0	-0.20

Fig. 14.21 List of the number sequences and resulting voltages

represent the magnitude. When using a four-bit straight binary counter, the circuit in Fig. 14.20 is obtained. The number sequence generated is listed in Fig. 14.21. With a 3-bit number at the input of the analog multiplexer, four positive steps $+0, 1, 2, 3$ of the sine function and correspondingly, four negative steps $-0, -1, -2, -3$ are selected. If the steps are distributed as in Fig. 14.19, the function values in Fig. 14.21 are obtained, and the appropriate resistors can be determined. As the chosen quantization is rather coarse, it is sufficient to select the nearest standard resistance value.

Since in a complete period each step occurs twice, the sine wave is approximated by a total of 16 steps. Correspondingly, the input frequency f_i of the counter must be 16 times that of the sine wave.

14.4 Basic principles of A/D conversion

The purpose of an A/D converter (ADC) is to transform an analog input voltage into a proportional digital number. There are three basically different conversion methods:

the parallel method	(word at a time),
the weighing method	(digit at a time),
the counter method	(level at a time).

The parallel method compares the input voltage with n reference voltages simultaneously and determines between which two reference levels the value of the input voltage lies. The resulting number is thus obtained in a single operation. However, the circuitry involved is very extensive as a separate comparator is required for each possible number. For a range of measurement from 0 to 100 in unit steps, $n = 100$ comparators are needed.

With the weighing (successive approximation) method, the end result is not obtained in a single operation; instead, one bit of the corresponding straight binary number is determined at a time. The input voltage is first checked against the most significant bit, i.e. whether it is larger or smaller than the reference voltage for this bit. If it is larger, the bit is 1 and the reference voltage is subtracted from the input voltage. The remainder is compared with the next lower bit, etc. The number of conversion operations and reference voltages needed is the same as the number of bits in the result.

The simplest method is the counter method. It involves counting how often the reference voltage of the least significant bit must be added to arrive at the input voltage. The number of operations is the required result. If the maximum number to be represented is n, a maximum of n operation steps must be performed to obtain the result.

For a comparison of the individual methods, we have listed their most important characteristics in Fig. 14.22. The different conversion techniques are often combined.

approach	number of steps	number of reference voltages	characteristics
parallel method	1	n	extensive circuit, fast
weighing method	ld n	ld n	
counter method	n	1	simple, slow

Fig. 14.22   Comparison of different approaches to A/D conversion

14.5   Accuracy of A/D converters

14.5.1   Static errors

When converting an analog signal to a digital quantity having a finite number of bits, a systematic error is incurred, due to the limited resolution, which is known as the quantization error. According to Fig. 14.23, it is $\pm \frac{1}{2} U_{\text{LSB}}$, i.e. it corresponds to half the input voltage step required to change the least significant bit.

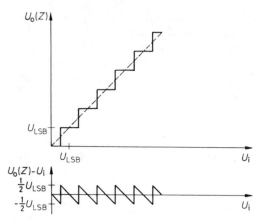

Fig. 14.23 Quantization error. Voltage $U_o(Z)$ is obtained by D/A conversion of the number Z at the A/D converter output

If the number sequence generated is reconverted by a D/A converter to a voltage, the quantization error gives rise to superimposed noise, the r.m.s. value of which is given by [14.2]

$$U_{n\,\text{r.m.s.}} = \frac{U_{\text{LSB}}}{\sqrt{12}}. \qquad (14.18)$$

For full sinusoidal swing, the r.m.s. output signal voltage for an N-bit converter is determined by

$$U_{s\,\text{r.m.s.}} = \frac{1}{\sqrt{2}} \cdot \frac{1}{2} \cdot 2^N \cdot U_{\text{LSB}}.$$

Hence, the signal-to-noise ratio

$$S = 20\,\text{dB}\,\lg\frac{U_{s\,\text{r.m.s.}}}{U_{n\,\text{r.m.s.}}} = N \cdot 6\,\text{dB} + 1.8\,\text{dB}. \qquad (14.19)$$

In addition to the systematic quantization error, errors arise from non-ideal circuitry. When the mid-values of the steps are connected, as shown in Fig. 14.23 by a broken line, a straight line is obtained which passes through the origin and has the slope 1. For a real A/D converter, this line misses the origin (offset error) and its slope is different from 1 (gain error). The gain error gives rise to a *relative* deviation of the output quantity from the desired value, which is constant over the whole range of operation. The offset error gives rise to a constant *absolute* deviation. Both errors can usually be eliminated by adjusting the output at zero and full-range operation. The remaining deviation is then due to drift and nonlinearity only.

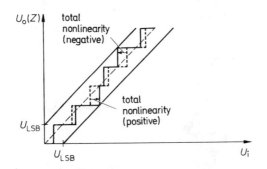

Fig. 14.24 Transfer characteristic of an A/D converter having a linearity error

A nonlinearity error exceeding the quantization error is incurred whenever the steps are of different heights. For a determination of the nonlinearity error, offset and gain are adjusted and the maximum deviation of the input voltage from the ideal straight line is measured. This value, reduced by the systematic quantization error of $\frac{1}{2}U_{\mathrm{LSB}}$, is the *total nonlinearity*. It is usually quoted as a percentage of the LSB voltage unit. For the example in Fig. 14.24, the total nonlinearity is $\pm\frac{1}{2}U_{\mathrm{LSB}}$.

A further measure of the linearity error is the *differential nonlinearity* which indicates by how much the widths of the individual steps deviate from the desired value U_{LSB}. If this deviation is larger than U_{LSB}, a number is skipped (missing code). For even larger deviations, the number Z may even decrease for an increasing input voltage (monotonicity error).

14.5.2   Dynamic errors

A/D converter applications fall into two categories: those in digital voltmeters and those in signal-processing circuits. Their use in digital voltmeters is based on the assumption that the input voltage remains constant during the conversion process. For signal-processing applications, however, the input voltage changes continually. For digital processing, samples must be taken from the alternating voltage by means of a sample-and-hold circuit. The samples are then A/D converted. It has been shown in Section 12.1 that the resulting number sequence $\{Z\}$ represents the continuous input signal without loss of information only if the *sampling theorem* is satisfied. The sampling frequency f_s must therefore be at least twice the highest signal frequency f_{max}. This requires that the A/D conversion time is smaller than $1/2f_{\mathrm{max}}$.

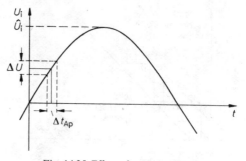

Fig. 14.25 Effect of aperture jitter

In order to judge the attainable accuracy, the properties of the A/D converter and of the sample-and-hold circuit (see Section 7.4) must, in this application, be considered together. For example, it is pointless to operate a 12-bit A/D converter in conjunction with a sample-and-hold circuit which does not settle to $1/4096 \approx 0.025\,\%$ of full scale within the allotted time.

Another dynamic error is incurred due to *aperture jitter*. As can be seen from Fig. 14.25, an uncertainty Δt_{Ap} of the sampling time results in an uncertainty ΔU of the sampled voltage. The aperture time itself is not critical as it only produces a constant delay. For the calculation of the maximum error ΔU, the input signal is taken to be a sine wave with the highest frequency f_{max} for which the system is designed. The maximum slope is at the origin

$$\left.\frac{dU}{dt}\right|_{t=0} = \hat{U}\,\omega_{max}.$$

Hence, the amplitude error

$$\Delta U = \hat{U}\,\omega_{max} \cdot \Delta t_{Ap}.$$

If the error is to be smaller than the smallest conversion voltage U_{LSB} of the A/D converter, the condition for the aperture jitter must be as follows:

$$\Delta t_{Ap} < \frac{U_{LSB}}{\hat{U}\,\omega_{max}} = \frac{U_{LSB}}{\frac{1}{2}\,U_{max}\,\omega_{max}}. \tag{14.20}$$

It is extremely difficult to fulfill this condition for high signal frequencies, as the following example illustrates. For an 8-bit converter, $U_{LSB}/U_{max} = 1/255$. For a maximum signal frequency of $10\,\mathrm{MHz}$, the aperture jitter must, according to Eq. (14.20), be smaller than $125\,\mathrm{ps}$.

14.6   Design of A/D converters

14.6.1   Parallel method

Figure 14.26 illustrates the application of the parallel method for 3-bit numbers. With three bits, 8 different numbers including zero can be represented, for which 7 comparators are required. The appropriate 7 equidistant reference voltages can be generated by means of a voltage divider.

For an input voltage having a value between, for instance, $\frac{5}{2}U_{LSB}$ and $\frac{7}{2}U_{LSB}$, the comparators 1 to 3 produce ones, the comparators 4 to 7, zeros. A logic circuit is therefore needed which converts these comparator states to the number "3". Figure 14.27 lists the comparator output states and the corresponding straight binary numbers. A comparison with Fig. 9.7 shows that the required conversion can be carried out by the priority encoder described in Section 9.1.1.

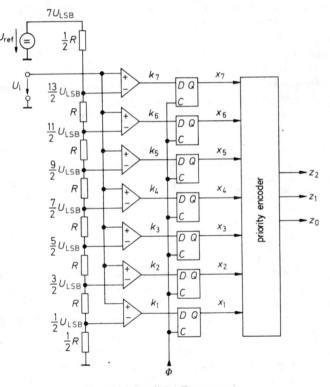

Fig. 14.26   Parallel A/D converter.

$$Z = \frac{U_i}{U_{LSB}} = 7\,\frac{U_i}{U_{ref}}$$

input voltage	comparator states								straight binary number			decimal equivalent
U_i/U_{LSB}	k_7	k_6	k_5	k_4	k_3	k_2	k_1		z_2	z_1	z_0	Z
0	0	0	0	0	0	0	0		0	0	0	0
1	0	0	0	0	0	0	1		0	0	1	1
2	0	0	0	0	0	1	1		0	1	0	2
3	0	0	0	0	1	1	1		0	1	1	3
4	0	0	0	1	1	1	1		1	0	0	4
5	0	0	1	1	1	1	1		1	0	1	5
6	0	1	1	1	1	1	1		1	1	0	6
7	1	1	1	1	1	1	1		1	1	1	7

Fig. 14.27 States of variables in the parallel A/D converter, dependent on the input voltage

However, the priority encoder must not be connected directly to the outputs of the comparators since totally erroneous straight binary numbers may arise if the input voltage is not constant. The example of a change from 3 to 4, i.e. in straight binary code from 011 to 100, illustrates this. If the most significant bit changes before the two other bits because of a shorter propagation delay, the number 111 (i.e. 7_{dec}) occurs temporarily. This is equivalent to an error of half the range. The result of an A/D conversion is usually transferred to a memory, and there is therefore a certain probability that this wrong number would be carried over and stored. The use of a sample-and-hold circuit can prevent this effect as it holds the input voltage constant during the conversion process. However, since the sample-and-hold circuit requires an acquisition time, this measure limits the range of input frequencies. Nevertheless, spurious transitions of the comparators cannot be entirely avoided since fast sample-and-hold circuits have a considerably drift.

These drawbacks are not incurred if the analog storage is replaced by edge-triggered flip-flops connected to the comparators as in Fig. 14.26, and operated by the clock of the following sequential logic circuit. This method ensures that the output signals of the priority encoder are stationary at the trigger edge.

It can be seen from Fig. 14.27 that all the comparator outputs below the highest one at ON, are at logic 1. This property is not guaranteed for steep signal edges since differing delay times of the comparators cause them to change state in the wrong sequence. This intermediate condition may be stored accidentally by the flip-flops if trigger edge end signal edge nearly coincide. However, this has no effect on the priority encoding as the less-significant 1's are not taken into account (see Fig. 9.7).

The instant of conversion is essentially determined by the trigger edge of the clock pulse, although the actual instant occurs a little earlier because of the comparator delay time. The difference in the delays therefore defines the aperture jitter. In order that the low values demanded in the last section can be obtained, comparators having shortest possible delay times must be chosen.

Because of the parallel processing of the bits, this method is by far the fastest approach to conversion. When employing ECL circuits, signal frequencies of over 50 MHz can be dealt with.

Suitable IC types:

	Comparator	Storage	Priority Encoder
TTL:	NE 521	SN 74S273	SN 74148
ECL:	Am 687	included	MC 10165

Complete arrangement with memory and priority encoder:

4 bits: MC 10331 (Motorola),
8 bits: TDC 1007 J (TRW).

14.6.2   Extended parallel method

A disadvantage of the parallel conversion method is that the number of comparators required rises exponentially with the word length. For an 8-bit converter for example, a total of 255 converters is needed. This number can be considerably reduced by a small sacrifice in conversion speed. Such a compromise involves the combination of the parallel method with the weighing method.

An 8-bit converter for the extended parallel conversion method is realized in the following manner; initially, the four most significant bits are parallel-converted, as can be seen in the block diagram of Fig. 14.28. The result is the coarsely quantized value of the input

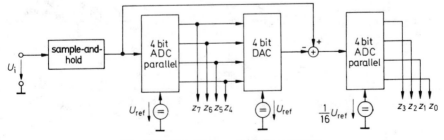

Fig. 14.28 Extended parallel A/D converter.

$$Z = \frac{U_i}{U_{LSB}} = 255\,\frac{U_i}{U_{ref}}$$

voltage. A D/A converter is used to produce the appropriate analog voltage which is then subtracted from the input voltage. The remainder is digitized by a second 4-bit A/D converter.

If the difference between the coarse value and the input voltage is amplified by a factor of 16, two A/D converters with the same input voltage range can be employed. There are, however, different requirements for the accuracy of the two converters: the accuracy of the first 4-bit converter must be as high as that of an 8-bit converter, otherwise the calculated difference is meaningless.

It is obvious that the coarse and fine values must be evaluated from the same input voltage $U_i(t_j)$. However, there is a delay due to the switching time of the first stage. The input voltage must therefore be kept constant by a sample-and-hold circuit until the whole of the number has been converted.

A list of available modules and other equipment operating on this method, can be found in ref. [14.3].

14.6.3 Weighing method

Figure 14.29 shows an A/D converter, based on the weighing method. The control logic (e.g. a microcomputer [14.4]) resets the register to zero at the start of the conversion. Afterwards, the most significant bit, i.e. z_7, is set to one. Hence, the voltage

$$U(Z) = 2^7 U_{LSB}$$

appears at the output of the D/A converter. This is half the input range. If the input voltage U_i is higher than this voltage, z_7 must remain 1; if it is smaller it must become zero. The control logic must therefore reset z_7 to zero if the output variable k of the comparator is zero. Subsequently,

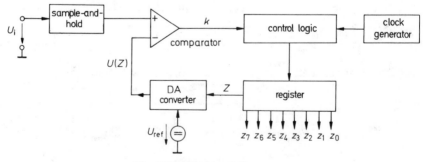

Fig. 14.29 Weighing A/D converter.

$$Z = \frac{U_i}{U_{LSB}}.$$

the remaining voltage
$$U_i - z_7 \cdot 2^7 \cdot U_{LSB}$$

is compared in the same way with the next lower bit, etc. After eight approximation steps of this kind, the register holds a straight binary number Z which, when converted by the DAC, produces a voltage coinciding with U_i within the resolution U_{LSB}. Hence,

$$U_i = Z \cdot U_{LSB}$$

or

$$Z = \frac{U_i}{U_{LSB}}. \qquad (14.21)$$

If the input voltage changes during conversion, a sample-and-hold circuit must be provided for temporary storage of the sampled input voltage to ensure that all bits are evaluated from the same voltage $U_i(t_j)$.

A/D converters operating on the above principle are available as monolithic ICs.

8 bits in 40 µs: ADC 0800 (National);
8 bits in 20 µs: AD 7570J (Analog Devices);
10 bits in 20 µs: AD 7570L (Analog Devices);
12 bits in 25 µs: ADC 1210 (National).

14.6.4 Counter method

A/D conversion by the counter method requires the least components and high accuracy can be attained with relatively simple circuits. However, conversion time is considerably longer than with the other methods. It is usually between 1 and 100 ms, although this is quite sufficient for many applications. The counter method is therefore the most widely used and the variety of different circuits is extensive. The most important ones are described below.

Compensation method

The compensating A/D converter in Fig. 14.30 is closely related to the circuit described in the previous section. There is, however, the essential difference that the register is replaced by a counter. A large part of the control logic can therefore be omitted.

The subtracting circuit enables comparison of the input voltage U_i with the compensating voltage $U(Z)$. If the difference $U_i - U(Z) > \frac{1}{2}U_{LSB}$, the counter counts upwards. $U(Z)$ thereby approaches the input voltage. If the difference $U_i - U(Z) < -\frac{1}{2}U_{LSB}$, the counter counts

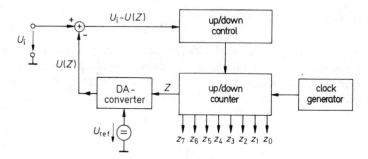

Fig. 14.30 Compensating A/D converter.

$$Z = \frac{U_i}{U_{LSB}}$$

downwards. In this manner, the compensating voltage is always able to follow the input voltage, and the circuit is therefore also known as a *tracking A/D converter.*

When the magnitude of the difference becomes smaller than $\frac{1}{2}U_{LSB}$, the counter is stopped, thereby avoiding the least significant bit of the counter continually changing state.

In contrast to the weighing method, a BCD-format output can be readily obtained with this approach. The straight-binary counter is simply replaced by a BCD-counter.

The reduction in control logic circuitry is offset by a considerable increase in conversion time, as the compensating voltage can only change in steps of U_{LSB}. However, if the input voltage changes only slowly, there is also a short conversion time with this solution. This is because the approximation operates continually due to the tracking property and need not, as in the weighing method, always re-start at zero.

Single-slope method

The single-slope A/D converter in Fig. 14.31 does not require a D/A converter. The principle of the circuit is based on the conversion of the input voltage to a proportional time interval. The sawtooth generator and the comparators CA 1, CA 2 serve this purpose.

The sawtooth voltage rises from negative to positive values according to the expression

$$V_S = \frac{U_{ref}}{\tau} t + V(0).$$

The output of the exclusive-OR gate is at logic 1 for as long as the sawtooth voltage is between the limits 0 and U_i. The corresponding

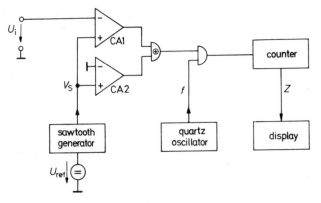

Fig. 14.31 Single-slope A/D converter.

$$Z = \frac{\tau \cdot f}{U_{\text{ref}}} \cdot U_i$$

time interval is therefore given as

$$\Delta t = \frac{\tau}{U_{\text{ref}}} U_i.$$

It is measured by counting the pulses of a quartz oscillator. If the counter is reset to zero at the beginning of each conversion cycle, the counter state is at

$$Z = \frac{\Delta t}{T} = \frac{\tau f}{U_{\text{ref}}} U_i \qquad (14.22)$$

when the voltage exceeds the upper trigger level of the comparator circuit.

For a negative input voltage, the sawtooth voltage will first cross the input voltage and then through zero. This sequence can therefore be used to determine the polarity of the input voltage. The conversion time is independent of the sequence and depends only on the magnitude of the input voltage. After each conversion, the counter must be reset to zero and the sawtooth voltage to its negative initial value $V(0)$. To avoid a change in the display during the new conversion, the old result is usually stored. For the continual tracking compensation method, this is not necessary as the counter state no longer changes when the system is balanced and U_i constant.

As is apparent from Eq. (14.22), the tolerance of the time constant τ directly determines the accuracy of the result. It is defined by an RC element and is thus subject to the temperature drift and long-term stability of the capacitor. For this reason, an accuracy of better than 0.1% is very difficult to attain.

Dual-slope method

There is another method, in which not only the reference voltage but also the input voltage is integrated, and this is illustrated by the block diagram in Fig. 14.32. At rest, the switches S_1 and S_2 are open; S_3 is closed. The output voltage of the integrator is zero.

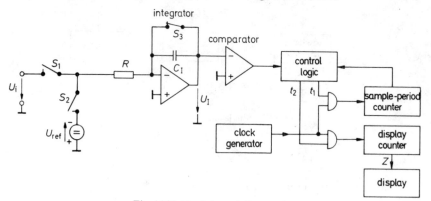

Fig. 14.32 Dual-slope A/D converter.

$$Z = \frac{\overline{U}_i}{U_{ref}} n_1$$

At the beginning of the conversion, switch S_3 is opened and S_1 closed, and the input voltage is integrated, this integration being carried out for a constant time interval defined by the sample-period counter. At the end of the integration interval t_1, the output voltage of the integrator is given by

$$U_1(t_1) = -\frac{1}{\tau} \int_0^{t_1} U_i \, dt = -\frac{\overline{U}_i n_1 T}{\tau}, \tag{14.23}$$

where n_1 is the number of clock pulses predetermined by the sample-period counter and T the period of the clock generator.

After measuring, evaluation is started by opening switch S_1 and applying the reference voltage to the integrator via switch S_2. The polarity of the reference voltage is opposite to that of the input voltage so that the magnitude of the integrator output voltage decreases, as can be seen in Fig. 14.33.

The comparator and the display counter are now used to determine the time interval required for the integrator voltage to reach zero again:

$$t_2 = n_2 T = \frac{\tau}{U_{ref}} |U_1(t_1)|. \tag{14.24}$$

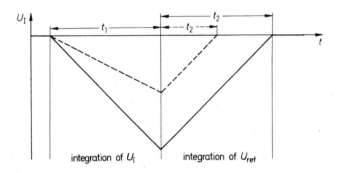

Fig. 14.33 Time function of the integrator output voltage for different input voltages

Hence, with Eq. (14.23), the end result is given by

$$Z = n_2 = \frac{\overline{U}_i}{U_{ref}} n_1. \tag{14.25}$$

This equation shows the most prominent feature of the dual-slope method: neither the clock frequency $1/T$ nor the integration time constant $\tau = RC_I$ influence the result. The only condition is that the clock frequency remains constant during the interval $t_1 + t_2$. This short-term stability presents no problem, even for simple clock generators. For these reasons, the method can produce an accuracy of better than $0.01\% = 100$ ppm [14.5].

As is shown by Eq. (14.23), the result is influenced not by the instantaneous input voltage but by its mean value taken over the measuring time t_1. Alternating voltages are therefore more attenuated the higher their frequency. Alternating voltages, the frequencies of which are multiples of $1/t_1$, are completely rejected. It is therefore sensible to control the clock frequency so that t_1 equals the period of the mains supply voltage or a multiple thereof. In this manner, all hum interference is eliminated.

The dual-slope method results in a simple circuit, produces very accurate results and is therefore preferable for use in digital voltmeters. The relatively long conversion time is quite acceptable for such applications.

Automatic zero-balancing

It has been shown for the dual-slope method that the time constant $\tau = RC_I$ and the clock frequency $f = 1/T$ do not influence the result. The accuracy is therefore defined essentially by the tolerance of the reference voltage and the offset errors of integrator and comparator. These offset errors can be eliminated to a large extent by an automatic zero

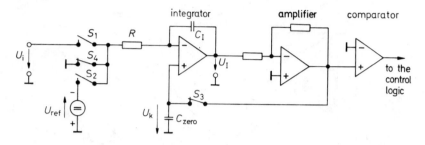

Fig. 14.34 Dual-slope method with automatic zero balance

balance. To implement this, the short-circuiting switch S_3 in Fig. 14.32 is replaced by a control circuit as in Fig. 14.34, which sets the integrator to suitable initial conditions.

At rest, switch S_3 is closed. The integrator and the comparator preamplifier constitute a voltage follower, the output voltage U_k of which charges the zeroing capacitor C_{zero}. For offset correction, the integrator input is simultaneously grounded via switch S_4. The voltage U_k therefore assumes the correction value $U_{OI} - I_B R$, where U_{OI} is the integrator offset voltage and I_B the integrator input bias current. At steady state, this compensation effects zero current through C_I, as for an ideal integrator.

For the integration of the input voltage, the switches S_3 and S_4 are opened and S_1 closed. As the voltage U_k remains stored at the capacitor C_{zero}, the offset is balanced during the integration interval. The zero drift is then only influenced by the short-term stability.

This method also largely eliminates the offset error of the comparator. The reason for this is that with the converter inactive, the output voltage U_I of the integrator does not settle at zero as for the previous circuit, but at the offset voltage of the comparator preamplifier, i.e. at the trigger level of the total arrangement.

Oscillation may arise due to the two series-connected amplifiers in the compensation loop. For damping, a resistor is connected in series with C_{zero}, and it is also advisable to limit the gain of the comparator preamplifier to values below 100. This also effects the short delay times necessary for comparator operation during the integration mode.

Integrating A/D converters are available as monolithic ICs in CMOS technology. One must differentiate between two major groups: those with parallel output for general applications (particularly for data processing by microcomputers) and those with multiplex BCD output for the control of display units.

IC types with straight-binary coded parallel outputs:

   13 bits: AD 7550 (Analog Devices),
8...12 bits: ADC-EK 8B...12B (Datel).

IC types with parallel BCD outputs:

3-digit BCD: ADC-EK 12D (Datel).

$3\frac{1}{2}$-digit 7-segment: 7107 (Intersil).

IC types with $3\frac{1}{2}$-digit BCD multiplex outputs:

LD 131 (Siliconix),   MC14433 (Motorola),

MN 2301 (Analogic).

15 Measurement circuits

In the previous chapters, a number of methods for the processing of analog and digital signals are described. Many applications, however, require that even electrical signals are first adapted before they can be processed in analog computing circuits or A/D converters. In such cases, measurement circuits are needed which have a low-resistance single-ended output, i.e. produce a ground-referenced output voltage.

15.1 Measurement of voltage

15.1.1 Impedance converter

If the signal voltage of a high-impedance source is to be measured without affecting the load conditions, the non-inverting amplifier (follower-with-gain, electrometer amplifier) in Fig. 2.3 can be employed for impedance conversion. It must be noted, however, that the high-impedance input is very sensitive to noise arising from capacitive stray currents in the input lead. Therefore the lead is usually screened, but this results in considerable capacitive loading of the source to ground (30...100 pF/m). For an internal source resistance of 1 GΩ, for instance, and a lead capacitance of 100 pF, an upper cutoff frequency of only 1.6 Hz is obtained.

The capacitance is not constant but can change, for example when the lead is moved. This gives rise to an additional problem in that this variation produces very large noise voltages. If, for instance, the lead is charged-up to 10 V, a change of capacitance by 1 % results in a voltage step of 100 mV!

These disadvantages can be avoided if a non-inverting amplifier is employed to keep the voltage low between the inner conductor and the shield. The shield is then connected not to ground but to the amplifier output, as shown in Fig. 15.1. In this manner, the lead capacitance is

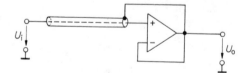

Fig. 15.1 Reduction of shield capacitance and of stray current noise by allowing the shield potential to follow the measuring potential

virtually reduced by the open-loop gain of the operational amplifier. Since now only the offset voltage of the amplifier appears across the lead capacitance, the noise also is eliminated to a large extent.

Increase of the output voltage swing

The maximum permissible supply voltage of standard integrated operational amplifiers is usually ± 18 V. The attainable output voltage swing is thereby limited to values of about ± 15 V. This limit can be raised by allowing the operational amplifier supply potentials to follow the input voltage. This is achieved by the bootstrap circuit in Fig. 15.2, using two emitter followers to stabilize the potential differences $V_1 - U_o$ and $U_o - V_2$ to a value $U_z - 0.7$ V. The maximum output voltage swing is thus no longer determined by the operational amplifier, but by the permissible voltage across the emitter followers and the current sources.

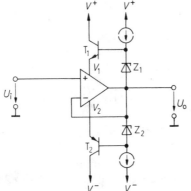

Fig. 15.2 Voltage follower for large input voltages

15.1.2   Measurement of potential differences

The measurement of potential differences can, in principle, be carried out by the subtracting amplifier in Fig. 1.3. As is shown in Section 1.2.2, the common-mode rejection ratio is defined mainly by the matching tolerance of the resistance ratios α_N and α_P. The resistance ratios are influenced by the internal resistance of the signal source. In analog computing circuitry, the signal source is usually an operational amplifier in feedback, with a very low output resistance so that this influence can be neglected. For the application in measuring equipment, however, a defined low source resistance must be established by means of voltage followers. This leads to the universal subtracting circuit in Fig. 15.3 (instrumentation amplifier). For $R_1 = \infty$, OA 1 and OA 2 operate as voltage followers.

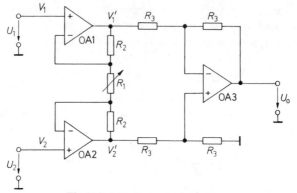

Fig. 15.3 Instrumentation amplifier.

Output voltage: $U_o = \left(1 + \dfrac{2R_2}{R_1}\right)(V_2 - V_1)$

The circuit has the advantage that the differential gain can be adjusted by variation of a single resistor. It can be seen in Fig. 15.3 that the potential difference $V_2 - V_1$ appears across resistor R_1. Hence,

$$V_2' - V_1' = \left(1 + \frac{2R_2}{R_1}\right)(V_2 - V_1).$$

This difference is transferred to the single-ended output by the subtracting amplifier OA 3.

For a purely common-mode input drive ($V_1 = V_2 = V_{CM}$), $V_1' = V_2' = V_{CM}$. The common-mode gain of OA 1 and OA 2 is therefore unity, independent of the differential gain chosen. With Eq. (1.6), we thus obtain the common-mode rejection ratio as

$$G = \left(1 + \frac{2R_2}{R_1}\right)\frac{2\alpha}{\Delta\alpha},$$

where $\Delta\alpha/\alpha$ is the relative matching tolerance of the resistors R_3.

15.1.3   Isolation amplifier

The common-mode voltage of the subtracting circuit described in the previous section is limited to values within the supply voltage range. With the method of Fig. 15.2, it can be raised to voltages of about $\pm 100\,\mathrm{V}$.

There are many applications, however, in which the voltage to be measured is superimposed on a considerably higher common-mode voltage of perhaps several kV. To deal with such high potentials, the

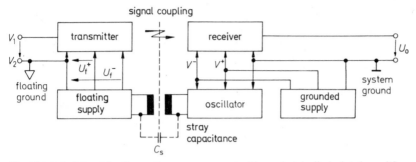

Fig. 15.4 Principle of floating voltage measurement with an electrically isolated amplifier

measuring circuit is split, as in Fig. 15.4, into two electrically isolated units. The transmitter unit operates at the measuring potential, the receiver unit at system ground. For such an operation, the transmitter must have its own floating supply, the common of which is used as one of the two differential inputs (floating ground). Although this terminal is electrically isolated from system ground, it must not be forgotten that there is still some capacitive coupling due mainly to the capacitance C_s of the supply transformer. This is indicated in Fig. 15.4. To keep the coupling low, it is sensible not to use a mains supply transformer but a high-frequency transformer for about 100 kHz, fed from a separate oscillator.

If both test points have a high internal resistance, even the reduced stray capacitance current may produce a considerable voltage between test point 2 and floating ground. In such a case, it is advisable to connect the floating ground terminal to a third point and to determine the potential difference between the two measuring points with the instrumentation amplifier of Fig. 15.3. The measuring leads then carry no current; the instrumentation amplifier is connected to the floating supply. The remaining common-mode voltage with respect to floating ground can usually be kept below 10 V if the floating ground is connected to a suitable potential within the circuit being tested.

There is still the question of electrically isolated transmission of the measured voltage to the receiver unit. Two methods exist: transformers and opto-couplers. For the transmission by transformers, the signal must be modulated on a carrier of sufficiently high frequency (amplitude modulation or pulse-width modulation), whereas opto-couplers enable the direct transmission of d.c. voltages. When high accuracy is required, the analog signal can be digitized at floating-ground potential, and the digital values subsequently transmitted by opto-couplers to the receiver unit, as shown in Fig. 13.4. The non-linearity of the opto-coupler is then no longer significant.

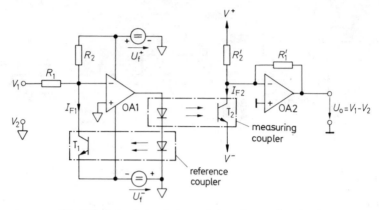

Fig. 15.5 Opto-electronic transmission of an analog signal

A method of optical analog transmission is presented in Fig. 15.5. In order to compensate for the linearity error of the opto-coupler, the operational amplifier OA1 controls the current through the light-emitting diodes in such a way that the photo-electric current in the reference receiver T_1 equals a desired value. The feedback loop is closed via the reference coupler, so that

$$I_{F1} = \frac{U_f^+}{R_2} + \frac{V_1 - V_2}{R_1}.$$

As the photo-electric current cannot change polarity, a constant current U_f^+/R_2 is superimposed to enable transmission of bipolar input

type	3650	3456	275
manufacturer	Burr-Brown	Burr-Brown	Analog Devices
coupling	optical	pulse-width modulation	amplitude modulation
bandwidth	15 kHz	2.5 kHz	1.5 kHz
nonlinearity	0.1 %	0.01 %	0.1 %
common-mode rejection at 60 Hz	100 dB	130 dB	120 dB
input circuit	summing point	instrumentation amplifier	voltage follower
insulation voltage	2 kV	2 kV	2 kV
floating supply	external	internal	internal

Fig. 15.6 Examples of isolation amplifiers

signals. If the two opto-couplers are well matched, we obtain at the receiver side the current $I_{F2} = I_{F1}$, and thus the output voltage

$$U_o = \frac{R'_1}{R_1}(V_1 - V_2) \quad \text{for} \quad \frac{U_f^+}{R_2} = \frac{V^+}{R'_2}.$$

Isolation amplifiers with transformer or optical coupling are available as modules. At the primary side, inverting or non-inverting amplifiers as well as instrumentation amplifiers are employed. Most types comprise a d.c./d.c. converter for the floating supply of the transmitter unit so that only the usual grounded supply is required externally. The permissible potential difference between the transmitter and receiver unit is usually several kV. Figure 15.6 lists data of some typical modules.

15.2   Measurement of current

15.2.1   Floating zero-resistance ammeter

Section 2.2 describes a current-to-voltage converter which is almost ideally suited for the measurement of currents because of its extremely low input resistance. However, as the input represents virtual ground, only currents to ground can be measured.

Floating ammeters can be realized by the instrumentation amplifier in Fig. 15.3 if its two inputs are connected to a measuring shunt. The advantage of the low input resistance is then lost, but if the shunt is incorporated in the feedback loop of the input amplifiers, as illustrated in Fig. 15.7, a floating ammeter is obtained having virtually zero input resistance.

Due to the feedback via the resistors R_2 and R'_2, the potential V_N assumes a value V_i such that the potential difference across the inputs 1

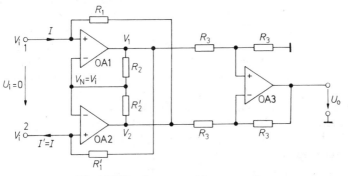

Fig. 15.7 Floating zero-resistance ammeter

Output voltage: $U_o = 2RI$ for $R_1 = R'_1 = R_2 = R'_2 = R$

and *2* becomes zero. If a current I flows into terminal *1*, the feedback causes the output potential of OA 2 to have the value

$$V_2 = V_i - IR_1. \tag{15.1}$$

With $V_N = V_i$, we obtain

$$V_1 = V_2 + \left(1 + \frac{R_2}{R_2'}\right)(V_i - V_2) = V_i + \frac{R_1 R_2}{R_2'} I. \tag{15.2}$$

Hence, the current leaving terminal *2* is given by

$$I' = \frac{V_1 - V_i}{R_1'} = \frac{R_1 R_2}{R_1' R_2'} I. \tag{15.3}$$

If both inputs are to behave as those of a floating circuit, I' must be the same as I, or the current difference $\Delta I = I' - I$ will flow through the operational amplifier outputs to ground. Hence the condition that

$$\frac{R_1}{R_1'} = \frac{R_2'}{R_2}. \tag{15.4}$$

The subtracting amplifier OA 3 computes the difference $V_1 - V_2$. Its output voltage, with Eqs. (15.1) and (15.2), is therefore

$$U_o = R_1 \left(1 + \frac{R_2}{R_2'}\right) I, \tag{15.5}$$

i.e. it is proportional to the current I.

15.2.2 Measurement of current at high potentials

The permissible common-mode voltage of the previous circuit is limited to values between the supply voltages. For measurement of currents at higher potentials, the simple circuit of Fig. 2.5 can be employed if it is connected to the floating ground terminal of an isolation amplifier rather than to system ground. Its output voltage is referenced to system ground with the help of the isolation amplifier.

The required circuitry may be reduced quite considerably if a voltage drop of 1 V to 2 V can be tolerated for the measurement of current (e.g. in the anode circuit of high-voltage tubes). In such cases, the current to be measured is made to flow through a light-emitting diode of an opto-coupler, so that a floating supply is no longer needed. For linearization of the transfer characteristic, a reference opto-coupler may be used on the receiver side, as demonstrated in Fig. 15.8. Its input current I_2 is controlled by the operational amplifier in such a manner that the photo-electric currents of the reference and the measuring

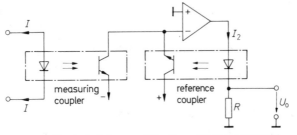

Fig. 15.8 Simple isolation amplifier for current measurement.

Output voltage: $U_o = R \cdot I$

coupler cancel. If the couplers are well matched,

$$I_2 = I.$$

This current can be measured as a voltage across the grounded resistor R.

15.3   A.C./D.C. converters

Several quantities are used for the definition of alternating voltages: the arithmetic mean absolute value, the root-mean-square (r.m.s.) value and the positive and negative peak value.

15.3.1   Measurement of the mean absolute value

To obtain the absolute values of an alternating voltage, a circuit is required in which the sign of the gain changes with the polarity of the input voltage; i.e. its transfer characteristic must have the shape represented in Fig. 15.9.

Such a full-wave rectifier can be realized by a diode bridge. However, the accuracy of such a circuit is limited owing to the forward voltages of the diodes. This effect can be avoided if the bridge rectifier is operated from a controlled current source; a simple solution is

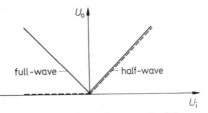

Fig. 15.9 Characteristics of a half-wave and a full-wave rectifier

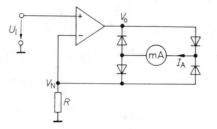

Fig. 15.10 Full-wave rectifier for floating meters.

Meter current: $I_A = |U_i|/R$

shown in Fig. 15.10. The operational amplifier is employed as a voltage-controlled current source, in accordance with Fig. 2.8. Hence, independently of the diode forward voltage,

$$I_A = \frac{|U_i|}{R}.$$

For the display of the mean value of this current, a moving-coil ammeter can be used for instance, and this method is therefore often made use of in analog multimeters.

For output potentials in the range $-2U_D < V_o < 2U_D$, the amplifier has no feedback, as none of the diodes conduct. While V_o changes from $2U_D$ to $-2U_D$, V_N remains constant, this causing a delay time within the control loop. Because of this, any phase shift can be incurred in the control loop, depending on the frequencies involved so that stabilization of the operational amplifier is particularly difficult. To reduce the delay time, amplifiers having a fast slew rate must be chosen and the frequency compensation must be stronger than for linear feedback.

Full-wave rectifier with single-ended output

In the previous circuit, the load (i.e. the moving-coil instrument) had to be floating. However, if the signal is to be processed further, e.g. digitized, a ground-referenced output voltage is needed; this can be derived from the current I_A with the help of, for instance, a floating current-to-voltage converter. Figure 15.11 shows a more simple method.

We deal first with the operation of OA1. For positive input voltages, it operates as an inverting amplifier since V_2 is negative, i.e. diode D_1 is forward and D_2 reverse biased. Hence, $V_1 = -U_i$. For negative input voltages, V_2 is positive; D_1 is OFF and D_2 is conducting, thereby applying feedback to the amplifier and preventing OA1 from saturating so that the summing point remains as zero voltage. Since D_1

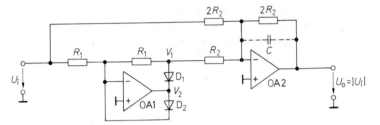

Fig. 15.11 Full-wave rectifier with single-ended output

is reverse biased, V_1 is also zero. Therefore

$$V_1 = \begin{cases} -U_i & \text{for } U_i \geq 0, \\ 0 & \text{for } U_i \leq 0. \end{cases} \tag{15.6}$$

Amplifier OA1 thus operates as an inverting half-wave rectifier.

The extension to the full-wave rectifier is effected by amplifier OA2. It computes the expression

$$U_o = -(U_i + 2V_1), \tag{15.7}$$

which, with Eq. (15.6), becomes

$$U_o = \begin{cases} U_i & \text{for } U_i \geq 0, \\ -U_i & \text{for } U_i \leq 0. \end{cases} \tag{15.8}$$

This is the desired characteristic of a full-wave rectifier. The operation is illustrated by Fig. 15.12.

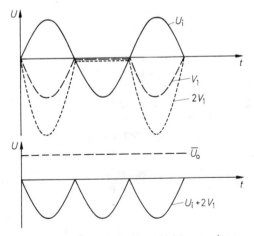

Fig. 15.12 Operation for sinusoidal input voltage

Amplifier OA2 can be extended by the addition of capacitor C to become a first order lowpass filter. If the filter cutoff frequency is chosen small compared to the lowest signal frequency, a smooth direct voltage is obtained at the output, having the value

$$U_o = \overline{|U_i|}.$$

As for the previous circuit, amplifier OA1 must have a fast slew rate to keep the delay time of the change-over from one diode to the other as short as possible.

Broadband full-wave rectifier

A differential amplifier has available an inverting and a non-inverting output and can therefore be used as a fast full-wave rectifier. In this application, two parallel-connected emitter followers T_3/T_4 are used, as in Fig. 15.13, to connect the more positive collector potential to the output. The Zener diode compensates for the collector quiescent potential to make the output quiescent potential zero.

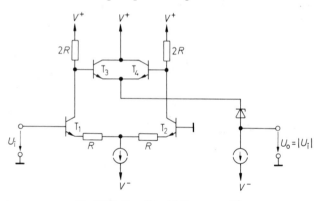

Fig. 15.13 Broadband full-wave rectifier

This method can be used to rectify signal voltages of up to 100 MHz with good linearity. For the detailed design, the same considerations apply as for the broadband differential amplifiers in Section 4.5.

15.3.2   Measurement of the r.m.s. value

Whereas the arithmetic mean absolute value (mean modulus) is defined as

$$\overline{|U|} = \frac{1}{T} \int_0^T |U| \, \mathrm{d}t, \tag{15.9}$$

the definition of the root-mean-square value (r.m.s. value) is given as

$$U_{\text{r.m.s.}} = \sqrt{\overline{(U^2)}} = \sqrt{\frac{1}{T}\int_0^T U^2\,\mathrm{d}t}, \qquad (15.10)$$

where T is the measuring interval which must be large compared to the longest period contained in the signal spectrum. In this manner, a reading is obtained which is independent of T. For strictly periodic functions, averaging over one period is sufficient to get the correct result.

For sinusoidal voltages,

$$U_{\text{r.m.s.}} = \hat{U}/\sqrt{2}$$

so that the r.m.s. value can be determined simply by measuring the peak value. For other waveshapes, this method would produce errors, particularly for highly peaked voltages, i.e. for waveshapes having large *crest factors* $\hat{U}/U_{\text{r.m.s.}}$.

The errors become smaller if the measurement of r.m.s. values is reduced to that of the mean absolute values. For a sinusoidal voltage,

$$\overline{|U|} = \frac{\hat{U}}{T}\int_0^T |\sin\omega t|\,\mathrm{d}t = \frac{2}{\pi}\hat{U} \qquad (15.11)$$

and, with $U_{\text{r.m.s.}} = \hat{U}/2$,

$$U_{\text{r.m.s.}} = \frac{\pi}{2\sqrt{2}}\overline{|U|} \approx 1.11\cdot\overline{|U|}. \qquad (15.12)$$

The relative magnitude of the individual values is demonstrated in Fig. 15.14.

The *form factor* of 1.11 is incorporated in the calibration of most available mean-absolute-value meters; for sinusoidal signals therefore, they show the r.m.s. value although they actually measure the mean-absolute value. For other waveshapes, this modified reading produces varying deviations from the true r.m.s. value. For a triangular wave-

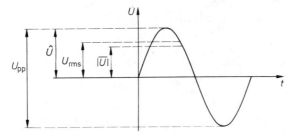

Fig. 15.14 Relative magnitude of peak value, r.m.s. value and mean absolute value, of a sinusoidal signal

shape, $U_{\text{r.m.s.}} = (2/\sqrt{3})\,\overline{|U|}$; for white noise, $U_{\text{r.m.s.}} = \sqrt{\pi/2}\,\overline{|U|}$, whereas for a d.c. voltage, $U_{\text{r.m.s.}} = \overline{|U|}$. The following deviations in the readings are obtained for different waveshapes:

d.c., square wave: reading too large by 11 %,
triangular wave: reading too small by 4 %,
white noise: reading too small by 11 %.

True measurement of r.m.s. values

For a true waveshape-independent measurement of an r.m.s. value, either the definition Eq. (15.10) can be employed, or the power can be measured.

The operation of the circuit in Fig. 15.15 is based on Eq. (15.10). In order to evaluate the mean value of the squared input voltage, a simple first order lowpass filter is used, the cutoff frequency of which is low in comparison with the lowest signal frequency.

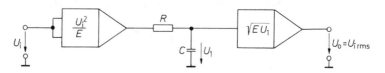

Fig. 15.15   Measurement of r.m.s. values by analog computing circuits

One drawback of the circuit is that its minimum input signal must be relatively large; if, for instance, a voltage of 10 mV is applied to the input and the computing unit E is 10 V as usual, a voltage of only 10 μV is obtained at the output of the squarer. This, however, will be drowned by the noise of the square-rooter.

In this respect, the circuit of Fig. 15.16 is preferable since the square-rooting operation at the output is replaced by a division at the input. The voltage at the output of the lowpass filter is thus

$$U_o = \overline{\left(\dfrac{U_i^2}{U_o}\right)}.$$

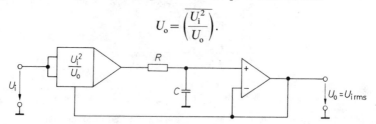

Fig. 15.16   Circuit for r.m.s. measurement, having an improved input voltage range

At steady state, $U_o = $ const; hence,

$$U_o = \frac{\overline{(U_i^2)}}{U_o},$$

i.e.

$$U_o = \sqrt{(\overline{U_i^2})} = U_{r.m.s.}.$$

An advantage of this method is that the input voltage U_i is not multiplied by the factor U_i/E which, for low input voltages, is small against unity, but by the factor U_i/U_o which is close to unity. The available input range is therefore considerably larger. However, the precondition for this is that the division U_i/U_o can be carried out sufficiently accurately even for small signals. Divider circuits based on logarithmic operations are best suited, e.g. the circuit in Fig. 1.42. The multiplier/divider of Fig. 1.39 is also suitable but has the disadvantage that it can operate only for positive input signals. It must therefore be connected to a precision full-wave rectifier. The monolithic integrated r.m.s. converter AD 536 from Analog Devices is based on this principle. Because of the logarithmic calculus involved, an additional output is available which is calibrated in dB. The accuracy is 0.2% for frequencies of up to 20 kHz and 1% for up to 100 kHz.

Thermal conversion

The problem of r.m.s. measurement can be reduced to that of the determination of power; this is carried out by a thermocouple, used to measure the temperature of a resistance wire. However, measurement of the very small thermal voltages presents difficulties.

Another possibility is to use the input voltage U_i as in Fig. 15.17, to heat a resistor R_1, the rise in temperature of which is then detected by means of a change in U_{BE} of transistor T_1. In order that the influences

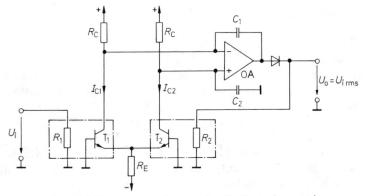

Fig. 15.17 Measurement of r.m.s. values by thermal conversion

of ambient temperature and junction characteristic are eliminated, a second pair R_2, T_2, well-matched with the first, is inserted in the circuit. Resistor R_2 is heated by the positive d.c. voltage U_o supplied by the amplifier OA. The thermal coupling with T_2 results in a rise of its collector current, thereby effecting a reduction of U_o. Hence, the circuit is stabilized by thermal feedback. The output voltage now assumes such a level that

$$I_{C2} = I_{C1}, \quad \text{i.e.} \ U_{BE2} = U_{BE1}.$$

The temperature and hence, the thermal power of the two heating resistors is the same, and one obtains

$$U_o = U_{i\,\text{r.m.s.}}.$$

The two pairs R_1, T_1 and R_2, T_2 must be thermally well isolated to guarantee that T_1 is heated only by R_1 and T_2 only by R_2.

The diode at the amplifier output prevents resistor R_2 being heated by a negative voltage, as this would result in latch-up of the circuit due to positive thermal feedback.

The capacitors C_1 and C_2 effect an additional frequency compensation in order to match the controller circuit to the thermal time constant. With this external circuitry, the amplifier integrates the difference of the two collector currents I_{C1} and I_{C2} and thus operates as an integral controller.

Since the thermal power is proportional to the square of U_o, the loop gain is also proportional to U_o^2, giving a nonlinear step response: the switch-off time constant is considerably larger than the switch-on time constant. Substantial improvement is achieved with an additional square-function a.c. feedback circuit [5.1].

The operation of the r.m.s. converter 4130 from Burr-Brown is based on the above principle. Its accuracy is 0.05% for frequencies of up to 100 kHz and 2% for up to 10 MHz.

15.3.3   Measurement of the peak value

The peak value can be measured very simply by charging a capacitor via a diode. For elimination of its forward voltage, the diode is inserted in the feedback loop of a voltage follower, as in Fig. 15.18. For input voltages $U_i < V_C$, the diode is reverse biased. For $U_i > V_C$, the diode conducts and, due to the feedback, $V_C = U_i$. The capacitor therefore charges to the peak input voltage. The voltage follower OA 2 draws only very little current from the capacitor so that the peak value can be stored over a long period. Push-button switch S discharges the capacitor to prepare it for the next measurement.

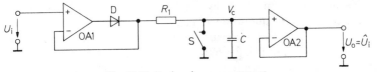

Fig. 15.18 Peak value measurement

The capacitive load of amplifier OA1 may give rise to oscillations; this effect is eliminated by resistor R_1. However, the charging time is increased as the capacitor voltage now only approaches steady-state conditions asymptotically. A further disadvantage of the circuit is that OA1 has no feedback if $U_i < V_C$, i.e. is saturated. The resulting recovery time limits the use of this circuit to low-frequency applications.

Both these drawbacks are avoided with the peak detector in Fig. 15.19. Here, OA1 is operated in the inverting mode. If U_i exceeds the value of $-V_C$, V_1 becomes negative and diode D_1 conducting. Because of the feedback across the two amplifiers, V_1 assumes such a value that $U_o = -U_i$. In this way the diode forward voltage, as well as the offset voltage of the impedance converter OA2, is eliminated. If the input voltage falls, V_1 rises, thus reverse biasing diode D_1. The feedback via R_2 is opened, but V_1 can only rise until diode D_2 becomes conducting and provides feedback for OA1. In this manner, saturation of OA1 is avoided.

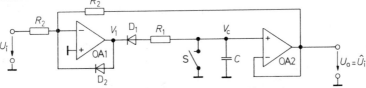

Fig. 15.19 Improved method of peak value measurement

The inverted positive peak value of U_i remains stored at capacitor C as no leakage current can flow either through D_1 or through the impedance converter OA2. Before a new reading, capacitor C must be discharged by switch S. For measurement of negative peak voltages, the diodes must be reversed.

Continuous measurement of the peak value

The method described can be adapted for continuous peak voltage measurements if switch S is replaced by a resistor, the value of which is chosen so that the discharge of the capacitor C between two voltage crests is negligible. However, this method has the disadvantage that a decrease in peak amplitude is registered only very slowly.

It is important for many applications, particularly in control circuits, that the peak amplitude is measured with the shortest possible delay. With the methods described so far, the measuring time is at least one period of the input signal. However, for sinusoidal signals, the peak amplitude can be computed at any instant with the trigonometric equation

$$\hat{U}=\sqrt{\hat{U}^2\sin^2\omega t+\hat{U}\cos^2\omega t}. \qquad (15.13)$$

We have already used this relationship for the amplitude control of the oscillator in Fig. 8.24. There it was particularly practicable since both the sine and the cosine function were available.

For the measurement of an unknown sinusoidal voltage, we must establish the $\cos\omega t$ function from the input signal; a differentiator can be employed for this purpose. We obtain its output voltage as

$$V_1(t)=-RC\frac{\mathrm{d}U_i(t)}{\mathrm{d}t}=-\hat{U}_iRC\frac{\mathrm{d}\sin\omega t}{\mathrm{d}t}=-\hat{U}_i\omega RC\cos\omega t. \qquad (15.14)$$

If the frequency is known, the coefficient ωRC can be adjusted to 1, so that the required term for Eq. (15.13) is obtained. By squaring and adding $U_i(t)$ and $V_1(t)$, a continuous voltage for the amplitude is obtained which needs no filtering.

For variable frequencies, the circuit must be expanded as in Fig. 15.20, by inserting an integrator to enable provision of the $\cos^2\omega t$ term with frequency-independent amplitude. The output potential of the integrator is

$$V_2(t)=-\frac{1}{RC}\int U_i(t)\,\mathrm{d}t=-\frac{1}{RC}\int\hat{U}_i\sin\omega t\,\mathrm{d}t=\frac{\hat{U}_i}{\omega RC}\cos\omega t, \qquad (15.15)$$

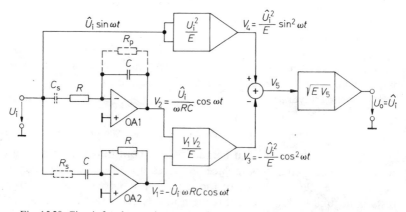

Fig. 15.20 Circuit for the continuous amplitude measurement of sinusoidal signals

where the constant of integration is zero at steady state, because of the resistor R_p. By the multiplication of V_1 and V_2, the required expression

$$V_3(t) = -\frac{\hat{U}_i^2}{E}\cos^2\omega t$$

is obtained. By subtracting this from V_4 and square-rooting the result, the output voltage

$$U_o = \hat{U}_i$$

is measured which equals the amplitude of the input voltage at any instant. For steep changes in the amplitude, a temporary departure from the correct value is incurred until the integrator output resettles to zero mean voltage. However, the output voltage will instantly change to the right direction so that a connected controller for example, obtains correct information about the trend without delay.

15.3.4 Synchronous demodulator

In a synchronous demodulator (synchronous detector, phase-sensitive rectifier), the sign of the gain is not influenced by the polarity of the input voltage, but rather by an external control voltage $U_{CS}(t)$. For such a circuit, the switches for polarity change described in Sections 7.3.2 and 7.3.3 can be employed.

A synchronous demodulator can be used in the arrangement of Fig. 15.21 to separate a sine wave from a noisy signal and to determine its amplitude. The sine wave selected has a frequency which equals that of the control signal, and a phase shift φ which is constant with respect to the control signal. The special case when $f_i = f_{CS}$ and $\varphi = 0$ is illustrated in Fig. 15.22. It can be seen that under these conditions, the synchronous demodulator operates as a full-wave rectifier. If $\varphi \neq 0$ or $f_i \neq f_{CS}$, negative voltage-time areas occur as well as the positive areas and reduce the mean value of the output voltage so that it is always lower than that of the example.

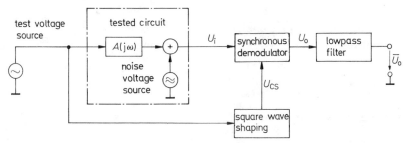

Fig. 15.21 Application of a synchronous demodulator to the measurement of noisy signals

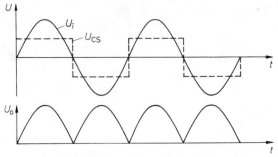

Fig. 15.22 Operation of a synchronous demodulator

The output voltage will now be determined as a function of frequency and phase. The input voltage U_i is multiplied by $+1$ or by -1 in the rhythm of the control frequency f_{CS}, this effect being represented mathematically as

$$U_o = U_i(t) \cdot S(t), \qquad (15.16)$$

where

$$S(t) = \begin{cases} 1 & \text{for } U_{CS} > 0 \\ -1 & \text{for } U_{CS} < 0. \end{cases}$$

By rewriting this in Fourier series form, we obtain

$$S(t) = \frac{4}{\pi} \sum_{n=0}^{\infty} \frac{1}{2n+1} \sin(2n+1)\omega_{CS}t. \qquad (15.17)$$

Let us now assume the input voltage to be a sinusoidal voltage having the frequency $f_i = m \cdot f_{CS}$ and the phase angle φ_m. With Eqs. (15.16) and (15.17), we then have the output voltage

$$U_o(t) = \hat{U}_i \sin(m\omega_{CS}t + \varphi_m) \cdot \frac{4}{\pi} \sum_{n=0}^{\infty} \frac{1}{2n+1} \sin(2n+1)\omega_{CS}t. \quad (15.18)$$

The arithmetic mean value of this voltage is evaluated by the subsequent lowpass filter. With the auxiliary equation

$$\frac{1}{T} \int_0^T \sin(m\omega_{CS}t + \varphi_m) = 0$$

and the orthogonal property of sinusoidal functions, i.e.

$$\frac{1}{T} \int_0^T \sin(m\omega_{CS}t + \varphi_m) \sin l\omega_{CS}t \, dt = \begin{cases} 0 & \text{for } m \neq l \\ \frac{1}{2}\cos\varphi_m & \text{for } m = l \end{cases}$$

and with Eq. (15.18), we obtain the final result

$$\overline{U}_o = \begin{cases} \dfrac{2}{\pi m}\, \hat{U}_i \cdot \cos \varphi_m & \text{for } m = 2n+1 \\[2mm] 0 & \text{for } m \neq 2n+1, \end{cases} \tag{15.19}$$

where $n = 0, 1, 2, 3, \ldots$.

If the input voltage is a mixture of frequencies, only those components contribute to the mean value of the output voltage, the frequencies of which equal the control frequency or are an odd multiple thereof. This explains why the synchronous demodulator is particularly suitable for selective amplitude measurements. The synchronous demodulator is also known as the *phase-sensitive rectifier* since the mean output voltage is dependent on the phase angle between the appropriate component of the input voltage and the control voltage.

For $\varphi_m = 90°$, $\overline{U}_o$ is zero even if the frequency condition is fulfilled. For our example in Fig. 15.22, we have $\varphi_m = 0$ and $m = 1$. In this case, Eq. (15.19) yields

$$\overline{U}_o = \frac{2}{\pi}\, \hat{U}_i,$$

which is the arithmetic mean of a full-wave rectified sinusoidal voltage; a result which could have been deduced directly from Fig. 15.22.

Equation (15.19) has shown that only those components contribute to the ouptput voltage, the frequencies of which are equal to the control frequency or are odd multiples thereof. However, this holds only if the time constant of the lowpass filter is infinitely large. In practice, this is not possible and is not even desirable as the upper cutoff frequency would be zero, i.e. the output voltage could not change at all. If $f_c > 0$, the synchronous demodulator no longer picks out discrete frequencies, but individual frequency bands. The 3 dB bandwidth of these frequency bands is $2f_c$, and Fig. 15.23 shows the resulting filter characteristic.

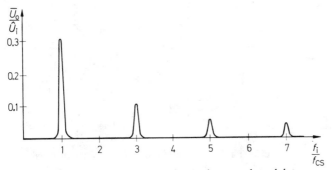

Fig. 15.23 Filter characteristic of a synchronous demodulator

The mostly unwanted contribution of the odd-order harmonics can be avoided by using an *analog multiplier* for synchronous demodulation instead of the polarity changing switch. The input voltage is then multiplied by a sinusoidal function $U_{cs} = \hat{U}_{cs} \sin \omega t$ rather than by the square wave function $S(t)$. As this sine wave no longer contains harmonics, Eq. (15.19) holds only for $n = 0$. If the amplitude of the control voltage is chosen to be equal to the computing unit voltage E of the multiplier, we obtain, instead of Eq. (15.19), the result

$$\overline{U}_{o} = \begin{cases} \frac{1}{2}\hat{U}_{i} \cos \varphi & \text{for } f_{i} = f_{cs} \\ 0 & \text{for } f_{i} \neq f_{cs}. \end{cases} \tag{15.20}$$

According to Eq. (15.20), the synchronous demodulator does not produce the amplitude $\hat{U}_{i}$ directly, but gives the real part $\hat{U}_{i} \cos \varphi$ of the complex amplitude $\underline{U}_{i}$. For the determination of the magnitude $|\underline{U}_{i}| = \hat{U}_{i}$, the phase angle of the control voltage can be adjusted by a suitable phase shifting network so that the output voltage of the demodulator is at maximum. The signal $U_{i}(t)$ and the control voltage $U_{cs}(t)$ are then in phase, and we obtain from Eq. (15.20)

$$\overline{U}_{o} = \tfrac{1}{2}\hat{U}_{i} = \tfrac{1}{2}|\underline{U}_{i}|\big|_{f_{i} = f_{cs}}.$$

If a calibrated phase-shifter is employed, the phase shift φ of the tested circuit can be read directly.

One is often interested only in the amplitude of a spectral input component and not in its phase angle. In such a case, synchronization of the control voltage is no longer necessary if two synchronous demodulators are used as in Fig. 15.24 and operated by two quadrature control voltages:

$$V_{1} = E \sin \omega_{cs} t \quad \text{and} \quad V_{2} = E \cos \omega_{cs} t$$

where E is the computing unit voltage of the demodulating multipliers. The oscillator in Fig. 8.24 is particularly well suited to the generation of these two voltages.

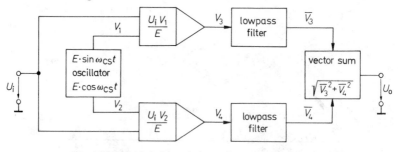

Fig. 15.24 Phase-independent synchronous demodulation.

Output voltage: $U_{o} = \tfrac{1}{2}\hat{U}_{i}$ for $f_{cs} = f_{i}$

Only the spectral component of the input voltage having the frequency f_{CS} contributes to the output voltages of the two demodulators. If it has the phase angle φ with respect to V_1, it is of the form

$$U_i = \hat{U}_i \sin(\omega_{CS} t + \varphi).$$

According to Eq. (15.20), the upper demodulator produces the output voltage

$$\overline{V}_3 = \tfrac{1}{2} \hat{U}_i \cos \varphi, \qquad (15.21)$$

whereas the lower demodulator gives

$$\overline{V}_4 = \tfrac{1}{2} \hat{U}_i \sin \varphi. \qquad (15.22)$$

By squaring and adding, we obtain the output voltage

$$U_o = \tfrac{1}{2} \hat{U}_i \sqrt{\sin^2 \varphi + \cos^2 \varphi} = \tfrac{1}{2} \hat{U}_i, \qquad (15.23)$$

which is independent of the phase angle φ. The circuit can therefore be used as a tunable selective voltmeter. Its bandwith is constant and equal to twice the cutoff frequency of the lowpass filters. The attainable Q-factor is considerably higher than that of conventional active filters: a 1 MHz signal can be filtered with a bandwidth of 1 Hz without any trouble. This corresponds to a Q-factor of 10^6.

If the control frequency is made to sweep through a given range, the circuit operates as a Fourier analyzer.

16   Electronic controllers

16.1   Underlying principles

The purpose of the controller is to bring a physical quantity (the controlled variable X) to a predetermined value (the reference variable W) and to hold it at this value. To achieve this goal the controller must counteract in a suitable way the influence of disturbances.

The basic arrangement of a simple control circuit is shown in Fig. 16.1. The controller influences the controlled variable X by means of the correcting variable Y so that the error signal $W-X$ is as small as possible. The disturbances acting on the controlled system (plant) are represented formally by the disturbance variable Z which is superimposed on the correcting variable. Below, we shall assume that the controlled variable is a voltage and the system is electrically controlled. Electronic controllers can then be employed.

In the simplest case, such a controller is a circuit amplifying the error signal $W-X$. If the controlled variable X rises above the reference signal W, the difference $W-X$ becomes negative. The correcting variable Y is thereby reduced by a factor defined by the amplifier gain. This reduction counteracts the increase in the controlled variable, i.e. there is negative feedback. At steady state, the remaining error signal is smaller, the larger the gain A_C of the controller. It can be seen from Fig. 16.1 that for linear systems

$$Y = A_C(W-X) \quad \text{and} \quad X = A_S(Y+Z), \tag{16.1}$$

where A_S is the gain of the controlled system. Hence, the controlled variable X

$$X = \frac{A_C A_S}{1 + A_C A_S} W + \frac{A_S}{1 + A_C A_S} Z. \tag{16.2}$$

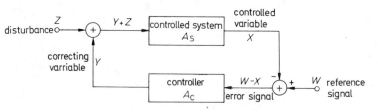

Fig. 16.1 Block diagram of a feedback control loop

It is now obvious that the response of the control system to a reference input, $\partial X/\partial W$, approaches unity more closely as the loop gain

$$g = A_C A_S = \frac{\partial X}{\partial(W - X)} \qquad (16.3)$$

becomes larger. The response to a disturbance, $\partial X/\partial Z$, approaches zero more closely the larger the gain A_C of the controller.

It must be pointed out, however, that there is a limit to the value of the loop gain g since, for too large a gain, the unavoidable phase shifts within the control loop give rise to oscillations. This problem has already been discussed for the frequency compensation of operational amplifiers. The objective of control engineering is to obtain, despite this restriction, the smallest possible error signal and good transient behaviour. For this reason, an integrator and a differentiator are added to the proportional amplifier and the P-controller is thus turned into one showing PI or even a PID action. The electronic realization of such circuits is dealt with below.

16.2   Controller types

16.2.1   P-controller

A P-controller (controller with proportional action) is a linear amplifier. Its phase shift must be negligibly small within the frequency range in which the loop gain g of the control system is larger than unity. Such a P-controller is, for example, an operational amplifier with resistive feedback.

For the determination of the maximum possible proportional gain A_P, we consider the Bode plot of a typical controlled system, represented in Fig. 16.2. The phase lag is $180°$ at the frequency $f = 3.3$ kHz. The negative feedback then becomes a positive feedback. In other words, the phase condition of Eq. (8.3) for a self-sustaining oscillation is fulfilled. The value of the proportional gain A_P determines whether or not the amplitude condition of Eq. (8.2) is also fulfilled. For the example in Fig. 16.2, the gain A_S of the system at 3.3 kHz is about $0.01 \cong -40$ dB. If we choose A_P to be $A_P = 100 \cong +40$ dB, the loop gain at this frequency would then be $|g| = |\underline{A}_S| \cdot A_P = 1$, i.e. the amplitude condition of an oscillator would also be fulfilled and the system would oscillate permanently at $f = 3.3$ kHz. If $A_P > 100$ is chosen, an oscillation with exponentially rising amplitude is obtained; and for $A_P < 100$, a damped oscillation occurs.

There is now the question of how much A_P must be reduced for an optimum transient behaviour. An indication of the damping of the

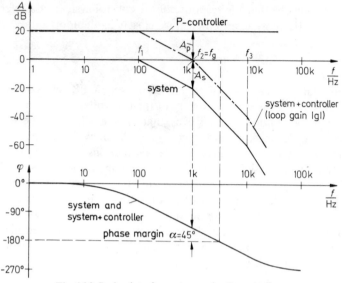

Fig. 16.2 Bode plot of a system and a P-controller

transient response can be taken directly from the Bode diagram: the *phase margin* is the phase angle required at the *gain crossover frequency* f_g to make a phase lag of 180°. The gain crossover frequency f_g is the frequency for which the loop gain is unity and is therefore often called the unity loop-gain frequency. Hence, the phase margin

$$\alpha = 180° - |\varphi_{\text{loop}}(f_g)| = 180° - |\varphi_S(f_g) + \varphi_C(f_g)|, \qquad (16.4)$$

where φ_S is the phase shift of the system and φ_C that of the controller. For the case of a P-controller, by definition $\varphi_C(f_g) = 0$, so that

$$\alpha = 180° - |\varphi_S(f_g)|. \qquad (16.5)$$

A phase margin of $\alpha = 0°$ results in an undamped oscillation, since the amplitude condition and the phase condition for an oscillator are simultaneously fulfilled; $\alpha = 90°$ represents the case of critical damping. For $\alpha \approx 65°$, the step response of the closed control loop shows an overshoot of about 4%. The settling time is then minimum. This phase margin is optimum for most cases. The oscillogram in Fig. 16.3 enables a comparison of the step responses obtained for different phase margins.

For the example in Fig. 16.2, the system's phase shift φ_S at frequency $f_2 = 1\,\text{kHz}$ is about $-135°$, and its gain $A_S = 0.1 \triangleq -20\,\text{dB}$. If the gain A_P of the P-controller is adjusted to $+20\,\text{dB}$, the loop gain is unity at this frequency, and the phase margin 45°. $A_P = 10$ is thus the highest

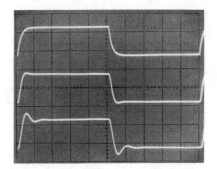

Fig. 16.3 Step response as a function of the phase margin, for constant gain crossover frequency f_g. Upper trace: $\alpha = 90°$, middle trace: $\alpha = 65°$, and lower trace: $\alpha = 45°$

gain for which an acceptable transient behaviour is obtained. The low-frequency limit of the loop gain is therefore

$$g = A_S A_P = 1 \cdot 10.$$

From Eq. (16.2), we obtain the relative error at steady state as

$$\frac{W - X}{W} = \frac{1}{1 + g} \approx 9\%.$$

Only for systems having a first order lowpass filter characteristic can the value of the proportional gain A_P be unlimited, since for such circuits the phase margin is larger than 90° at all frequencies.

16.2.2   PI-controller

The previous section has shown that for reasons of stability, the gain of a P-controller must not be too large. One possibility of improving the control accuracy is to raise the loop gain at low frequencies, as Fig. 16.4 illustrates. In the vicinity of the critical frequency f_g, the frequency response of the loop gain thereby remains unchanged so that the transient behaviour is not affected. However, the remaining error signal is now zero since

$$\lim_{f \to 0} |g| = \infty.$$

For the implementation of such a frequency response, an integrator is parallel-connected to the P-controller, as is shown in Fig. 16.5. The Bode plot of the resulting *PI-controller* (controller with proportional-integral action) is presented in Fig. 16.6. It can be seen that for low frequencies, the PI-controller behaves as an integrator and for high

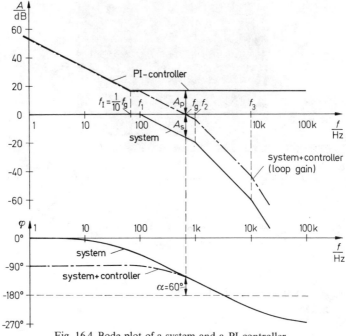

Fig. 16.4 Bode plot of a system and a PI-controller

frequencies as a proportional amplifier. The change-over between the two regions is characterized by the cutoff frequency f_i of the PI-controller. At this frequency the phase shift is $-45°$, and the controller gain $|\underline{A}_C|$ is 3 dB above A_P.

For a determination of the cutoff frequency f_I, we calculate from Fig. 16.5 the complex controller gain

$$\underline{A}_C = A_P + \frac{1}{j\omega\tau_I} = A_P\left(1 + \frac{1}{j\omega\tau_I A_P}\right).$$

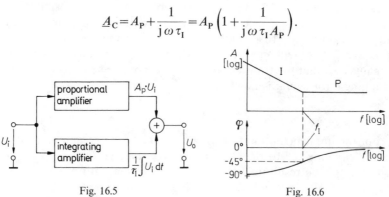

Fig. 16.5
Block diagram of the PI-controller

Fig. 16.6
Bode plot of the PI-controller

Hence,

$$\underline{A}_C = A_P\left(1 + \frac{\omega_I}{j\omega}\right) \quad \text{where} \quad \omega_I = 2\pi f_I = \frac{1}{\tau_I A_P}. \tag{16.6}$$

A PI-controller can also be realized with a single operational amplifier; the appropriate circuit being shown in Fig. 16.7. Its complex gain is

$$\underline{A}_C = -\frac{R_2 + 1/j\omega C_1}{R_1} = -\frac{R_2}{R_1}\left(1 + \frac{1}{j\omega C_1 R_2}\right). \tag{16.7}$$

By comparing coefficients with Eq. (16.6), we obtain the controller parameters as

$$A_P = -\frac{R_2}{R_1} \quad \text{and} \quad f_I = \frac{1}{2\pi C_1 R_2}. \tag{16.8}$$

On the basis of Fig. 16.4, we now deal in more detail with the choice of the optimum integration cutoff frequency f_I. In the first step of controller optimization, the integral term is omitted and the proportional gain A_P raised by as much as is permissible for the desired damping. According to Fig. 16.4, the phase shift of our system at 700 Hz is about $-120°$ and the corresponding gain $|\underline{A}_S| = 0.14 \triangleq -17\,\text{dB}$. If the phase margin is to be 60°, the proportional gain must be $A_P = 7 \triangleq +17\,\text{dB}$. This value is used in Fig. 16.4. The gain crossover frequency for these parameters is thus $f_g \approx 700\,\text{Hz}$.

As mentioned earlier, the integration cutoff frequency f_I must be small in comparison with f_g so as not to increase the phase lag at f_g. On the other hand, it is not sensible to choose f_I smaller than necessary as this would increase the time taken by the integrator to adjust for zero error signal. The upper limit for f_I is at about $0.1 f_g$, this value being used in Fig. 16.4. The appropriate transient behaviour of the error signal is shown by the oscillogram of Fig. 16.8. It is obvious from the lower trace that, with these optimum parameters, the PI-controller settles to zero error signal in the same time interval that a purely P-action controller requires to adjust to a relative error signal of $1/(1+g)$ $= 1/8 = 12.5\%$.

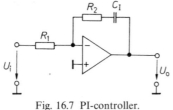

Fig. 16.7 PI-controller.

$$A_P = -\frac{R_2}{R_1}, \quad f_I = \frac{1}{2\pi C_1 R_2}$$

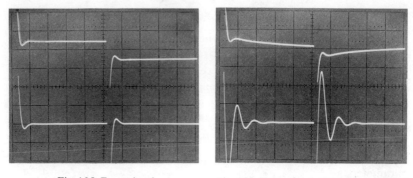

Fig. 16.8 Error signal.
Upper trace: P-controller.
Lower trace: PI-controller with optimum f_I

Fig. 16.9 Error signal of the PI-controller.
Upper trace: f_I too small.
Lower trace: f_I too large

The effect of a not quite optimum f_I is demonstrated by the oscillogram in Fig. 16.9. For the upper trace, f_I is too small: the settling time is increased. For the lower trace, f_I is too large: the phase margin is reduced.

16.2.3 PID-controller

By parallel-connecting a differentiator as in Fig. 16.10, a PI-controller can be extended to become a PID-controller (controller with proportional-integral-derivative action). Above the differentiation cut-off frequency f_D, the circuit behaves as a differentiator. The phase shift rises to $+90°$, as can be seen from the Bode plot in Fig. 16.11. This phase lead at high frequencies can be used to partially compensate for the phase lag of the controlled system in the vicinity of f_g, allowing a higher proportional gain, so that a higher gain crossover frequency f_g is obtained. The transient behaviour is thereby speeded up.

The selection of the controller parameters will again be illustrated for our example system. In the first step, we raise the proportional

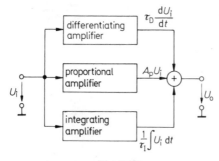

Fig. 16.10
Block diagram of the PID-controller

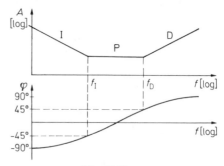

Fig. 16.11
Bode plot of the PID-controller

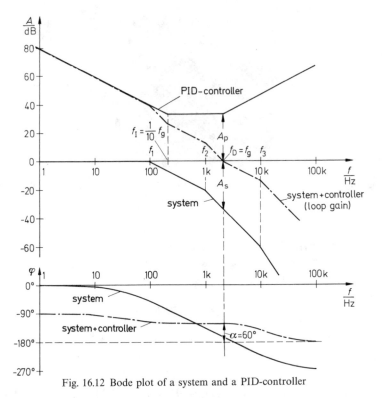

Fig. 16.12 Bode plot of a system and a PID-controller

gain A_P until the phase margin is only about 15°. In this case, we can infer from Fig. 16.12 that $A_P = 50 \hat{=} 34 \, \mathrm{dB}$ and $f_g \approx 2.2 \, \mathrm{kHz}$, as against 700 Hz for the PI-controller. If the differentiation cutoff frequency is now chosen to be $f_D \approx f_g$, the phase lag of the controller at frequency f_g is approx. $+45°$, i.e. phase margin is increased from 15° to 60°, and we obtain the desired transient behaviour.

For the determination of the integration cutoff frequency f_I, the same principles apply as for the PI-controller, i.e. $f_I \approx \frac{1}{10} f_g$. This results in the frequency response of the loop gain shown in Fig. 16.12.

The reduction in settling time is illustrated by a comparison of the oscillograms for a PI- and a PID-controller, shown in Fig. 16.13.

In order to implement a PID-controller we must first determine the complex gain from the block diagram in Fig. 16.10.

$$\underline{A}_C = A_P + j\,\omega\,\tau_D + \frac{1}{j\,\omega\,\tau_I} = A_P \left[1 + j\left(\frac{\omega}{\omega_D} - \frac{\omega_I}{\omega}\right)\right], \qquad (16.9)$$

where

$$f_D = \frac{A_P}{2\pi\,\tau_D} \quad \text{and} \quad f_I = \frac{1}{2\pi A_P\,\tau_I}. \qquad (16.10)$$

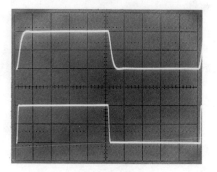

Fig. 16.13 Comparison of the transient behaviour of a system with PI-controller (upper trace) and with PID-controller (lower trace)

A circuit having the frequency response of Eq. (16.9) can also be realized with a single operational amplifier; this is shown in Fig. 16.14. Its complex gain is given by

$$\underline{A}_C = -\left[\frac{R_2}{R_1} + \frac{C_D}{C_I} + j\omega C_D R_2 + \frac{1}{j\omega C_I R_1}\right].$$

Hence, with $\dfrac{C_D}{C_I} \ll \dfrac{R_2}{R_1}$,

$$\underline{A}_C = -\frac{R_2}{R_1}\left[1 + j\left(\omega C_D R_1 - \frac{1}{\omega C_I R_2}\right)\right]. \qquad (16.11)$$

A comparison of the coefficients with those in Eq. (16.9) yields the controller parameters

$$A_P = -\frac{R_2}{R_1}, \qquad f_D = \frac{1}{2\pi C_D R_1}, \qquad f_I = \frac{1}{2\pi C_I R_2}. \qquad (16.12)$$

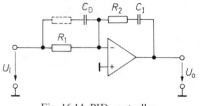

Fig. 16.14 PID-controller.

$$A_P = -\frac{R_2}{R_1}, \qquad f_I = \frac{1}{2\pi C_I R_2}, \qquad f_D = \frac{1}{2\pi C_D R_1}$$

16.2.4 PID-controller with adjustable parameters

For the determination of the different controller parameters, we assumed that the parameters of the controlled system were known. However, these data are often difficult to measure, particularly for very slow systems. It is therefore usually better to establish the optimum controller parameters by experiment. A circuit is then required which allows *independent* adjustment of the controller parameters A_P, f_I and f_D. It can be seen from Eqs. (16.12) and (16.10) that this condition can be fulfilled neither by the circuit in Fig. 16.14 nor by that in Fig. 16.10 since, for a variation of A_P, the cutoff frequencies f_I and f_D will also change.

For the circuit in Fig. 16.15, however, the parameters are decoupled and their independent adjustment is therefore possible. The complex gain of the circuit is

$$\underline{A}_C = \frac{R_P}{R_1}\left[1 + j\left(\omega C_D R_D - \frac{1}{\omega C_I R_I}\right)\right]. \tag{16.13}$$

Comparing coefficients with Eq. (16.9) gives the controller parameters:

$$A_P = \frac{R_P}{R_1}, \quad f_D = \frac{1}{2\pi C_D R_D}, \quad f_I = \frac{1}{2\pi C_I R_I}. \tag{16.14}$$

The controller optimization is illustrated again for our example system. In the first step, switch S is closed to render the integrator inactive. Resistor R_D is made zero, i.e. the differentiator does not contribute to the output signal. The circuit is therefore a purely P-action controller.

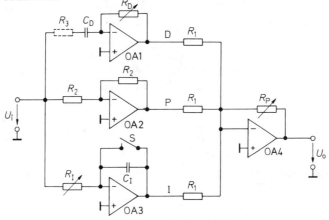

Fig. 16.15 PID-controller with decoupled parameters.

$$A_P = \frac{R_P}{R_1}, \quad f_I = \frac{1}{2\pi C_I R_I}, \quad f_D = \frac{1}{2\pi C_D R_D}$$

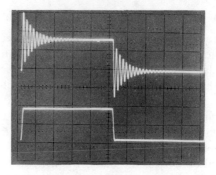

Fig. 16.16 Experimental determination of the proportional and derivative controller
parameters

We now apply a square wave to the reference signal input and
record the transient behaviour of the controlled variable X. Starting
from zero, A_P is increased until the step response is only slightly
damped, as in the upper trace of Fig. 16.16. The step response is then
that obtained for the system in Fig. 16.12, for a phase margin of 15°
and no derivative nor integral controller action.

In the second step, the differentiation cutoff frequency f_D is lowered
from infinity by increasing R_D, and is adjusted to a value for which the
desired damping is attained (see Fig. 16.16, lower trace).

In the third step, we consider the transient behaviour of the error
signal $W - X$. After opening switch S, the integration cutoff frequency f_I
is increased until the settling time is minimum. The appropriate oscillograms have already been shown in Figs. 16.8 and 16.9.

The great advantage of this optimization method is that the optimum controller parameters represented by Fig. 16.12 are obtained
immediately and without iteration. With the controller parameters
found in this way, the simple PID-controller in Fig. 16.14 can be
designed.

16.3   Control of non-linear systems

16.3.1   Static nonlinearity

We have assumed so far that the equation for the controlled system
is

$$X = A_S \cdot Y$$

i.e. that the controlled system is linear. For many systems, however, this
condition is not fulfilled so that in general

$$X = f(Y).$$

For small changes about a given operating point X_0, each system can be considered linear as long as its transfer characteristic is continuous and differentiable in the vicinity of this point. In such cases, the derivative

$$a_s = \frac{dX}{dY}$$

is used so that for small-signal operation

$$x \approx a_s y,$$

where $x = (X - X_0)$ and $y = (Y - Y_0)$. For a fixed point of operation, the controller can now be optimized as described above. However, a problem arises if larger changes in the reference variable are allowed: since the incremental system gain a_s is dependent on the actual point of operation, the transient behaviour varies depending on the magnitude of W.

This can be avoided by enforcing linearity, i.e. by connecting a function network of Section 1.7.5 in front of the controlled system. The appropriate block diagram is shown in Fig. 16.17. If the function network is used to implement the function $Y = f^{-1}(Y')$, we obtain the required linear system equation

$$X = f(Y) = f[f^{-1}(Y')] = Y'.$$

If, for instance, the system shows exponential behaviour, i.e.

$$X = A e^Y,$$

the function network must be a logarithmic one having the characteristic

$$Y = f^{-1}(Y') = \ln \frac{Y'}{A}.$$

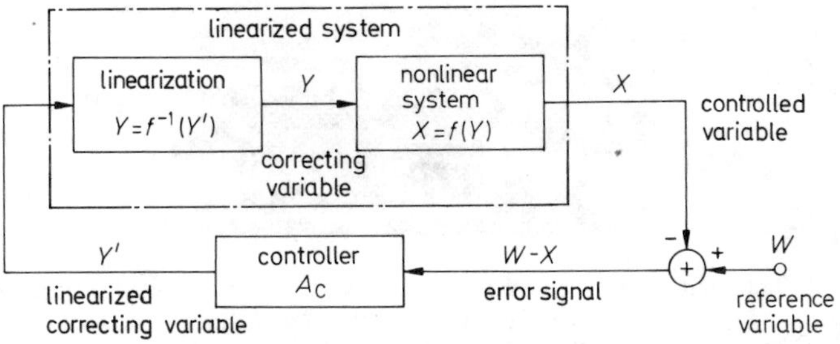

Fig. 16.17 Linearization of a system having static nonlinearity

16.3.2   Dynamic nonlinearity

A different kind of nonlinearity of a controlled system may be due to the rate of change of some quantity in the system being limited to a value which cannot be raised by increasing the correcting variable. We have encountered this effect in operational amplifiers as a limitation of the slew rate. For large input steps and controllers with integral action, the above effect leads to large overshoots which decay only slowly.

The reason for this is as follows. For optimized integral action of the controller and for a small voltage step, the integrator reaches its steady-state output voltage at the precise instant when the error signal becomes zero. For double the input step of a linear system, the rate of change in the system, as well as that of the integrator, would double. The increased reference value would therefore be reached within exactly the same settling time.

For a system having a limited slew rate, only the rate of change of the integrator is doubled whereas that of the system remains unchanged. This results in the controlled system reaching the reference value very much later and the integrator overshooting. The controlled variable therefore greatly exceeds the reference value and the decay takes longer, the farther away the integrator output voltage is from the steady-state value. The decay time constant for this non-linear operation therefore becomes larger for increasing input steps.

The effect is avoided by increasing the integration time constant (i.e. reducing f_I) until no overshoot is incurred for the largest possible input step. However, this results in considerably prolonged settling times for small-signal operation (see Fig. 16.9 lower trace).

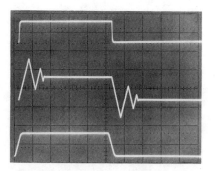

Fig. 16.18 Transient behaviour of the controlled variable for a limited system slew rate.
Upper trace: small-signal characteristic.
Middle trace: large-signal characteristic.
Lower trace: large-signal characteristic for a slew-rate limited reference variable

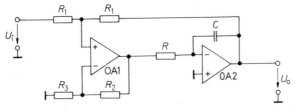

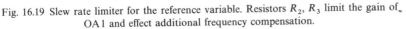

Fig. 16.19 Slew rate limiter for the reference variable. Resistors R_2, R_3 limit the gain of. OA1 and effect additional frequency compensation.

Steady-state output voltage: $U_o = -U_i$

Maximum slope of output voltage: $\dfrac{dU_o}{dt} = \dfrac{U_{max}}{RC}$

A much more effective measure is to limit the slew rate of the reference variable to the maximum slew rate of the controlled system. This ensures linear operation throughout, and the overshoot effect is thus avoided. The settling time for large reference signals is thereby not increased since the controlled variable cannot change any faster anyway. This is illustrated by the oscillogram in Fig. 16.18.

In principle, a lowpass filter could be used to limit the slew rate, but this would also reduce the small-signal bandwidth. A better solution is shown in Fig. 16.19. If a voltage step is applied to the circuit input, amplifier OA1 saturates at the output limit U_{max}. The output voltage of OA2 therefore rises at the rate

$$\frac{dU_o}{dt} = \frac{U_{max}}{RC}$$

until it reaches the values $-U_i$, determined by the over-all feedback. A square-wave voltage would therefore be shaped into the required trapezoidal voltage. The signal remains unchanged if the rate of change of the input voltage is smaller than the predetermined maximum. The small-signal bandwidth is therefore not affected.

16.4   Phase-locked loop

A particularly important application of feedback control in communication systems is the phase-locked loop (PLL). Its purpose is to control the frequency f_2 of an oscillator in such a manner that it is the same as the frequency f_1 of a reference oscillator, and to do this so

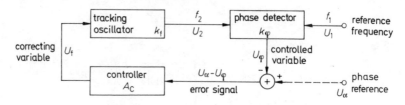

Fig. 16.20 Principle of the phase-locked loop (PLL)

accurately that the phase shift between the two signals remains constant. The basic arrangement of such a circuit is illustrated in Fig. 16.20.

The frequency of the tracking oscillators can be varied by means of the control voltage U_f, according to the relationship

$$f_2 = f_0 + k_f U_f. \tag{16.15}$$

Such voltage-controlled oscillators (VCOs) are described in Chapter 8. For low frequencies, the second-order differential equation circuit of Section 8.3.2 or the function generators of Section 8.4.2 can be employed. For higher frequencies, the emitter-coupled multivibrator of Fig. 8.43 is more suitable, or any LC oscillator if a varactor diode is parallel-connected to the capacitor of the oscillating circuit. However, the linear relationship of Eq.(16.15) then holds only for small variations around the point of operation, f_0, as the incremental control constant (VCO "gain") $k_f = df_2/dU_f$ is dependent on the point of operation.

The phase detector produces an output voltage which is defined by the phase angle φ between the tracking oscillator voltage U_2 and the reference alternating voltage U_1;

$$U_\varphi = k_\varphi \cdot \varphi.$$

The integrating property of the controlled system is thereby of particular interest. If frequency f_2 deviates from the reference frequency f_1, the phase shift φ will increase proportionally with time. The error signal in the closed loop therefore rises, even for a finite controller gain, until both frequencies are exactly the same. The remaining error signal of the *frequency* is thus zero.

The remaining error signal of the *phase shift*, however, does not usually become nil. From Fig. 16.20 we deduce that $U_\alpha - U_\varphi = U_f/A_C$; hence

$$\alpha - \varphi = \frac{f_1 - f_0}{A_C k_f k_\varphi}, \tag{16.16}$$

where f_0 is the VCO frequency for $U_f = 0$. If it is of importance not only to keep the phase shift constant but also hold it precisely at a predetermined value α, a PI-controller must be used. In many applications it is sufficient to control for a *constant* phase shift α so that the control input U_α can be omitted. Voltage U_φ is then the error signal.

For the determination of the controller parameters, the frequency response of the system must be known. As mentioned before, the phase detector shows integral behaviour so that the phase shift is given by

$$\varphi = \int_0^t \omega_2 \, d\tilde{t} - \int_0^t \omega_1 \, d\tilde{t} = \int_0^t \Delta\omega \, d\tilde{t}, \qquad (16.17)$$

where $\tilde{t}$ is a dummy time variable of integration. For the determination of the frequency response of the controlled system, we modulate frequency ω_2 sinusoidally with a modulating frequency ω_m around the centre frequency ω_1. Hence

$$\Delta\omega(t) = \widehat{\Delta\omega} \cos \omega_m t.$$

By inserting this in Eq. (16.17), we obtain

$$\varphi(t) = \frac{\widehat{\Delta\omega}}{\omega_m} \sin \omega_m t.$$

Taking into account the phase lag of 90°, we obtain in complex notation

$$\frac{\varphi}{\Delta\omega} = \frac{1}{j\omega_m}, \qquad (16.18)$$

which is the equation for an integrator. With the constants k_f and k_φ, the complex gain of the controlled system is then

$$\boxed{\underline{A}_s = \frac{\underline{U}_\varphi}{\underline{U}_f} = \frac{2\pi k_f k_\varphi}{j\omega_m} = \frac{k_f k_\varphi}{j f_m}.} \qquad (16.19)$$

As will be seen later, the measurement of the phase shift involves a certain delay and the factor k_φ is therefore complex, i.e. the order of the system is raised.

The behaviour of a phase-locked loop generally depends on the type of the phase detector used. The most important circuits are now dealt with.

16.4.1 Sample-and-hold circuit as phase detector

The phase angle φ between two voltages U_1 and U_2 can, for example, be measured by sampling with a sample-and-hold circuit the instantaneous value of U_1 when U_2 has a positive-going zero crossing.

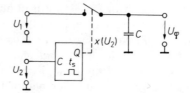

Fig. 16.21 Sample-and-hold circuit as a phase detector

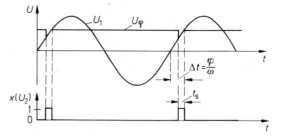

Fig. 16.22 Voltages of the phase detector. The dips in U_φ disappear to a great extent if the sampling time t_s is not much larger than the time constant of the sample-and-hold circuit

For this purpose, U_2 is used, as in Fig. 16.21, to activate an edge-triggered one-shot producing the sampling pulse for the sample-and-hold circuit. It can be seen in Fig. 16.22 that the output voltage of the circuit is given by

$$U_\varphi = \hat{U}_1 \sin \varphi. \tag{16.20}$$

Around the point of operation ($\varphi = 0$), the detector characteristic is approximately linear:

$$U_\varphi \approx \hat{U}_1 \varphi.$$

Hence, the factor of the phase detector

$$k_\varphi = \hat{U}_1. \tag{16.21}$$

It is obvious from Fig. 16.23 that a further possible operating point where $U_\varphi = 0$, would be at $\varphi = \pi$. Then, $k_\varphi = -\hat{U}_1$. The sign of the

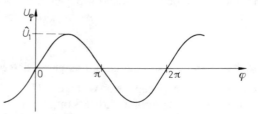

Fig. 16.23 Transfer characteristic of a sample-and-hold circuit as phase detector

controller gain defines which of the two operating points is assumed. Further stable points of operation occur at intervals of 2π. This indicates that the phase detector does not recognize a displacement by a whole number of periods.

If a triangular waveshape U_1 is employed rather than the sinusoidal one, a triangular detector characteristic is obtained. For rectangular input voltages U_1, the circuit is obviously not usable.

Dynamic behaviour

The phase detector described determines a new value for the phase shift only once during a period of the input waveform and thus has a dead time. The delay is between 0 and $T_2 = 1/f_2$, depending on the instant at which a change in phase occurs. The average delay is therefore $\frac{1}{2}T_2$. To allow for this, the "gain" k_φ of the phase detector for higher phase modulation frequencies f_m must be written in complex form so that

$$\underline{k}_\varphi = k_\varphi \, e^{-\,\mathrm{j}\omega_m \cdot \frac{1}{2} T_2} = \hat{U}_1 \, e^{-\mathrm{j}\pi f_m/f_2}. \tag{16.22}$$

With Eq. (16.19), we therefore obtain for the complex gain of the entire controlled system

$$\underline{A}_\mathrm{s} = \frac{k_\mathrm{f}\,\underline{k}_\varphi}{\mathrm{j} f_m} = \frac{k_\mathrm{f}\,\hat{U}_1}{\mathrm{j} f_m \, e^{\mathrm{j}\pi f_m/f_2}},$$

or

$$|\underline{A}_\mathrm{s}| = \frac{|U_\varphi|}{|U_\mathrm{f}|} = \frac{k_\mathrm{f}\,\hat{U}_1}{f_m} \quad \text{and} \quad \varphi_m = -\frac{\pi}{2} - \frac{\pi f_m}{f_2}. \tag{16.23}$$

Controller parameters

For the controller it is best to choose a circuit without derivative action since the output voltage of the sample-and-hold element changes only in steps. According to Eq. (16.23), the phase shift φ_m between $\underline{U}_\varphi$ and $\underline{U}_\mathrm{f}$ at frequency $f_m = \frac{1}{4} f_2$, is $-135°$. The phase margin is thus $45°$ if we adjust the proportional gain A_P in such a way that the gain crossover frequency $f_\mathrm{g} = \frac{1}{4} f_2$. By definition, at $f_m = f_\mathrm{g}$,

$$|\underline{g}| = |\underline{A}_\mathrm{s}| \cdot |\underline{A}_\mathrm{c}| = 1.$$

With $\underline{A}_\mathrm{c} = A_\mathrm{P}$ and Eq. (16.23), this results in

$$A_\mathrm{P} = \frac{f_\mathrm{g}}{k_\mathrm{f} k_\varphi} = \frac{f_2}{4 k_\mathrm{f} \hat{U}_1}.$$

A typical example is $f_2 = 10\,\mathrm{kHz}$, $k_\mathrm{f} = 5\,\mathrm{kHz/V}$ and $k_\varphi = \hat{U}_1 = 10\,\mathrm{V}$. Hence, $A_\mathrm{P} = 0.05$. The controller can in this case be a passive voltage divider.

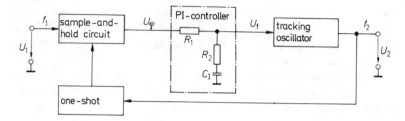

Fig. 16.24 PLL with sample-and-hold circuit as phase detector

For the reduction of the remaining phase error [see Eq. (16.16)], the gain at low frequencies can be raised by integral action ($f_i \approx \frac{1}{10} f_g = \frac{1}{40} f_2$). It is sensible, however, to limit the low-frequency value of the gain to a finite value A_1, since otherwise the integrator would drift and saturate for as long as the loop is unlocked. The VCO can thereby be detuned so much that the loop will not lock in.

The passive voltage divider can be extended in a simple way to become a PI-controller with limited gain A_1, by connecting a capacitor in series with the resistor R_2, as is illustrated in Fig. 16.24. The controller parameters are then

$$A_P = \frac{R_2}{R_1 + R_2}; \qquad f_I = \frac{1}{2\pi R_2 C_I}; \qquad A_1 = 1.$$

Pull-in

After switch-on there is usually a frequency offset $\Delta f = f_1 - f_0$. The phase shift therefore rises proportionally with time. According to Fig. 16.23, this produces an alternating voltage at the output of the phase detector, having the frequency Δf and the amplitude $\hat{U}_\varphi = \hat{U}_1$, so that the tracking oscillator is thus frequency modulated by the voltage

$$U_f = A_P \hat{U}_1 \sin \Delta \omega t.$$

There will therefore be an instant at which the frequencies are identical, and the loop will lock in. The precondition for this is that the frequency offset $\Delta f = f_1 - f_0$ is smaller than the sweep width

$$\Delta f_{2\,max} = \pm k_f A_P \hat{U}_1. \qquad (16.24)$$

This maximum permissible offset is known as the *capture range* and represents the normal range of operation of the loop. For our example, it is $\pm 2.5\,\text{kHz}$, i.e. 25% of f_0.

16.4.2   Synchronous demodulator as phase detector

Section 15.3.4 describes the application of the multiplier as a phase-sensitive rectifier. If two sinusoidally alternating voltages of approximately identical frequencies and of amplitude E, are applied to the inputs, the output voltage will be

$$U_\text{o}=\frac{E}{2}\cos\varphi\,\cos(\omega_2-\omega_1)t+\frac{E}{2}\cos[(\omega_1+\omega_2)t-\varphi].\quad(16.25)$$

For $\omega_1=\omega_2$, its mean absolute value is given by

$$U_\varphi=\overline{U}_\text{o}=\frac{E}{2}\cos\varphi,\quad(16.26)$$

in accordance with Eq. (15.20). This function is represented in Fig. 16.25. It becomes immediately obvious that the voltage cannot be used as control variable in the vicinity of $\varphi=0$ as the sign of the error signal cannot be identified. The two points of operation $\varphi=\pm\pi$ are better suited because voltage U_φ crosses zero. The sign of the gain determines at which of the two points lock-in occurs. Further stable points of operation occur at intervals of 2π. This implies that this kind of phase detector also cannot recognize a phase displacement by a whole number of periods, i.e. by a multiple of 2π.

Within a range of about $\pm\pi/4$ around the stable point of operation, φ_0, the characteristic of the phase detector is approximately linear, so that, with $\varphi=\varphi_0+\vartheta$,

$$U_\varphi=\frac{E}{2}\cos(\varphi_0+\vartheta)=\pm\frac{E}{2}\sin\vartheta\approx\pm\frac{E}{2}\vartheta.\quad(16.27)$$

The sensitivity (phase detector "gain") is therefore

$$k_\varphi=\frac{U_\varphi}{\vartheta}=\pm\frac{E}{2}.\quad(16.28)$$

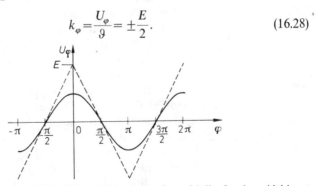

Fig. 16.25 Mean absolute value of the output voltage of a multiplier for sinusoidal input voltages of the amplitude E. Broken curve: multiplier mean output voltage for a square-wave input signal with the levels $\pm E$

If instead of the two sinusoidal voltages, two square-waves with the magnitude $\pm E$ are applied, the detector characteristic becomes triangular, as shown by the broken lines in Fig. 16.25. The stable points of operation are again at $\varphi_0 = \pm\dfrac{\pi}{2} \pm n \cdot 2\pi$, the sensitivity in this case being

$$k_\varphi = \pm \frac{2E}{\pi}. \tag{16.29}$$

For square-wave input voltages, an analog multiplier is obviously no longer required. In this case, considerably higher frequencies are attained with the transistor modulator in Fig. 7.16.

If the ripple of U_φ is to be kept small, a lowpass filter must be connected to the multiplier output, the cutoff frequency f_c of which is small against $2f_1$, according to Eq. (16.25). This is a definite disadvantage since, unlike for the previous circuit, the proportional gain of the controller must now be chosen so low that the gain crossover frequency $f_g \approx f_c$. At this frequency, the phase shift of the controlled system together with that of the lowpass filter is already $-135°$. If $f_g \approx f_c \ll f_1$, a control loop would be obtained which is impractical because of its very slow response. In principle, it could be speeded up by derivative action of the controller, but this would cancel the effect of the lowpass filter i.e. increase the ripple.

An increase of the control system bandwidth at the expense of the ripple of U_φ can be attained very simply by using a proportional action controller and omitting the lowpass filter. A phase margin of 90° is then available for any proportional gain chosen, i.e. the control loop is critically damped.

Because of the feedback, the ripple of U_φ causes the tracking oscillator to be frequency modulated with twice the signal frequency. This results in a distortion of the output sine wave. For square-waves, the mark-space ratio is changed. The proportional gain must not be too high so as to keep the distortion within tolerable limits. The condition $f_g \leq \frac{1}{3}f_1$ can be used as a rule of thumb.

The resulting arrangement is shown in Fig. 16.26 and is available as an integrated PLL. Usually, the multiplier is simplified and reduced to a modulator such as that of Fig. 7.16. The types NE 560...566 from Signetics, for instance, are based on this principle.

When operated without a lowpass filter, the circuit is usable only in those applications where it is important to have frequency f_2 identical to f_1 and where the shape and phase shift of the output signal are not significant; for example, as a discriminator for frequency demodulation. The reference oscillation is used as input signal. If the VCO frequency f_2 is linearly dependent on U_f, this voltage is proportional to the

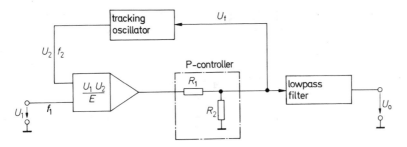

Fig. 16.26 PLL with a multiplier as phase detector for frequency demodulation

frequency deviation Δf_1. The superimposed ripple can be filtered out by a steep lowpass filter outside the control loop.

16.4.3   Frequency-sensitive phase detector

One drawback of the phase detectors described above is that they possess only a limited capture range, i.e. they cannot pull in if the initial frequency offset exceeds a certain limit. The reason for this is that the phase-equivalent signal U_φ for a frequency deviation is an alternating voltage symmetrical about zero. The voltage U_f therefore effects a periodic frequency modulation of the tracking oscillator, but never a systematic tuning in the right direction, i.e. towards the lock-in frequency.

The phase detector in Fig. 16.27, however, is different in this respect in that it produces a signal having the correct sign information for any given frequency offset. The circuit consists essentially of two RS flip-flops. It is operated with short L-pulses x_1, x_2 which can be generated from the input voltages $U_1(t)$, $U_2(t)$ at each positive-going zero crossing.

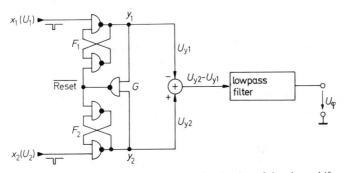

Fig. 16.27 Phase detector with memory for the sign of the phase shift

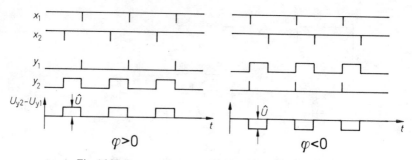

Fig. 16.28 Input and output signals of the phase detector

We now assume that both flip-flops are reset. If voltage U_2 leads voltage U_1, i.e. $\varphi > 0$, we first obtain a pulse x_2 which sets flip-flop F_2. It remains set until the following pulse x_1 sets flip-flop F_2. The state in which both flip-flops are set, exists only during the propagation delay time since they are jointly reset afterwards by gate G. It can be seen from Fig. 16.28 that the output of the subtractor shows a sequence of positive rectangular pulses. Correspondingly, a sequence of negative pulses is obtained if pulse x_2 occurs *after* pulse x_1, i.e. if $\varphi < 0$. This behaviour is summarized in the state diagram of Fig. 16.29.

The duration of the output pulses is equal to the time interval between the positive-going zero crossings of $U_1(t)$ and those of $U_2(t)$. Hence, the mean value of the output voltage

$$U_\varphi = \hat{U}\,\frac{\Delta t}{T} = \hat{U}\,\frac{\varphi}{2\pi}. \tag{16.30}$$

As the value of the time interval increases proportionally with φ until the limits $\pm 360°$ are reached, a range of linear phase measurement of $360°$ is obtained. When exceeding this limit, the output voltage jumps to zero and increases again, still having the same polarity. The result is the sawtooth characteristic in Fig. 16.30.

The basic difference between this characteristic and all previous ones is that, for $\varphi > 0$, U_φ is always positive and, for $\varphi < 0$, always negative. This is the reason for the frequency sensitivity of the detector.

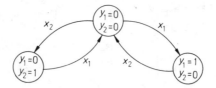

Fig. 16.29 State diagram of the phase detector

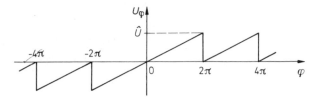

Fig. 16.30 Transfer characteristic of the phase detector

If frequency f_2 is, for instance, larger than f_1, the phase shift increases continuously and proportionally with time. According to Fig. 16.30, we then obtain for U_φ a sawtooth voltage having a positive mean value. If this detector is used in a phase-locked loop, it always indicates a leading phase. For a controller with integral action, the tracking frequency f_2 is therefore reduced until it coincides with f_1. The capture range is thus theoretically infinite and in practice limited only by the input voltage range of the VCO.

We have described in Section 16.4.2 that the averaging lowpass filter has a very unfavourable effect on the transient behaviour. For this reason, it is usually omitted in this circuit too. If one wishes to have $\varphi = 0$ (with the help of an I-controller), no phase distortions are incurred, since in this case $U_\varphi = 0$ even without filtering. The flip-flops then produce no output pulses.

One drawback of the circuit is that very small deviations in phase are not recognized, since the flip-flops would then have to produce extremely short output pulses which would be lost due to the limited rise times within the circuit. This is the reason why the phase jitter (phase noise) is somewhat larger than that with the sample-and-hold detector.

If a PLL with a large capture range and small phase jitter is required, this circuit can be combined with a sample-and-hold detector. After lock-in has been accomplished, the sample-and-hold detector is switched into the loop instead of the frequency-sensitive phase detector.

The frequency-sensitive phase detector is available as a monolithic integrated digital circuit. However, instead of pulse triggering, edge triggering is employed so that square-wave signals can be applied directly to the input. For this purpose, the circuit is extended, as in Fig. 16.31, by two RS flip-flops. They also suppress the undesired spikes shown in Fig. 16.28.

IC types:

MC 4044 (TTL) for up to 10 MHz,
MC 12040 (ECL) for up to 80 MHz.

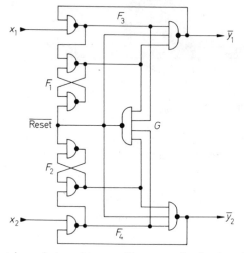

Fig. 16.31 Edge-triggered phase detector with memory for the sign of the phase shift

16.4.4   Phase detector with extensible measuring range

With the phase detectors described so far it is not possible to detect a phase shift of more than one oscillation period, as the phase measuring range is limited to values below 2π. There are applications, however, for which a phase delay of several oscillations must be recovered. The phase detector in Fig. 16.32 suits this purpose. It is based on the straight binary up-clock/down-clock counter as illustrated in Fig. 10.11, which is insensitive to coinciding clock pulses. In order that the output can produce positive and negative differences,

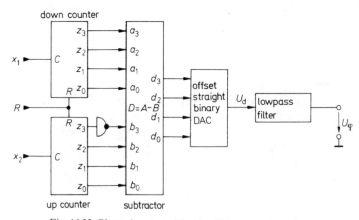

Fig. 16.32 Phase detector with extensible measuring range

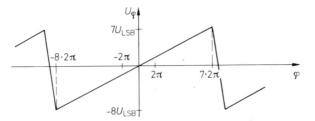

Fig. 16.33 Detection characteristic of the phase detector

the offset binary representation is chosen (see Chapter 10). The most significant bit of the up-counter is thus complemented, so that at phase shift zero, the difference D is $D = 1000_2$. In this representation, the difference is directly applicable to an offset binary D/A converter (see Section 14.3.1).

Near zero phase shift, the detector behaves in the same way as the previous circuit; if x_2 leads x_1, positive pulses of the magnitude U_{LSB} arise, the duration of which equals the interval between the zero crossings of the input voltages. For a phase lag, negative pulses occur. The mean value of the pulses is

$$U_{\varphi} = \overline{U}_D = U_{LSB} \frac{\Delta t}{T} = U_{LSB} \frac{\varphi}{2\pi}.$$

If the phase displacement reaches the values $\pm 2\pi$, the value for the interval Δt jumps from T to 0. In contrast to the previous circuit, however, the output does not assume zero voltage, but is increased by the value U_{LSB} by action of the counter. The resulting detector characteristic is illustrated in Fig. 16.33, for 4 bits. The range of phase measurement can be extended to any multiple of 2π by a suitable extension of the counter.

16.4.5   PLL as a frequency multiplier

A particularly important application of the PLL is that of frequency multiplication. A frequency divider is connected to each of the two inputs of the phase detector, as illustrated by Fig. 16.34. The frequency of the tracking oscillator assumes such a value that

$$\frac{f_1}{n_1} = \frac{f_2}{n_2}.$$

In this manner, the frequency of the tracking oscillator

$$f_2 = \frac{n_2}{n_1} f_1$$

can be adjusted to any rational multiple of the reference frequency f_1.

In this application the phase detector may operate at a frequency considerably lower than that of the tracking oscillator. It must therefore be ensured that the control voltage U_φ contains no ripple. An undesired frequency modulation would otherwise occur instead of simply the distortion of the output waveform, as describe in Section 16.4.2.

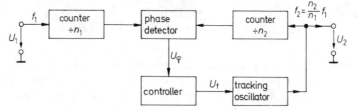

Fig. 16.34 Frequency multiplication with a PLL

17 Appendix: Definitions and nomenclature

We hope that the following list of definitions will help to avoid confusion and enable a better understanding. Where possible, the definitions are based on IEC recommendations.

Voltage. A voltage between two points x and y is denoted by U_{xy}. It is defined as being positive if point x is positive with respect to point y, and negative, if point x is negative with respect to point y. Therefore, $U_{xy} = -U_{yx}$. The statement

$$U_{BE} = -5\,V$$

or

$$-U_{BE} = 5\,V$$

or

$$U_{EB} = 5\,V$$

thus indicates that there is a voltage of 5 V between E and B where E is positive with respect to B. In a circuit diagram, the double indices are often omitted and the notation U_{xy} is replaced by a voltage arrow U pointing from node x to node y.

Potential. The potential V is the voltage of a node with relation to a common reference node 0:

$$V_x = U_{x0}.$$

In electrical circuits, the reference potential is denoted by a ground symbol. For the voltage between two nodes, x and y,

$$U_{xy} = V_x - V_y.$$

Current. The current is indicated by a current arrow on the connecting line. One defines the current I as being positive if the current in its conventional sense, i.e. the transport of positive charge, flows in the direction of the arrow. I is thus positive if the arrow of the current flowing through a load points from the larger to the smaller potential. The directions of the current and voltage arrows in a circuit diagram are not important as long as the actual values of U and I are given the correct signs.

If current and voltage arrow of a circuit element have the same direction, Ohm's law, with the above definitions, is $R = U/I$. If they have opposite directions, it changes to $R = -U/I$. This fact is illustrated in Fig. 17.1.

$$R = \frac{U}{I} \qquad\qquad R = -\frac{U}{I}$$

Fig. 17.1 Ohm's law

Resistance. If the resistance is voltage- or current-dependent, a static resistance $R = U/I$ and an incremental resistance $r = \partial U/\partial I \approx \Delta U/\Delta I$ can be defined. These formulae are valid if the voltage and current arrows point in the same direction. If the directions are opposed, a negative sign must be inserted, as in Fig. 17.1.

Voltage and current source. A real voltage source can be described by the equation

$$U_o = U_0 - R_{int} I_o, \tag{17.1}$$

where U_0 is the no-load voltage and $R_{int} = -dU_o/dI_o$ the internal resistance. This is represented by the equivalent circuit in Fig. 17.2. An ideal voltage source is characterized by the property $R_{int} = 0$, i.e. the output voltage is independent of the current.

A different equivalent circuit for a real voltage source can be deduced by rewriting Eq. (17.1):

$$I_o = \frac{U_0 - U_o}{R_{int}} = I_{sc} - \frac{U_o}{R_{int}}, \tag{17.2}$$

where $I_{sc} = U_0/R_{int}$ is the short-circuit current. The appropriate circuit is shown in Fig. 17.3. It is obvious that, the larger R_{int}, the less the output current depends on the output voltage. For $R_{int} \to \infty$, one obtains an ideal current source.

According to Figs. 17.2 and 17.3, a real voltage source can be represented either by an ideal voltage source or by an ideal current source. Which representation is chosen depends on whether the internal resistance R_{int} is small or large in comparison with the load resistance R_L.

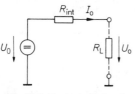

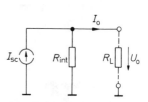

Fig. 17.2 Equivalent circuit   Fig. 17.3 Equivalent circuit
  of a real voltage source   of a real current source

Kirchhoff's current law (KCL). For the calculation of the parameters of many electronic circuits, we use Kirchhoff's current law. It states that the sum of all currents flowing into a node, is zero. Currents flowing towards the node are counted as being positive, and currents flowing from the node are negative. Figure 17.4 demonstrates this fact. It can be seen that, for node N,

$$\sum_i I_i = I_1 + I_2 - I_3 = 0.$$

Kirchhoff's voltage law (KVL). Kirchhoff's voltage law states that the sum of all voltages around any loop in an electrical network is zero. The voltage is entered in the appropriate equation with a positive sign if its arrow points in the direction in which one proceeds around the loop; the voltage is entered with a negative sign if the voltage arrow points against this direction. For the example in Fig. 17.5,

$$\sum_i U_i = U_1 + U_4 - U_2 - U_3 = 0.$$

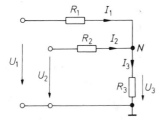
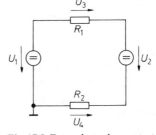
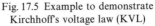

Fig. 17.4 Example to demonstrate
Kirchhoff's current law (KCL)

Fig. 17.5 Example to demonstrate
Kirchhoff's voltage law (KVL)

A.C. circuits (alternating current circuits). If the circuit can be described by a d.c. (direct current) transfer characteristic $U_o = f(U_i)$, this relationship necessarily also holds for any time-dependent voltage, i.e. $U_o(t) = f[U_i(t)]$, as long as the changes in the input voltage are quasi-stationary, i.e. not too fast. For this reason, we use capital letters for d.c. as well as time-dependent quantities, e.g. $U = U(t)$.

However, cases exist where a transfer characteristic is only valid for alternating voltages without d.c. components and it is therefore sensible to have a special symbol to distinguish such alternating voltages. We use the lower-case letter u to denote their instantaneous values.

A particularly important special case is that of sinusoidally alternating voltages, i.e.

$$u = \hat{U} \sin(\omega t + \varphi_u), \tag{17.3}$$

where $\hat{U}$ is the peak value (amplitude). Other values for the characterization of the voltage magnitude are the root-mean-square value $U_{\text{r.m.s.}} = \hat{U}/\sqrt{2}$ or the peak-to-peak value $U_{\text{pp}} = 2\hat{U}$.

The calculus for trigonometric functions is rather involved, but that for exponential functions fairly simple. Euler's theorem

$$e^{j\alpha} = \cos\alpha + j\sin\alpha \tag{17.4}$$

enables the expression of a sine function by the imaginary part of the complex exponential function:

$$\sin\alpha = \text{Im}\{e^{j\alpha}\}.$$

Equation (17.3) can therefore also be written as

$$u = \hat{U} \cdot \text{Im}\{e^{j(\omega t + \varphi_u)}\} = \text{Im}\{\hat{U}\,e^{j\varphi_u} \cdot e^{j\omega t}\} = \text{Im}\{\underline{U}\,e^{j\omega t}\},$$

where $\underline{U} = \hat{U}\,e^{j\varphi_u}$ is the complex amplitude. Its magnitude is given by

$$|\underline{U}| = \hat{U} \cdot |e^{j\varphi_u}| = \hat{U}[\cos^2\varphi_u + \sin^2\varphi_u] = \hat{U};$$

i.e. it is equal to the peak value of the sine wave. Time-dependent currents are treated in an analogous way. The corresponding symbols are then

$$I, \quad I(t), \quad i, \quad \hat{I}, \quad \underline{I}.$$

Arrows can also be assigned to alternating voltages and currents. Of course, the direction of the arrow then no longer indicates polarity but denotes the mathematical sign with which the values must be entered in the formulae. The rule illustrated in Fig. 17.1 for d.c. voltages also applies in this case.

In analogy to the resistance in a d.c. circuit, a complex resistance is defined and called the impedance $\underline{Z}$:

$$\underline{Z} = \frac{\underline{U}}{\underline{I}} = \frac{\hat{U}\,e^{j\varphi_u}}{\hat{I}\,e^{j\varphi_i}} = \frac{\hat{U}}{\hat{I}}\,e^{j(\varphi_u - \varphi_i)} = |\underline{Z}|\,e^{j\varphi},$$

where φ is the phase angle between current and voltage. If the voltage is leading with respect to the current, φ is positive. For a purely ohmic resistance, $\underline{Z} = R$; for a capacitance

$$\underline{Z} = \frac{1}{j\omega C} = -\frac{j}{\omega C}$$

and for an inductance $\underline{Z} = j\omega L$. The laws for the d.c. circuit quantities can be applied to complex quantities as well [17.1, 17.2].

One can also define a complex gain

$$\underline{A} = \frac{\underline{U}_o}{\underline{U}_i} = \frac{\hat{U}_o\,e^{j\varphi_o}}{\hat{U}_i\,e^{j\varphi_i}} = \frac{\hat{U}_o}{\hat{U}_i}\,e^{j(\varphi_o - \varphi_i)} = |\underline{A}|\,e^{j\varphi},$$

where φ is the phase angle between input and output voltage. If the output voltage is leading with respect to the input voltage, φ is positive; for a lagging output voltage, it is negative.

Logarithmic voltage ratio. In electrical engineering, a logarithmic value $|\underline{A}|^{\#}$ is often used to express the voltage ratio $|\underline{A}| = \hat{U}_o/\hat{U}_i$, i.e. the gain of a circuit. The relationship between $|\underline{A}|^{\#}$ and $|\underline{A}|$ is given by

$$|\underline{A}|^{\#} = 20\,\text{dB} \lg \frac{\hat{U}_o}{\hat{U}_i} = 20\,\text{dB} \lg |\underline{A}|.$$

The table in Fig. 17.6 lists some values.

linear voltage ratio $|A|$	logarithmic voltage ratio $|A|^{\#}$
0.5	$-6\,\text{dB}$
$1/\sqrt{2} \approx 0.7$	$-3\,\text{dB}$
1	$0\,\text{dB}$
$\sqrt{2} \approx 1.4$	$3\,\text{dB}$
2	$6\,\text{dB}$
10	$20\,\text{dB}$
100	$40\,\text{dB}$
1000	$60\,\text{dB}$

Fig. 17.6 Table for conversion of voltage ratios

Logarithms. The logarithm of a denominate number (e.g. $\lg 1.0\,\text{Hz}$) is not defined. We therefore write, for example, not $\lg f$ but $\lg(f/\text{Hz})$. The difference of logarithms is a different matter: the expression $\Delta \lg f = \lg f_2 - \lg f_1$ is well defined since it can be written in the form $\lg(f_2/f_1)$.

Mathematic symbols. We often use a shortened notation for the time derivatives:

$$\frac{dU}{dt} = \dot{U} \quad \text{and} \quad \frac{d^2 U}{dt^2} = \ddot{U}.$$

The symbol $\sim$ represents a *proportional* relationship; the symbol $\approx$ stands for *approximately equal to*; the symbol $\hat{=}$ means *corresponding to*. The symbol $\|$ means *parallel*. We use it to indicate that resistors are connected in parallel, i.e.

$$R_1 \| R_2 = \frac{R_1 R_2}{R_1 + R_2}.$$

List of the most important symbols

U	any time-dependent voltage, also d.c. voltages
u	alternating voltage without d.c. component
$\hat{U}$	amplitude (peak value) of a voltage
$\underline{U}$	complex voltage amplitude
$U_{\text{r.m.s.}}$	root-mean-square value of a voltage
E	computing unit voltage
U_{T}	thermal voltage kT/e_0 (k = Boltzmann's constant, T = absolute temperature, e_0 = charge of electron)
U_{b}	supply voltage
V^+	positive supply potential; in circuit diagrams indicated by $(+)$
V^-	negative supply potential; in circuit diagrams indicated by $(-)$
I	any time-dependent current, also direct currents
i	alternating current without d.c. component
$\hat{I}$	amplitude (peak value) of a current
$\underline{I}$	complex current amplitude
$I_{\text{r.m.s.}}$	root-mean-square value of a current
R	ohmic resistance
r	incremental resistance
$\underline{Z}$	complex resistance (impedance)
t	time
τ	time constant
T	period, cycle time
$f = 1/T$	frequency
f_c	3 dB cutoff frequency
f_{cA}	3 dB cutoff frequency of the open-loop gain $\underline{A}_D$ of an operational amplifier
f_T	gain-bandwidth product; unity-gain bandwidth
B	3 dB bandwidth
$\omega = 2\pi f$	angular frequency
$\Omega = \omega/\omega_0$	normalized angular frequency
$p = j\omega + \sigma$	complex angular frequency
$P = p/\omega_0$	normalized complex angular frequency
$A = \partial U_o/\partial U_i$	small-signal voltage gain for low frequencies
$\underline{A}(j\omega) = \underline{U}_o/\underline{U}_i$	complex voltage gain; frequency response
$A(p)$	general transfer function
A_D	differential gain, difference-mode gain, open-loop gain of an operational amplifier
g	loop gain
G	common-mode rejection ratio, CMRR

k	feedback factor
$\beta = \partial I_2 / \partial I_1$	small-signal current gain
$g_m = \partial I_C / \partial U_{BE}$	forward transconductance of a bipolar transistor
$g_{fs} = \partial I_D / \partial U_{GS}$	forward transconductance of a FET
ϑ	temperature on the Celsius scale
T	temperature on the Kelvin scale; absolute temperature
$\dot{x}$	first derivative of x with respect to time
$\ddot{x}$	second derivative of x with respect to time
$^a\log x$	logarithm to the base a
$\lg x$	logarithm to the base 10
$\ln x$	logarithm to the base e
$\operatorname{ld} x$	logarithm to the base 2

Logic symbols

$$x_1, x_2 \rightarrow y = x_1 \cdot x_2 \qquad \text{logic AND operation}$$

$$x_1, x_2 \rightarrow y = x_1 + x_2 \qquad \text{logic OR operation}$$

$$x \rightarrow y = \bar{x} \qquad \text{logic NOT operation}$$

$$x_1, x_2 \rightarrow y = x_1 \oplus x_2 \qquad \text{logic exclusive-OR operation}$$

Bibliography

1   Tietze, U.; Schenk, Ch.: Halbleiter-Schaltungstechnik. 4. Aufl. Berlin, Heidelberg, New York: Springer 1978.

1.1 Paterson, W.-L.: Multiplication and Logarithmic Conversion by Operational-Amplifier-Transistor Circuits. The Review of Scientific Instruments. 34 (1963) 1311–1316.

1.2 Roberge, J.K.: Operational Amplifiers. New York, London, Sydney, Toronto: J. Wiley 1975.

1.3 Sheingold, D.H. (Editor): Nonlinear Circuits Handbook. Analog Devices, Norwood, Mass., 1974.

1.4 Graeme, J.G.: Applications of Operational Amplifiers. New York: McGraw-Hill 1973.

2.1 Schenk, Ch.: Ein neues Schaltungskonzept für eine bipolare, spannungsgesteuerte Präzisions-Stromquelle. Nachrichtentechn. Z. 27 (1974) 102–104.

2.2 Antoniou, A.: 3-Terminal Gyrator Circuits Using Operational Amplifiers. Electronics Letters 4 (1968) 591.

2.3 Schenk, Ch.: Neue Schaltungen spannungsgesteuerter Stromquellen und ihre Anwendung in elektronischen Y-Gyratoren. Dissertation Universität Erlangen-Nürnberg 1976.

2.4 Rollett, J.M.: Greenaway, P.E.: Direct Coupled Active Circulators. Electronics Letters 4 (1968) 579.

3.1 Ghausi, M.S.: Principles and Design of Linear Active Circuits. New York: McGraw-Hill 1965, p. 84.

3.2 Weinberg, L.: Network Analysis and Synthesis. New York: McGraw-Hill 1962, p. 494.

3.3 ibid., p. 518.

3.4 Saal, R.: Handbuch zum Filterentwurf. Berlin: Elitera 1976.

3.5 Storch, L.: Synthesis of Constant-Delay Ladder-Networks Using Bessel Polynomials. Proc. IRE 42 (1954) 1666.

3.6 Schaumann, R.: A Low-Sensitivity, High-Frequency, Tunable Active Filter without External Capacitors. Proc. IEEE Int. Symp. on Circuits and Systems 1974, p. 438.

3.7 Unbehauen, R.: Synthese elektrischer Netzwerke. München, Wien: Oldenbourg 1972.

3.8 Heinlein, W.E.; Holmes, W.H.: Active Filters for Integrated Circuits. München, Wien: Oldenbourg 1974.

4.1 Müller, R.: Bauelemente der Halbleiter-Elektronik. Halbleiter-Elektronik, Bd. 2. Hrsg. v. W. Heywang u. R. Müller. Berlin, Heidelberg, New York: Springer 1973.

4.2 Cherry, E.M.; Hooper, D.E.: The Design of Wide-Band Transistor Feedback Amplifiers. Proc. I.E.E. 110 (1963) 375–389.

4.3 DeVilbiss, A.J.: A Wideband Oscilloscope Amplifier. Hewlett-Packard Journal 21 (1970) No. 5, p. 11.

4.4 Tietze, U.: Untersuchung eines Daten-reduzierenden Multiplex-Verfahrens für die digitale Bildübertragung. Dissertation Universität Erlangen-Nürnberg, 1975.

6.1 Kühn, R.: Der Kleintransformator. Prien: C.F. Winter 1964.

6.2 Transistor power supply transformers. Manual GT.25 by Gardners Transformers, Christchurch, England.

6.3 Emmermann, G.: Voltage Precision and High Current Capability. – Both in One Power Supply. Hewlett-Packard Journal 24 (1972) No. 3, p. 16.

6.4 Müller, R.: Grundlagen der Halbleiter-Elektronik. 2. Aufl. Halbleiter-Elektronik, Bd. 1. Hrsg. v. W. Heywang u. R. Müller. Heidelberg, New York: Springer 1975, p. 135.

6.5 Dudley, B.W.; Peck, R.D.: High Efficiency Modulator Power Supplies Using Switching Regulators. Hewlett-Packard Journal 25 (1973) No. 4, pp. 15–19.

6.6 Hnatek, E.R.: Applications of Linear Integrated Circuits. New York, London, Sydney, Toronto: J. Wiley 1975.

7.1 Digital-To-Analog-Converter Handbook. 2nd edition. Hybrid Systems Corporation, 1970.

8.1 Boyd, H.: Destroy your Microwave Transistors. Electronic Design 15 (1967) No. 20, p. 98.

8.2 Riedel, R.J.; Danielson, D.D.: The Dual Function Generator: A Source of a Wide Variety of Test Signals. Hewlett-Packard Journal 26 (1975) No. 7, pp. 18–24.

8.3 Smith, J.I.: Modern Operational Circuit Design. New York: Wiley-Interscience 1971.

9.1 Spaniol, O.: Arithmetik in Rechenanlagen. Stuttgart: Teubner 1976.

9.2 Barna, A.; Porat, D.I.: Integrated Circuits in Digital Electronics. New York, London, Sydney, Toronto: J. Wiley 1973.

9.3 Schmid, H.: Decimal Computation. New York, London, Sydney, Toronto: J. Wiley 1974.

9.4 Flores, I.: The Logic of Computer Arithmetic. Englewood Cliffs, N.J.: Prentice-Hall 1963.

9.5 Peatman, J.B.: The Design of Digital Systems. Tokyo: McGraw-Hill Kogakusha 1972.

10.1 Schaltbeispiele. Manual by Intermetall, Freiburg 1967, p. 40.

10.2 Martin, A.J.: PRBS Can Fool the System. Electronics 42 (1969) No. 8, p. 82.

10.3 Donn, E.S.: Manipulating Digital Patterns with a New Binary Sequence Generator. Hewlett-Packard Journal 22 (1971) No. 8, pp. 2–8.

10.4 Peterson, W.W.; Weldon, E.J.: Error-Correcting Codes. 2nd edition. Cambridge, Mass.: The MIT-Press 1972.

10.5 Anderson, G.C.; Finnie, B.W.; Roberts, G.T.: Pseudo-Random and Random Test Signals. Hewlett-Packard Journal 19 (1967) No. 1, pp. 2–16.

10.6 Wendt, S.: Entwurf komplexer Schaltwerke. Berlin, Heidelberg, New York: Springer 1974.

10.7 Clare, C.R.: Designing Logic Systems Using State Machines. New York: McGraw-Hill 1973.

11.1 Martin, D.P.: Microcomputer Design. Martin Research, Northbrook, Ill., 1976.

11.2 M6800 Microcomputer. System Design Data. Data book by Motorola, Phoenix, Arizona.

12.1 Unbehauen, R.: Systemtheorie. München, Wien: Oldenbourg 1971.

12.2 Schüßler, H.W.: Digitale Systeme zur Signalverarbeitung. Berlin, Heidelberg, New York: Springer 1973, pp. 19–29.

12.3 Daniels, R.W.: Approximation Methods for Electronic Filter Design. New York: McGraw-Hill 1974.

12.4 Lüke, H.D.: Signalübertragung. Berlin, Heidelberg, New York: Springer 1975.

13.1 Blood, W.R.: MECL System Design Handbook. 2nd edition. Handbook by Motorola, Phoenix, Arizona, 1972.

13.2 Peterson, W.W.; Weldon, E.J.: Error-Correction Codes. Cambridge, Mass.: The MIT-Press 1972.

13.3 Berlekamp, E.R.: Algebraic Coding Theory. New York: McGraw-Hill 1968.

13.4 Toschi, E.A.; Watanabe, T.: An All-Semiconductor Memory with Fault Detection, Correction, and Logging. Hewlett-Packard Journal 27 (1976) No. 12, pp. 8–13.

14.1   McGuire, P.L.: Digital Pulses Synthesize Audio Sine Waves. Electronics 48 (1975) No. 20, pp. 104, 105.

14.2   Hölzler, E.; Holzwarth, H.: Pulstechnik. Bd. 1 u. 2. Berlin, Heidelberg, New York: Springer 1975, 1976.

14.3   Pretzl, G.: Schnelle Analog-Digital-Umsetzer. Elektronik 25 (1976) No. 12, pp. 36–42.

14.4   Aldridge, D.: Analog-To-Digital Conversion Techniques with the M6800 Micro-processor System. Application note AN-757 by Motorola, Phoenix, Arizona.

14.5   Kay, B.; Harmon, J.L.: Twelve Functions in a New Digital Meter. Hewlett-Packard Journal 20 (1969) No. 7, pp. 2–13.

14.6   Seitzer, D.: Elektronische Analog-Digital-Umsetzer. Berlin, Heidelberg, New York: Springer 1977.

14.7   Sheingold, D.H. (Editor): Analog-Digital Conversion Handbook. Analog Devices. Norwood, Mass., 1973.

14.8   Schmid, H.: Electronic Analog-Digital Conversion. New York: Van Nostrand Reinhold 1970.

14.9   Hnatek, E.R.: A User's Handbook of D/A and A/D Converters. New York, London, Sydney, Toronto: J. Wiley 1976.

15.1   Ott, W.E.: A New Technique of Thermal RMS Measurement. IEEE Journal of Solid-State Circuits 9 (1974) No. 6, pp. 374–380.

16.1   Oppelt, W.: Kleines Handbuch technischer Regelvorgänge. Weinheim/Bergstraße: Verlag Chemie 1965.

16.2   Gardner, F.M.: Phaselock Techniques. New York, London, Sydney: J. Wiley 1966.

17.1   Weyh, U.; Benzinger, H.: Die Grundlagen der Wechselstromlehre. München, Wien: Oldenbourg 1967.

17.2   Brophy, J.J.: Basic Electronics for Scientists. New York: McGraw-Hill 1977.

Index